Risk Analysis and the Security Survey

Risk Analysis and the Security Survey

Fourth Edition

James F. Broder

Eugene Tucker

AMSTERDAM • BOSTON • HEIDELBERG • LONDON • NEW YORK
OXFORD • PARIS • SAN DIEGO • SAN FRANCISCO
SINGAPORE • SYDNEY • TOKYO
Butterworth-Heinemann is an imprint of Elsevier

Butterworth-Heinemann is an imprint of Elsevier
225 Wyman Street, Waltham, MA 02451, USA
The Boulevard, Langford Lane, Kidlington, Oxford, OX5 1GB, UK

Notices
Knowledge and best practice in this field are constantly changing. As new research and experience
broaden our understanding, changes in research methods, professional practices, or medical
treatment may become necessary.

Practitioners and researchers must always rely on their own experience and knowledge in evaluating
and using any information, methods, compounds, or experiments described herein. In using such
information or methods they should be mindful of their own safety and the safety of others,
including parties for whom they have a professional responsibility.

To the fullest extent of the law, neither the Publisher nor the authors, contributors, or editors,
assume any liability for any injury and/or damage to persons or property as a matter of products
liability, negligence or otherwise, or from any use or operation of any methods, products,
instructions, or ideas contained in the material herein.

Library of Congress Cataloging-in-Publication Data
 Broder, James F.
  Risk analysis and the security survey / James F. Broder, Eugene Tucker. – 4th ed.
  p. cm.
  ISBN 978-0-12-382233-8 (hardback) 1. Industries–Security measures. 2. Risk management. 3. Crime
prevention surveys. 4. Assistance in emergencies–Planning. I. Tucker, Eugene. II. Title.
  HV8290.B664 2012
 658.4'7–dc23
                                                                                            2011035312

British Library Cataloguing-in-Publication Data
A catalogue record for this book is available from the British Library

For information on all Butterworth-Heinemann publications
visit our Web site at www.elsevierdirect.com

Printed and bound by CPI Group (UK) Ltd, Croydon, CR0 4YY

Transferred to digital print 2012

Dedication

To the memory of Mr. Murphy, who, God willing, waits for me at the Rainbow Bridge.

"If you don't know where you're going, any road will get you there."
—Lewis Carroll, *Alice's Adventures in Wonderland*

Contents

About the Authors   xv

Acknowledgments   xvii

Introduction   xix

Part 1. The Treatment and Analysis of Risk	1
1. Risk	3
What Is Risk?	3
What Is Risk Analysis?	4
Risk Assessment	4
Risk Exposure Assessment	6
2. Vulnerability and Threat Identification	9
Risk Identification	9
Examples of the Problems of Identification	11
Security Checklist	12
3. Risk Measurement	21
Cost Valuation and Frequency of Occurrence	21
Principles of Probability	23
Probability, Risk, and Security	25
Estimating Frequency of Occurrence	27
4. Quantifying and Prioritizing Loss Potential	29
Assessing Criticality or Severity	30
The Decision Matrix	31

5. Cost/Benefit Analysis   33
    System Design Engineering   33
    Building Redundancy into the System   36
    A Security Countermeasure   37

6. Other Risk Analysis Methodologies   39
    National Infrastructure Protection Plan   40
    Review   44

7. The Security Survey: An Overview   45
    Why Are Security Surveys Needed?   45
    Who Needs Security Surveys?   46
    Attitude of Business Toward Security   48
    What Can a Security Survey Accomplish?   49
    Why the Need for a Security Professional?   50
    How Do You Sell Security?   50

8. Management Audit Techniques and the Preliminary Survey   53
    Audit Guide and Procedures   53
    The Preliminary Survey   58
    Summary   63

9. The Survey Report   69
    "I Must Write, Therefore I Shall"   69
    Five Criteria of Good Reporting   72
    Format   75
    Summary   78

10. Crime Prediction   79
    Analysis of Internal Crime   80
    Analysis of External Crime   81
    Inadequate Security   85

How to Establish Notice   87

Review   89

11. Determining Insurance Requirements   91

Risk Management Defined   91

Risk Control   92

Crime Insurance   92

K & R (Kidnap and Ransom) Coverage   94

Part 2. Emergency Management and Business Continuity
         Planning   99

12. Emergency Management – A Brief Introduction   101

Comprehensive Emergency Management   101

Standards   103

Private Sector Preparedness Accreditation and Certification Program   104

National Incident Management System (NIMS)   104

The Incident Command System (ICS)   104

Unified Command   109

Emergency Operations Center   110

Summary   112

13. Mitigation and Preparedness   113

Mitigation   113

Preparedness   130

Summary   133

14. Response Planning   135

Emergency Response Planning and Response Plans   136

Emergency Response Team   139

Emergency Procedures   141

Summary   199

15. Business Impact Analysis   201

Risk Analysis versus Business Impact Analysis   203

Business Impact Analysis Methodology   204

Other Questions for the Impact Analysis   214

Resource Questionnaires and Forms   215

Summary   221

16. Business Continuity Planning   223

Why Plan?   224

The Planning Process   225

Project Management   226

Summary   245

17. Plan Documentation   247

Required Elements of the Plan   247

Multihazard Functional Planning   248

Plan Organization and Structure   249

Summary   257

18. Crisis Management Planning for Kidnap, Ransom, and Extortion   259

Threat Identification   261

Plan Documentation   262

Plan Activation   263

Crisis Management Team   264

Handling the Initial Contact   266

Ransom Considerations   267

Preventive Security 268

Suggestions for Kidnapped Individuals 269

Media Control 270

Summary 271

Bibliography 271

19. Monitoring Safeguards 273

Monitoring or Testing the Existing System 273

The Scientific Method 274

Five Basic Types of Testing 274

Avoid Predictable Failure 275

Some Audit Guidelines 276

Develop a Plan of Action 277

20. The Security Consultant 279

In-House versus Outside Advice 279

Why Use Outside Security Consultants? 281

Security Proposals (Writing and Costing) 283

Summary 288

Evaluation of Proposals and Reports 288

Appendices 289

Appendix A   Security Survey Work Sheets 291

General Questions before Starting a Survey 291

Number of Employees 291

Cafeteria 292

Credit Union 292

Custodial Service 292

Company Store 292

Petty Cash or Funds on Hand   293

Classified Operations   293

Theft Experience   293

Some Reference Materials   303

Annex A: Hospital Surveys   303

Annex B: University and College Surveys   306

Appendix B   Sample Kidnap and Ransom Contingency Plan   309

   I.   Introduction   309

   II.   Basic Plan   309

Appendix C   Security Systems Specifications   323

Introduction   323

Example: Requirements Specification for an Integrated
          Electronic Security System   325

Conclusion   327

Index   329

About the Authors

James F. Broder, CFE, CPP, FACFE, has more than 40 years experience in security and law enforcement. He has worked as a security executive, instructor, and consultant as well as having served in Vietnam as a Police Advisor in the Counter Insurgency Directorate, Vietnamese National Police. A former FBI Special Agent and employee for the US State Department, U.S. House of Representatives, Washington D.C. Mr. Broder is considered to be one of the most highly respected security authorities in the United States.

Eugene Tucker, CPP, CFE, CBCP, CHST is the head of Praetorian Protective Service®, LLC, and a past member of the board of directors of the Business Recovery Managers Association. He has served as the coordinator of the Emergency Management Program at a major University, and his professional interests and experience range from Security and Safety Management to Business Continuity Planning.

Acknowledgments

The first edition of *Risk Analysis and the Security Survey* was published in 1984. The book continues to be widely accepted within both the security profession and the academic community worldwide. Originally written for security and risk management professionals, it has become widely accepted as a textbook in Security Management degree programs in universities throughout the English-speaking world. Accordingly, we have tried to continue to meet the needs of our principal reading audiences and at the same time expand the text to keep pace with current trends in the world as viewed by security and law enforcement professionals. In this regard, the second half of this text, written by coauthor Gene Tucker, addresses the subjects associated with recovery. It is important to point this out because when security fails, as it occasionally does, recovery becomes paramount and security professionals must understand the vital role they play if they are to fully meet their responsibilities.

Security, once regarded by many in management and government as a necessary evil, has become recognized as a means necessary to combat evil. The events since September 11, 2001, have changed our outlook regarding the vital role security plays in protecting our national interest. We are no longer complacent. We recognize that an attack upon our facilities and infrastructures can occur at any time, in any place. It will be a long time before we are ever again allowed the dubious luxury of resting on our laurels.

One of the little-known ironies of the attack on the World Trade Center (WTC) was the recognition by the New York Port Authority Management that the WTC was to remain a prime target for another terrorist attack. The first attack occurred in 1993, which put WTC management on notice that security had to become a prime concern. Among other things, the security department was given an almost unlimited budget to upgrade the existing security countermeasures and "harden the target" to avoid a repeat of the earlier attack. One month before the 9/11 attack, the former Assistant Director of the FBI for Counter Terrorism, John McGuire, retired from the FBI to head the security staff at the WTC. John McGuire had personally led the investigation into the terrorist bombings of the three U.S. embassies in Africa and the attack against the Navy destroyer U.S.S. *Cole* in the harbor at Yeman. John McGuire was the Bureau's leading expert on international terrorism. John McGuire and his staff of security professionals were all on duty that fateful morning in September when two aircraft were flown into the twin towers of the WTC. He, as well as all the members of his security staff and thousands of others, died on that unforgettable day!

So we would like to take this opportunity to acknowledge the great sacrifice made by John McGuire and the countless other security and law enforcement professionals who have dedicated their lives to the cause of freedom in an effort to provide us with a more secure future. For many years before 9/11, John McGuire and others tried to warn us about the threat posed by international terrorists. How sad that he and so many others had to die to really get our attention.

The events of September 11, 2001 have taught us some valuable lessons:

We in security are about probabilities and not guarantees.

The enhanced security countermeasures put in place at the World Trade Center as a result of the 1993 terrorist bombing were more than adequate! The failures that occurred that fateful morning in security were at locations hundreds of miles from the WTC.

John McGuire and the security professionals from the New York Port Authority should rest comfortably knowing that they more than met their responsibilities, because the failures in the system that occurred did not occur at their location on their watch.

James F. Broder
San Marino, CA

Introduction

In the early 1980s, we wrote the first edition of this book for the combined audiences of risk managers and security professionals. At that time, this author was employed as a security consultant for one of the largest insurance brokerage firms in the world. One of my early challenges was to explain to insurance brokers and their clients, risk managers for Fortune 500 companies, exactly what security professionals could do to reduce their risk regarding criminal and security issues. Risk management professionals were long accustomed to working with fire protection (property) consultants and safety (casualty) consultants. Few risk managers, however, had ever employed the services of a security professional. What gave rise to the need for property and casualty consultants was the necessity for clients to meet strict code requirements in order to be eligible for insurance coverage. Those requirements then, and now, were set forth in the form of "standards." The National Fire Protection Association (NFPA) and the Occupational Health & Safety Administration (OSHA), among others, have published standards that are accepted by the insurance industry as the minimum standards required by an insured in order to receive consideration to become insured against the risk inherent in fires and accidents.

In California, fire protection and safety consultants are required to be licensed as Professional Engineers (PEs). We are not aware that security consultants have to be licensed in any state.

The two obstacles that had to be overcome were (1) the absence of standards in the security profession and (2) the fact that anyone could be a security consultant without regard to licensing requirements. The lack of published standards in the security industry exists to this day. While no state requires a security consultant to be licensed as a professional engineer, the security industry does have the "Certification" by the American Society for Industrial Security (ASIS), which has educational and testing requirements, generally accepted by risk managers and other insurance professionals as proof of professional competence.

The issue of the absence of "standards" in the security profession is paramount! Standards are the hallmark of most professional associations and societies. Why is this important for security professionals? Because standards set forth minimum requirements (benchmarks), which can then be used in auditing one's security organization and program. As an example, this author has been engaged as an "expert witness" in many litigation matters (lawsuits regarding third-party liability). Invariably, the question I am often asked is, "What are the standards involved?" My usual answer is that we don't have many standards in the security profession; what we have, instead, are "acceptable practices." Another answer is, "Standards are whatever the client says they are!" Lawsuits involving the actions of security guards are very common. California, among others, licenses security guards. Regulatory requirements for licensing are often mistaken for standards by otherwise educated and experienced attorneys and judges. And in many cases the client or end users of security manpower and hardware do not have a clue about what does or does not constitute standards.

In 1980 we wrote, "The *Risk Management Journal* reported that 85 percent of those polled (Risk Managers) indicated that risk identification and evaluation was their number one priority." We suspect that this figure has not changed much over the past 30 years. Risk Analysis in this time has evolved as a methodology commonly used in the security industry. While we have not conducted a poll, there have been many articles written in security journals over these past years suggesting that the first order of business for security professionals is to fully identify their problems.

Readers of the earlier editions of this text will recognize that the material in Chapters 1–6, while updated, has not been significantly changed. The methodology we suggested in these chapters 30 years ago has stood the test of time. And, while the principles we suggest be used to perform a risk analysis are not categorized as "standards," they have nevertheless gained wide approval as "acceptable practices."

Accordingly, coauthor Gene Tucker and I once again set out here in the fourth edition to make a small contribution to the field of security. Granted, much has changed over the past 30 years. For one thing, the security profession has finally gained its rightful place, along with fire and safety, in the field of protection. Nevertheless, there is still much to be learned, and there will be continuing challenges for us in this profession to face. In this regard we urge the readers of this text to remember the basics. Spending time and using precious resources on the more glamorous but least likely harmful events that may occur is not serving the needs of our clients. International terrorism is a serious issue. However, the fact of the matter is that most of our clients have a greater chance of being struck by lightning than being attacked by terrorists.

James F. Broder, CFE, CPP, FACFE.

Similar to the security profession, the management of emergency and disaster situations is often guided by beliefs that are rooted in misunderstanding, misdirection, and mythology. Too often we believe and accept the mistakes or misconceptions repeated over and over by colleagues or, worse, by the media. Instead, we should turn to the research of social scientists and skeptically examine what we have done in the past or are encouraged to do in the future.

Careful attention to our text should point the reader in a direction that allows an effective analysis of potential events to reveal their true risk, highlight interdependencies, and understand that a single event can cause a cascade of failures or disasters. Consider as illustration the recent earthquake in Japan that caused a tsunami that interrupted power to a major nuclear power station and consequently affected the supply chain for the remainder of the world, not to mention the well-being of the environment and the surrounding population.

We have encountered more than one senior manager of an organization who has said that risk and business impact analysis is a useless exercise because a true manager is already aware of the issues, and, if they know their business, they don't need risk analysis or business continuity planning. Those who think a security program or a business continuity plan, once in place, is sufficient without keeping it alive should speak with the survivors of a coastal community in Japan that was completely swept away by the 2011 tsunami. The great majority of the population survived. Many believe this was because they identified the risk and analyzed it until they had a firm understanding of the problems they faced. They practiced tsunami evacuation on a regular basis. The managers mentioned above have looked at us with a "deer in the headlights" expression when the risk or business impact analysis revealed a serious exposure that they had not anticipated.

According to the British philosopher Herbert Spencer, "There is a principle which is a bar against all information, which is proof against all arguments and which cannot fail to keep a man in everlasting ignorance—that principle is contempt prior to investigation."

Gene Tucker, CPP, CFE, CBCP, CHST

The Treatment and Analysis of Risk

1

Risk

Risk assessment analysis is a rational and orderly approach as well as a comprehensive solution to problem identification and probability determination. . . . While it is not an exact science, it is, nevertheless...the art of defining probability in a fairly precise manner.

—James F. Broder, CFE, CPP, FACFE

What Is Risk?

Risk is associated with virtually every activity one can imagine, but for the purpose of this text, we limit the meaning of the word *risk* to the uncertainty of financial loss, the variations between actual and expected results, or the probability that a loss has occurred or will occur. In the insurance industry, risk is also used to mean "the thing insured"—for example, the XYZ Company is the risk. Additionally, risk can mean the possible occurrence of an undesirable event.

Risk should not be confused with perils, which are the causes of risk—such things as fire, flood, and earthquake. Nor should risk be confused with a hazard, which is a contributing factor to a peril. Almost anything can be a hazard—a loaded gun, a bottle of caustic acid, a bunch of oily rags, or a warehouse used for storing highly flammable products, as an example. The end result of risk is loss or a decrease in value.

Risks are generally classified as "speculative" (the difference between loss and gain—for example, the risk in gambling) and "pure risk," a loss/no-loss situation, to which insurance generally applies.

For the purposes of this text, the divisions of risk are limited to three common categories:

- Personal (people assets)
- Property (material assets)
- Liability (legal issues) that could affect both of the above. Here we include such problems as errors and omissions, wrongful discharge, workplace violence, sexual harassment, and last but not least, two issues which have become serious challenges to the business community: third-party and product liability. The first of these issues has given rise to a new profession: that of the "security expert witness." The subject is so important to the security professional that we have devoted a chapter in this text to it titled, "Crime Prediction" (see Chapter 10).

What Is Risk Analysis?

Risk analysis is a management tool, the standards for which are determined by whatever management decides it is willing to accept in terms of actual loss. To proceed in a logical manner to perform a risk analysis, it is first necessary to accomplish some fundamental tasks:

- Identify the asset(s) in need of protection (people, money, manufactured products, and industrial processes, to name a few).
- Identify the kinds of risks (or threats) that may affect the assets identified (kidnapping, extortion, internal theft, external theft, fire, and earthquake, for example).
- Determine the probability of the identified risk(s) occurring. Here one must keep in mind that the task of making such a determination is not an exact science but an art—the art of projecting probabilities. Remember this rule: "Nothing or no one can ever be made 100 percent secure; there is no such thing as a perfect security program."
- Determine the impact or effect on the organization in dollar values when possible, if a given loss does occur.

These subjects are given an in-depth treatment in the chapters that follow.

Risk Assessment

Risk assessment analysis is a rational and orderly approach, as well as a comprehensive solution, to problem identification and probability determination. It is also a method for estimating the expected loss from the occurrence of an adverse event. The key word here is "estimating" because risk analysis will never be precise methodology; remember, we are discussing probabilities. Nevertheless, the answer to most, if not all, questions regarding one's security exposures can be determined by a detailed risk assessment analysis.

What Can Risk Analysis Do for Management?

Risk analysis can provide management with vital information on which to base sound decisions. A thorough risk analysis can provide answers to many questions, such as: Is it always best to prevent the occurrence of a situation? Is it always possible? Should policy also consider how to contain the effect a hazardous situation may have? (This is what nuclear power plants prepare for.) Is it sufficient simply to recognize that an adverse potential exists and to do nothing for now but be aware of the hazard? (The principle here is called being self-insured.) The eventual goal of risk analysis is to strike an economic balance between the impact of the risk on the enterprise and the cost of implementing prevention and protective countermeasures.

A properly performed risk analysis can provide many benefits to management, a few of which are as follows:

- A properly performed analysis will show the current security posture (profile) of the organization—what it really is, not what management perceives it to be!

- It will highlight areas in which greater (or lesser) security is needed.
- It will help to assemble some of the facts needed for the development and justification of cost-effective countermeasures (safeguards).
- It will increase security awareness by assessing and then reporting the strengths and weaknesses of the security profile to all organizational levels from operations to top management. Risk analysis is not a task to be accomplished once and for all; it must be performed periodically if one is to stay abreast of changes in the environment and society as a whole. For example, in some industries "bomb threats" are seen to be on the increase. This is generally true for society as a whole. Bomb threats do not, however, fall into statistically predictable patterns. So it would behoove those of us who have any potential exposure to this problem to ensure our security plan is designed to deal with such anomalies. Also, because security measures designed at the inception of a program for building or expansion generally prove to be more effective than those superimposed later, risk analysis should have a place in the design or building phase of every new facility. This is being done increasingly as architects have begun to realize the importance of planning for security.

The major resource required for performing a risk analysis is trained manpower. For this reason, the first analysis is the most expensive. Subsequent analyses can be based in part on previous work history, and the time required to do a subsequent survey should decrease to some extent as experience and empirical knowledge are gained.

The time allocated to accomplish the risk analysis should be compatible with its objectives. Large facilities with complex, multi-shift operations and many files of historical data require more time to gather and review the necessary data than does a single-shift, limited-production location. If meaningful results are to be achieved, management must be willing to commit the resources necessary to accomplish the mission. It is best to delay or even abandon a security survey unless and until the necessary resources are made available to complete the tasks properly. In this regard, the security professional should be ever mindful of the legal risk associated with being "on notice" that serious threats exists which are not being addressed. After a situation is identified as being a "security risk" and is brought to management's attention, corrective action must be taken. Not doing so could expose the company to legal liability should an otherwise preventable event occur.

Role of Management in Risk Analysis

The success of any risk analysis undertaking is contingent on the role top management plays in the project. Management must support the project and express this support to all levels of the organization. Management must define the purpose and the scope of the risk assessment. It must select a qualified team and formally delegate the authority necessary to accomplish the mission. Finally, management must review the team's findings, decide which recommendations need to be implemented, and then decide on and establish the order of priorities for implementing the recommendations made in the survey report. Most of the professional security consultants the author has worked with list their recommendations in the order of priority. This takes the guesswork out of the equation for management.

Personnel who are not directly involved in the survey and the analysis process must be prepared to provide information and assistance to those who are. In addition, all employees must accept any inconvenience and the possible temporary interference of any activity that may result from actions of the survey team. Management should make it clear to all employees that it intends to rely on the final product and base its security decisions on the findings of the risk analysis team's report. The scope of the project should be well defined, and the statement of scope should spell out the parameters and depth of the analysis. It is often equally important to state specifically and in writing what the survey is not designed to accomplish or cover; this serves to eliminate any misunderstandings at the start of the project. An example might be the exclusion from a security survey of safety and evacuation procedures in the event of an emergency or disaster.

At this point, it may be helpful to define and explain two other terms that are sometimes used interchangeably with risk: threat—anything that could adversely affect the enterprise or the assets; and vulnerability—a weakness or flaw, such as holes in a fence, or virtually anything that may conceivably be exploited by a threat. Threats are most easily identified and organized by placing each in one of three classifications or categories: natural hazards (floods), accidents (chemical spills), or intentional acts (domestic or international terrorism). Vulnerabilities are most easily identified by interviewing long-term employees, supervisors, and managers in the facility; by field observation and inspection; and by reviewing incident reports. In the case of security hardware and electronics, tests can be designed and conducted to highlight vulnerabilities and expose weaknesses or flaws in the operation (see Chapter 18). Examples would be an out-of-date access control system or inadequate procedures to ensure operational procedures in the high-value cage of a storage warehouse.

Threat occurrence rates and probabilities are best developed from reports of occurrences or incident reports, whenever these historical data exist. If reports containing this information do not exist, it may be necessary to develop the information from other sources. This can be accomplished by conducting interviews with knowledgeable people or projecting data based on an educated guess, supported by studies in like industries at different locations.

Risk Exposure Assessment

Before any corrective action can be considered, it is necessary to make a thorough assessment of one's identifiable risk exposures. To accomplish this, it is essential that three factors be identified and evaluated in quantitative terms. The first is to determine (identify) the types of loss or risk that can affect the assets involved. Here, examples would be fire, flood, burglary, robbery, and kidnapping. If one of these were to occur (for now we will consider only single, not multiple, occurrences), what effect would the resulting disruption of operations have on the company? For example, if computers were destroyed by fire or flood, what would the effect be on the ability of the company to continue operating? There is a saying common to protection professionals: "One may well survive a burglary, but a major fire can put you out of business forever." If the chief executive officer, on an overseas trip, were to be killed in a terrorist bombing (or even suffer a serious heart attack), who would make the day-to-day operating decisions in his or her absence? What about unauthorized disclosure of trade secrets or other proprietary data? After all the risk exposures are identified (or as many as possible), one must proceed to evaluate those identified threats that, should they occur, would produce losses in quantitative terms—fire, power failure, flood, earthquake, and unethical or dishonest employees, to name a few of the more common risks necessary of consideration.

To do this, we proceed to the second factor: estimate the probability of occurrence. What are the chances that the identified risks may become actual events? For some risks, estimating probabilities can be fairly easy. This is especially true when we have documented historical data on identifiable problems. For example, how many internal and external theft cases have been investigated during the past year? Other risks are more difficult to predict. Workplace violence, embezzlement, industrial espionage, kidnapping, and civil disorder may have never occurred or may have occurred only once.

The third factor is quantifying (prioritizing) loss potential. This is measuring the impact or severity of the risk, if in fact a loss does occur or the risk becomes an actual event. This exercise is not complete until one develops dollar values for the assets previously identified. This part of the survey is necessary to set the stage for classification, evaluation, analysis, and for the comparisons necessary for the establishment of countermeasure (safeguards) priorities. Some events or kinds of risk with which business and industry are most commonly concerned are as follows:

- Natural catastrophe (tornado, hurricane, seismic activity)
- Industrial disaster (explosion, chemical spill, structural collapse, fire)
- Civil disturbance (sabotage, labor violence, bomb threats)
- International and domestic terrorism (the infamous 9/11 attack)
- Criminality (robbery, burglary, pilferage, embezzlement, fraud, industrial espionage, internal theft, hijacking)
- Conflict of interest (kickbacks, trading on inside information, commercial bribery, other unethical business practices)
- Nuclear accident (for example, Three Mile Island, Detroit Edison's Enrico Fermi #1, the 2011 major disaster in Japan). Some of these events (risks) have a low, or zero, probability of occurrence; also, some are less critical to an enterprise or community than others even if they do occur (fire versus burglary, for instance). Nevertheless, all of the identified possible events could occur and are thus deserving of consideration.

Examples include the nuclear accident at Chernobyl in the Soviet Union in 1987 and the chemical gas disaster ("breach of containment") at the Union Carbide plant in Bhopal, India, in 1984. Also, there are today in the United States chemical and nerve gas weapons stored in bunkers at military depots near populated areas. Do contingency plans exist to deal with these risks in the event of accidental fire, leak, or explosion? Are disaster drills and exercises conducted periodically to test the effectiveness of the contingency plans, if they exist? There are contingency plans for breach of containment and other industrial accidents in nuclear power generating plants in the United States, and these are strictly monitored by the Nuclear Regulatory Commission (NRC), which requires periodic drills to rehearse emergency plans.

In the following chapters, we continue to discuss vulnerabilities and threat identification, as well as risk measurement and quantification. The material presented above should give the reader some idea of what this textbook is all about. Risk analysis is a very powerful tool for management's use if properly applied to the problems unique to any given enterprise. To identify these unique problems, however, requires that an in-depth survey first be conducted.

2

Vulnerability and Threat Identification

Before the question of security can be addressed, it is first necessary to identify those harmful events which may befall any given enterprise.
—Charles A. Sennewald, CPP, Security Consultant, Author, and Lecturer

Risk Identification

In a systematic approach to the identification of threats, such as the one recommended in this text, the primary purpose of vulnerability identification or threat (exposure) determination is to make the task of risk analysis more manageable by establishing a base from which to proceed. When the risks associated with the various systems and subsystems within a given enterprise are known, the allocation of countermeasures (resources) can be more effectively planned. The need for such planning rests on the premise that security resources, like all resources, are limited and therefore must be allocated wisely.

Risk control begins, logically, with the identification and classification of the specific risks that exist in a given environment. To accomplish this task, it is necessary to examine by surveying all the activities and relationships of the enterprise in question and to develop answers to these basic considerations:

- Assets—What does the company own, operate, lease, control, have custody of or responsibility for, buy, sell, service, design, produce, manufacture, test, analyze, or maintain? It would be advisable to review the company's mission statement at this point in time.
- Exposure—What are the company's exposures? What could cause or contribute to damage, theft, or loss of property or other company assets, or what could cause or contribute to personal injury of company employees or others?
- Losses—What empirical evidence is available to establish the frequency, magnitude, and range of past losses experienced by this and other companies located nearby, performing a like service, or manufacturing the same or similar products?

Obviously, the answers to these questions and any additional questions that may be raised when conducting initial inquiries will be the basis for the identification of risks and eventually the quantification of all identifiable risks that may have a negative effect on the enterprise in question.

Security professionals can employ many techniques to develop the data for risk identification. They may review company policies, procedures (or take into account their absence), the structure of the organization, and its activities to ascertain what risks have previously been identified (if any) and to what extent they are perceived as management responsibilities. They may review insurance and risk-related files, including claims and loss records. Interviews with the heads of departments that have experienced loss exposures can develop vital information on the organization and the adequacy of loss-control procedures, if in fact any exist. Performing observation tours and inspections and interviewing management and other personnel in enough locations and activities will help develop a comprehensive picture of the company's risk exposures as a basis for later evaluation of existing loss control procedures and their effectiveness.

The tools necessary to accomplish the above tasks are the ability to conduct thorough interviews; the ability to conduct site inspections and field observations of operations, procedures, manpower utilization, the hardware, and electronics used in existing security systems; and the ability to identify, obtain, and analyze all pertinent records that one finds.

Another technique one may use is to develop asset data. To do this, one needs to completely identify all company assets, tangible and intangible, in terms of quantity and quality. One then locates all company assets and identifies obvious exposures that may exist at these locations. Next, one must determine the value of these assets in terms of actual dollars. This should be broken down into the following three categories:

- Owned assets $_____
- Leased assets $_____
- Facility losses $_____
  Total tangible assets $_____
  Total intangible assets $_____
  Grand total $_____

The identification of all company assets, coupled with a history of loss exposure for the company and other companies similarly located and engaged in a like or similar activity, will normally be sufficient to identify most of the major risks involved.

After this identification procedure is concluded, the security survey or inspection can be limited to those risks or exposures that specifically relate to the enterprise in question. These risks will usually include most, if not all, of the following:

- Crime losses, such as burglary, theft (internal and external), fraud, embezzlement, vandalism, arson, computer abuse, bomb threat, theft of trade secrets and intellectual property, industrial espionage, forgery, product forgery and trademark infringement, robbery, extortion, and kidnapping—to name the most common crime risks encountered by business and industry. Some others are:
- Cargo pilferage, theft, and damage
- Emergency and disaster planning
- Liability of officers and directors
- Environmental controls as directed by occupational safety and health codes

- Damage to property from fire, flood, earthquake, windstorm, explosion, building collapse, falling aircraft, and hazardous processes
- Comprehensive general liability arising from damage caused by any activity for which the entity can be held legally liable
- Business interruption and extra expense. An evaluation of this type of risk may require a detailed study of interdependencies connecting various segments of the entity and outside suppliers of goods and services. The study may indicate the need for extensive disaster recovery planning, for the reduction of risks not previously identified.
- Errors and omissions liability
- Professional liability
- Product liability

As can be seen from even a cursory review of these risks, the scope of risk identification alone, separate from risk evaluation and risk control, presupposes a degree of education and practical knowledge not often possessed by the average security manager. This implies that the person who is charged with this responsibility should have the education, training, and practical experience necessary to seek out, recognize, and thus identify not only the risks involved but also their impact upon the enterprise in question. The process of risk identification, evaluation, and control, in any dynamic organization, public or private, requires constant attention by professionals who possess the necessary knowledge and tools to accomplish these tasks. This knowledge is best acquired by study and practical experience. There is, however, no substitute for "hands-on" experience when it comes to the analysis of risks.

Examples of the Problems of Identification

We were given the assignment of conducting a security survey for a chain of fast-food restaurants. At the initial meeting with the management staff of this chain, reports from the company's internal audit division were furnished for our review. These reports showed all of their crime-related losses that had occurred for a 12-month period. They contained statistics that showed a disturbingly high incident rate for "robbery." One of the firm's top management people stated that he had a growing concern that a part-time high school cashier working in one of their restaurants might get shot in the course of a robbery because of the absence of procedures instructing employees what to do in case of an armed robbery.

Field inspections of a representative number of these restaurants produced evidence that the crime problem most often experienced was burglary (breaking and entering), followed closely by internal theft. "Armed robbery," or just plain robbery, as perceived by management, although admittedly always a dangerous situation, was not as frequent or as serious a problem as management had been led to believe by the statistics reported in the audit reports. Further inquiry revealed that the terms *robbery* and *burglary* were being used indiscriminately and, in many cases, synonymously. It was only after identifying the real problem that we could proceed to develop the proper procedures and allocate the necessary resources to address and then solve management's concern.

In another case, I met with the management of a national corporation that, among other things, printed negotiable instruments. The purpose of the meeting was to develop a mutual agreement regarding the scope for a security survey to be conducted at one of its West Coast plants. At the outset, one of the management representatives asked, "Have you ever conducted a security survey at a plant that prints negotiable instruments?" The simple fact at the time was that we had not. Nevertheless, the answer we gave was, "No, we haven't, but that really doesn't matter. It is immaterial to us if your plant manufactures widgets or prints negotiable instruments. You will either have a security program or you will not. If you do, it will either be functional, or it will not. In either event, it can be evaluated, and we can determine if the state of the security in existence at that facility is adequate, given the unique requirements for protection that this type of product requires. If we feel the security is not adequate, we will make recommendations to upgrade the quality and the quantity of the security safeguards necessary to accomplish this goal. If you have no program for security, we will design one for your consideration to meet the above requirements." When we arrived at the facility we found a total absence of even rudimentary security protection and had to design a basic program of safeguards and countermeasures for this operation. The managers at the site had been complaining to their superiors about the lack of protection for a number of years without success. The message here is "do not assume that management has any understanding of their security program or what they need to be protected."

In security, as with many other disciplines, we deal in acceptable practices and principles. These remain fairly constant, regardless of the problem involved or the environment encountered. The end result is loss prevention. This means that one either has or does not have an adequate security program in place. One way to find out is to conduct a security survey and identify those harmful events that may interfere with the company's objectives, as they are defined by the management of the enterprise in question.

Security Checklist

Until now we have been discussing some of the techniques and tools that the security professional needs to develop data for risk identification. Very often, security checklists are used to facilitate the gathering of pertinent information. These checklists take many forms. They can be simple lists of yes-or-no questions or open-ended questions requiring narrative responses. They may be brief and narrowly focused on the specific operation or activity in question, or they may be broader in scope and cover security concerns common to all the company's operations. No matter what its appearance, the purpose of a security checklist is to provide a logical recording of information and to ensure that no important question goes unasked.

The checklist is usually the backbone of the security survey or audit. This subject is covered extensively in Chapters 7, 8, and 9.

The following is a generic or introductory security survey checklist. It was designed for general use; therefore, it may include many items that are not appropriate in some situations. Additional checklists can be found in the appendices of this text. We do suggest, however, that one develop his or her own checklist tailored to the unique requirements of the operation being audited.

I. Policy and Program

1. Top management established a security policy?
  a. Policy published?
  b. Part of all managers' responsibility?
  c. Designated individual to establish and manage security program?
2. Top manager accessible to security supervisor?
3. Any regulations published? (Attach copy)
4. Disciplinary procedures?
  a. In writing?
  b. Specify offenses and penalties?
  c. Incidents recorded?
  d. Review by management?
  e. Uniformly enforced?
5. Any policy on criminal prosecution?
  a. Number of prosecutions attempted during past 5 years?
  b. Number of convictions?

II. Organization

1. Security supervisor full time?
  a. If part time, percentage of time spent on security?
  b. Describe chain of command from security supervisor to plant manager.
2. Number of full-time security personnel?
3. Number of personnel performing security duties each shift?
  a. Do they perform non security duties concurrently?
  b. Do security duties have first priority?
4. Have security personnel received security training?
5. Are written reports made of incidents?
6. Is there follow-up investigation of incidents?
7. Background investigation of security personnel?
8. Guards (security officers)?
  a. Number?
  b. Proprietary or contract service?
  c. If contracted service, does plant security supervisor interview and select?
  d. Is there a written contract for guard service?
    - Management's terms and conditions included?

    e. Written guard orders? (Attach copy)
    f. Weapons carried? (List type—for example, pistols, mace)
      - If yes, who inspects?
      - Company furnished?
    g. Make electronically recorded tours?
    h. Guards make written tour reports?
    i. Frequency of tours?
    j. Tour pattern varied?
    k. Number of report stations?
    l. Guards submit written report each shift? (Attach copy of form)
    m. Have guards received any formal training? (Describe on reverse side)
    n. Appearance of guards and uniforms?
9. Procedures
    a. Have security procedures been published? (Attach copy)
    b. Distributed to all those affected?
    c. Revised when conditions change?
    d. Used to conduct periodic audits and drills?
10. Does the security supervisor maintain contact with local law enforcement agencies to keep abreast of criminal activities and potential disorder in the community?

III.   Control of Entry and Movement

1. Is identification required of all persons entering?
2. Are there periodic 100 percent checks of identification?
3. How often? By whom?
4. Are all visitors registered?
5. How are employees distinguishable from visitors? (Explain on reverse)
6. Are all visitors escorted at all times?
7. Is there control of employee movement between areas within the plant? (Describe on reverse)
8. Are supervisors instructed to challenge strangers in their work areas?
    a. Do they? Nearly always? Sometimes? Never?
9. Are periodic traffic counts made at all points of entry as a means of detecting need for schedule change?
10. Are identification badges issued to all employees? Wearing enforced?

IV.   Barriers (Fences, Gates, Walls, etc.)

1. Is there a continuous barrier around the entire plant property?
    a. A major portion of it?
    b. Areas outside barrier? (List)
2. Fencing
    a. Eight feet high?
    b. Two-inch-square mesh?

      c. Eleven-gauge or heavier wire?
      d. Topped by three strands barbed wire or selvage?
      e. In good repair?
      f. Within 2 inches of firm ground at all points?
      g. Securely fastened to rigidly set posts?
      h. Metal posts set in concrete?
      i. Where attached to buildings, gaps not more than 4 inches?
      j. Gates in good repair? (How many?)
- Gates same height and construction as fence?
- Open only when required for operations?
- Locked other times?
- Equipped with alarm? (How many?)
- Guarded when open?
- Under surveillance when open? How?
- At least 10 feet of clear space both sides of fence?

3. Along embankments, fence on top or 20 feet from bottom?

4. Walls as perimeter barriers?

      a. At least 8 feet high?
      b. All doors equipped with alarm device or under surveillance?
      c. Means of surveillance?
      d. Windows:
- Permanently closed?
- Accessible for removal of property?
- Can be used for entry or exit?
- Protected by bars or heavy screen?
- Equipped with alarm?

      e. Final exit (perimeter) doors:
- Guarded?
- Alarm equipped? What type?
- Controlled by security personnel? Controlled by devices?
- Strong enough to resist heavy impact?
- Hinge pins concealed from outside or security type?

V. Lighting

1. Entire perimeter lighted?

2. Strip of light on both sides of fence?

3. Illumination sufficient to detect human movement easily at 100 yards?

4. Lights checked for operation daily before darkness?

5. Extra lighting at entry points and points of possible intrusion?

6. Lighting repairs made promptly?

7. Is the power supply for lights easily accessible (for tampering)?

8. Are lighting circuit drawings available to facilitate quick repairs?

9. Switches and controls
   a. Protected?
   b. Weatherproof and tamper resistant?
   c. Accessible to security personnel?
   d. Inaccessible from outside the perimeter barrier?
   e. Master switches centrally located?
10. Good illumination for guards on all routes inside the perimeter?
11. Materials and equipment in receiving, shipping, and storage areas adequately lighted?
12. Bodies of water on perimeter adequately lighted?
13. Auxiliary source of power for protective lighting?

VI.   Locks and Keys

1. Responsibility for control of locks and keys assigned to security supervisor?
2. Does he or she control locks and keys to all buildings?
3. Does he or she have overall authority and responsibility for issues, changes, and replacements?
4. Plant manager approve formula for issuing keys?
5. Managers approve issue of keys to their area?
6. Any keys issued to nonemployees?
7. Recipient sign receipt for key?
   a. Receipt show building and room number, date, name of authorizing manager?
   b. Receipt acknowledge obligation to turn in, report loss, not to duplicate?
8. Keys issued solely because of operational need for recipient to have key?
9. Lock and key control procedures and regulations in writing? (Attach copy)
10. All keys recovered from terminating employees?
11. Master keys not marked as such?
12. Spare keys stored under double lock or in combination-locked, fireproof cabinet?
13. Access to spare keys restricted to security supervisor and one other manager?
14. Locks changed immediately upon theft or loss of keys?
15. Locks on perimeter doors and gates changed annually?
16. Padlocks changed or rotated annually?
17. Manufacturer's serial number on padlocks obliterated and replaced by plant code number?
18. Padlock locked to hasp or staple when door or gate is open (to prevent substitution)?
19. Locks on inactive doors and gates checked regularly for evidence of tampering?
20. Door locks installed so that bolt extends 1/2 inch into jamb?
    a. Bolt covered by steel cover plate between door and jamb to prevent levering?
21. Combination locks
    a. Combination changed:
       - Annually?
       - When unauthorized person may have learned?
       - When knowledgeable person leaves or transfers?

   b. Combination memorized? (Not written!)
   c. Combination numbers: one odd, one even, one divisible by five?
      (Any sequence OK)
   d. Combination disclosed on basis of operational necessity
      (not convenience)?
22. Perimeter doors
   a. Lock installed without keyway in outside knob?
   b. Deadbolt locks in doors that must be unlocked from outside?
23. Safes
   a. Of substantial construction?
   b. Rated (labeled) for fire resistance?
   c. Rated (labeled) for burglary resistance?
   d. Lighted at night?
   e. Covered by proximity or motion detection alarm?

VII.   Alarms

1. Fire alarms
   a. Water flow? Is water pressure present?
   b. Valve condition?
   c. Water temperature?
   d. Particles of combustion detection?
   e. Heat or smoke sensing?
   f. Monitored continuously by:
      - Contract central station?
      - Proprietary station?
      - Direct connection to police or fire department?
   g. Tested regularly and tests recorded?
   h. Additional functions performed by alarm system (for example, shuts off computer
      power, lights, heating, air conditioning)?
2. Intrusion alarms
   a. Protects all of plant perimeter?
   b. Protects high-value storage areas?
   c. Protects other internal areas? (List)
   d. Types of sensors? (List)
   e. Proprietary or central station supervision?
   f. Regular recorded tests? How often?
3. Closed circuit television
   a. Used for surveillance only?
   b. Used for access control?
   c. Monitored continuously?

VIII.   Communications

1. Separate communications for security and emergency use?
   a. Telephone?
   b. Radio, pagers, cell phones?
2. If radio shared with other users, can security override?
3. Is there a means of contacting guard on patrol immediately? How?
4. Procedure for contacting local police and fire departments?
5. Means of alerting employees to emergency? How?

IX.   Property Control (Equipment, Material, Tools, Personal Property)

1. Covered by written procedures?
2. Specified form?
   a. Serial numbered?
   b. Multipart to provide separate audit trails?
3. Signed authorization required? Approved by higher level than beneficiary?
4. All transactions monitored at exit?
5. All exits controlled?
6. All transactions audited by third party (other than security)?
7. Follow-up on late returns (of borrowed items)?
8. Control points between work area and parking area?
9. Spot-checks of trucks and other vehicles?
10. All company tools (except small hand tools) marked with permanent company identification?
11. Employees sign receipt for tools and equipment issued?
12. Tools and desirable items secured in locked cages or rooms?
    a. Inventoried frequently?
13. All losses reported?
    a. Follow-up investigation?
    b. Written record?
    c. Statistics compiled? Reported to top management?
14. Shipping, receiving, and storage
    a. Guarded or within protected area?
    b. Under security of supervisor surveillance?
    c. Continuous spot-checks of complete shipments and receipts versus documents by other than shipping/receiving clerks?
    d. Outside drivers permitted inside plant?
    e. Vehicles locked?
    f. Tailgate check of existing vehicles?
    g. Storage areas under separate lock control when unattended?
    h. All withdrawals from stock recorded? Records provide separate audit trails?
    i. Continuous spot-checks of waste containers?

15. Scrap and salvage
   a. Written procedure for collection and disposal of scrap and salvage?
   b. Sealed bids required?
   c. Disposal action documented?
   d. Purchaser selected by management (other than administering employee)?
   e. Estimated annual sales of scrap and salvage?
   f. Any sold or given to employees? (Explain on reverse or attach procedure)
   g. Stored in a locked secure area?
   h. Waste spot-checked for saleable scrap and salvage?
   i. Classified and separated as to value?
   j. Spot-checked for "high grading"?
   k. Removed from premises under:
   - Signed authorization on specified form?
   - Surveillance of third-party employee—for example, accounting or security?
   - Verifies class and quantity?
   - Compares with purchaser's receipt?
   l. Auditors review all transactions?
   m. Are quantities within normal limits for this type of operation?

X.   Emergency Planning

1. Plans for reaction to imminent or actual:
   a. Fire?
   b. Explosion?
   c. Flood or tidal wave?
   d. Hurricane?
   e. Earthquake?
   f. Disorder?
   g. Aircraft accident?
   h. Bomb threats and bombs?
2. Responsibilities spelled out?
3. Responsible individuals designated?
4. Organization(s) completely staffed?
5. Periodic rehearsals of:
   a. All personnel?
   b. Key personnel?
6. Have critical features of plant and equipment been identified?
   a. Protected by barriers, access control, and lighting?
7. Coordinated with local public safety and disaster organizations?
8. Include plans for postdisaster recovery?
9. Identify resources available and required?

XI.   Personnel Screening

1.   Written, signed employment application required? Omissions not tolerated?
2.   All candidates interviewed?
3.   Investigation to verify?
    a. Previous employment
- Employers?
- Dates?
- Position and duties?
- Salary?
- Quality of performance?

    b. Education?
    c. Criminal record?
    d. Reputation?
    e. Medical record
- Illnesses?
- Physical handicaps and limitations?
- Work injuries?
- Occupational illnesses?

4.   Special screening of candidates for fiduciary positions? Describe.

XII.   Comments

Add any information you consider useful in arriving at a realistic assessment of the security in the facility. After all the significant loss potentials are identified, the next step is to evaluate the identified threats or vulnerabilities that may affect the enterprise and thus conceivably produce losses. The completed survey forms should be retained in the audit file for later use in the preparation of the final survey report. How to prepare a survey report is covered in Chapter 9 of this text.

3

Risk Measurement

Risk cannot always be eliminated. Properly identified, however,
it can usually be managed.
—Risk Assessment Guidelines, General Security, ASIS International, 2003

Risk measurement (quantification) is an essential element for later use in determining the impact (cost) of an unfavorable event on any operation or enterprise. It can also aid in predicting how often such an event may occur in a given period of time. Two necessities for performing risk measurement and quantification are a quantitative means of expressing potential cost and a logical expression of frequency of occurrence. Both must consider low as well as high frequencies of event occurrence.

There is no better way to state the impact of an adverse circumstance (unfavorable event)—whether the damage or cost is actual or abstract, or the victim a person, a piece of machinery, or the entire facility—than to assign it a monetary value. Ascertaining the cost of any adverse event is the logical way to equate value in our society. For a company that is concerned with cost (and which are not?), it is the only way! Because budgets and other financial matters are normally organized on a yearly basis, a year is obviously the most suitable time period to use in expressing the frequency of occurrence of threats. Of course, some threats may occur only once in a period of years, such as the 100-year flood. Others may occur daily or many times a day, such as internal theft. Each, however, can be measured in dollars as well as in frequency of occurrence.

Cost Valuation and Frequency of Occurrence

It is much more difficult to say that something happens every 1/73 of a year than that it happens, say, five times a day. It is also inconvenient to work with such fractions. For this reason, the transmutation of 1,000 days to 3 years, as shown below, has evolved. This method avoids unwieldy fractions yet maintains the flexibility of working with high-probability events in days and low-probability events in years.

In most cases, it is neither necessary nor desirable to make precise statements of impact and probability. The time needed for the analysis will be considerably reduced, and its usefulness will not be decreased, if impact (i) and frequency (f) correlations are given in factors of 10. It does not really matter to the overall estimation of threats whether the cost of the threat is valued at $110,000 or $130,000, or whether the anticipated frequency is 8 or 12 times a year. If at the time of deciding upon safeguards it becomes necessary to refine specific items, then by all means do so and end the argument! What is essential in the beginning is simplifying the measurement and quantification process, for reasons of efficiency and speed. This will facilitate the task by decreasing the amount of time spent on the analysis.

If the cost valuation (impact) of the event is:

$10, let i = 1
$100, let i = 2
$1,000, let i = 3
$10,000, let i = 4
$100,000, let i = 5
$1,000,000, let i = 6
$10,000,000, let i = 7
$100,000,000, let i = 8.

If the estimated frequency of occurrence is:

Once in 300 years, let f = 1
Once in 30 years, let f = 2
Once in 3 years, let f = 3
Once in 100 days, let f = 4
Once in 10 days, let f = 5
Once per day, let f = 6
Ten times per day, let f = 7
One hundred times per day, let f = 8.

Annual loss expectancy (ALE) is the product of impact and frequency. When using the values of f and i derived from the conversion tables (listed above), you can approximate the value of ALE by applying the following formula:

$$\text{ALE} = 10^{(f+i-3)}/3$$

No weighting factors have been introduced into the formula; the change is only for the purpose of accommodating the converted values. An even faster way to determine ALE is to use the matrix shown in Table 3-1, or alternatively, to develop one's own matrix. It would be impossible to list all the undesirable events that could plague any given security project. Most projects involving facilities, whether high-technology, refinery, manufacturing, or service, have more things in common than one would suspect at first glance (for example, fires on a cruise liner versus terrorism on the high seas). These common-origin problems have to be spliced or woven into the matrix along with events or occurrences peculiar to the particular analysis at hand. Terrorism and extortion on the high seas have, of late, become very serious problems for many U.S. ship owners, especially off the North African coast.

A thorough understanding of the elements that may affect frequency estimation is one of the keys to risk measurement. The following are some common elements that deserve consideration:

Table 3-1   Determination of Annual Loss Expectancy = ALE

Value of i	Value of f							
	1	2	3	4	5	6	7	8
1				$300	$3K	$30K	$30K	$300K
2			$300	3K	30K	300K	300K	3M
3		$300	3K	30K	300K	3M	3M	30M
4	$300	3K	30K	300K	3M	30M	30M	300M
5	$300	3K	30K	300K	3M	30M	300M	
6	3K	30K	300K	3M	30M	300M		
7	30K	300K	3M	30M	300M			

- Access—Is access difficult, limited, or open? Can an intruder gain access easily, or is it difficult? Can any employee do the same? What are the access criteria?
- Natural disasters—What kinds of natural disasters might realistically occur? To what degree would damage occur? How would it affect processing, storage, or supplies? How would loss of power or other utilities affect the entity?
- Environmental hazards—What special hazards are inherent in the operation? What is nearby? Are there any explosives, gasoline, or flammable objects in the area? Unused buildings next door? What can be the aftermath of fire? Water damage? Loss of stock, material? Proximity of fire and police departments?
- Facility housing—What protective devices are installed or can be installed? Anti-intrusion alarm systems, electronic access control systems? How is the building constructed? Type of roof? Sprinklers? What kind of flooring? What is flammable?
- Work environment—What is the relationship between personnel and management? (Loyal? Suspicious?) What are the aggravations of employees? Past labor history? How well do supervisors know employees? What is management's attitude toward employee dishonesty? (Condone? OK within bounds? Dismissal?) How open are lines of communication between employees and supervisors? Supervisors and upper management?
- Value—How much can an intruder profit? How much damage could result in the worst-case scenario? How much can a dishonest employee gain? How long before an intrusion will be detected? What is security response capability, time?

Principles of Probability

At this point, some statements about the nature of risk must be explained and considered. What we have stated so far is a simple approach to identifying and measuring risk. Risk is the possible happening of an undesirable event. An event is something that can occur, a definable occurrence. When an event happens, it can be described. Security counter-measures are designed to protect against harmful events occurring. For this reason, the question, "Is a system secure?" is meaningless. What should be asked is, "Is the system protected against all identified problems and incidents believed to be harmful?"

Any event can be described in at least two ways: it may be described in terms of the damage it will present if it occurs; or, it may be considered in terms of the probability of its occurrence. An identified risk, however, should be described in terms of its possibility of occurrence and its capacity for potential loss.

The study of the possibility of occurrence is known as *probability*. The principles that follow are based on philosophical (rather than mathematical) proofs derived in 1792 by the Marquis de Laplace in his *Théorie Analytique des Probabilités*. Excerpts from this classical treatise are reprinted below, in part. Laplace established 10 principles of probability, as quoted below.

1. Probability is defined as the ratio of the number of favorable cases to all possible cases.
2. If the cases are not equally possible, then the probability is the sum of the possibilities of each favorable case.
3. When the events are independent of each other, the probability of their simultaneous occurrence is the product of their separate probabilities.
4. If two events are dependent on each other, then the probability of the combined event is the product of the probability of the occurrence of the first event and the probability that the second event will occur given the occurrence of the first event.
5. If the probability of a combined event of the first phase and that of the second phase is determined, then the second probability divided by the first is the probability of the expected event drawn from an observed event.
6. When an observed event is linked to a cause, the probability of the existence of the cause is the probability of the event resulting from the cause divided by the sum of the probabilities of all causes.
7. The probability that the possibility of an event falls within given limits is the sum of the fractions [#6 above] falling within these limits.
8. The definition of *mathematical hope* is the product of the potential gain and the probability of obtaining it.
9. In a series of probable events, of which some produce a benefit and the others a loss, we shall have the advantage that results from it by making a sum of the products of the probability of each favorable event by the benefit that it procures, and subtracting from this sum that of the products of the probability of each unfavorable event by the loss that is attached to it. If the second sum is greater than the first, the benefit becomes a loss and hope is changed to fear.
10. Moral hope is defined as the relation between its absolute value divided by the total assets of the involved entity. This principle deals with the relation of potential gain to potential loss and describes the basis for not exposing all assets to the same risk.

Readers and students of this text have asked me to explain Laplace's theory in words of one syllable. Unfortunately, I am not astute enough to do so. The theory is merely set forth here for those readers who may desire a more precise methodology by which to arrive at probability determination in their unique environment or studies. For our purposes, however, the simpler the application, the better. Excessive refinement of the risk measurement process, in our opinion, would only serve to further delay the outcome of the project, without adding much benefit to the ultimate solution.

Probability, Risk, and Security

When security is defined as the implementation of a set of acceptable practices, procedures, and principles that, when taken as a whole, have the effect of altering the ratio of undesirable events to total events, the first principle and the importance of the probability theory become self-evident. The problem that security must constantly deal with is that all undesirable events are breaches of security! The goal of security design is to decrease the ratio of unfavorable events to total events. Obviously, some events are more likely to occur than others in the same environment. The risk of a flood that inundates a city would seem less likely than a transient power failure (tell this to the people of New Orleans, LA, and Nashville, TN). Both are undesirable events. Both can affect the operation of businesses. When the probability of each case is different, the ratios of favorable to unfavorable cases are altered accordingly.

Two events that have no relation to each other are considered to be independent events. If they are not linked in any way, the probability of their simultaneous occurrence is the product of their respective probabilities. An example: What is the probability of lightning striking a second time in the same spot? It is the same as the probability of lightning striking the first time: the two events are independent of each other and thus the probability ratio remains the same. In security, the penetration of a system and the simultaneous failure of the security system from causes other than penetration may be expressed as the product of the probabilities of the independent events. This fits the condition of Laplace's principle #3, above. Many security (and safety) systems, such as those employed at nuclear power facilities, are based on redundancy, such that if multiple failures occur, the redundant systems do not become operational until the preceding systems have failed. Principle #4 expresses the relation between dependent events: the probability of the first event is multiplied by the probability of the second event if the second event can happen only after the first event has occurred. Breaking and entering followed by theft, to produce a burglary, is an example of this theory.

The probability of security system failure may be expressed in terms of the lower-risk multiple or backup systems. As happens when two events are combined, principle #5 expresses the idea that when dealing with events, the past does not affect the future. If we assume the risk of a security breach is a given value and that it has occurred, we may not assume that it will not occur again. Probabilities of events are not guarantees. If an event has a probability of 1 in 100, the probability of that event happening again is still 1 in 100. For example, tossing a coin for heads or tails is a 50–50 proposition; that is, 1 time out of 2 it should come up heads. A coin toss could come up heads 10 times in succession;

however, as the past cannot affect the present or the future, the chance of heads coming up each time is still 50–50, and it will remain so on every toss of the coin.

Principle #6 deals with the attribution of causes to effects. It describes the relation between all causes and probable causes. This is effectively the expression of circumstantial evidence, as a probability leading to a conclusion but one less convincing than the conclusion reached using direct evidence. Principle #7 involves the basis of confidence limits. To illustrate, if a random sample of 100 variables is taken and is found to have a mean of 40 and a standard deviation of 11, it will not be possible to determine a precise mean. The best that can be established is limits within which the mean will fall with a specified probability or confidence, usually taken as 95 percent. Again we need to ask ourselves the question, "How precise a measurement do we need?" Security is, after all, an art and not a science.

The definition of mathematical hope is essential to the design of a secure system. This concept relates the potential gain to the probability of obtaining the gain. Principle #8 allows the utility of a procedure to be expressed in both monetary and probabilistic terms. If the potential gain from a security system was $1,000 and the probability of achieving this gain was 1 in 500, a value of 2 (in arbitrary units) could be assigned. Equivalent values could be assigned to other combinations to allow comparison among alternatives. But why go to all this trouble for so little gain?

Principle #9 allows for the fact that any solution to a problem introduces risk. Risk management solutions may fail, and this must be considered in the design stage. A backup system to provide redundancy is certainly to be considered, as well as the cost/benefit ratio for doing so. This principle is extensively used in the manufacture of commercial aircraft and the space shuttle, for obvious reasons. The cost of providing redundancy in a security system is generally more cost effective than buying an insurance policy to cover a failure.

The condition to be considered last is the situation in which one of the alternatives to positive action is simply to do nothing. In some cases the risks, upon analysis, become insignificant; the decision may be to accept the possibility of loss because the potential loss will not have a substantial effect on the client's assets. Principle #10 relates the amount and potential of risk to the wealth of the protected entity. A very profitable company may well afford to risk assets to maximize gains in profitability. The potential losses might be too great, however, for a less prosperous company, one that may be in greater need of relief from a catastrophe, such as a major hurricane or even an oil spill from a tanker at sea. In some instances, then, the most cost-effective security is simply not to implement a plan or solution; in some others, the most practical solution might be to cover the potential loss with some form of insurance.

To summarize, risk can be expressed in terms of probability of occurrence. The goal of security system design is to improve the ratio of favorable events to total events, or to reduce the ratio of unfavorable events. The basic technique used is to rate risks based upon their probability of occurrence and to establish economic values for potential risks and potential solutions. When possible (and cost-effective), redundant or backup systems may be designed to provide an added dimension of protection. Risk probability determination is not a guarantee that, because an event has a low probability and has occurred only one time, that it will not occur again or perhaps multiple times in the future.

Statistical analysis as it is used in many fields—astronomy, agriculture, engineering, or insurance—is approached by using the same procedures enumerated above.

Again, a word of caution: No statistical procedure can, in itself, ensure there will be no mistakes, inaccuracies, faulty reasoning, or incorrect conclusions. The data must be accurate, the methods properly applied, and the results interpreted by someone with a thorough understanding of the field in which the information obtained is being applied. That, after all, is the hallmark of the professional.

Estimating Frequency of Occurrence

When experience (history) has provided an adequate database, loss expectancy can be projected with a satisfactory degree of confidence. For example, if one leaves the keys in the ignition of an unlocked car on a downtown street in a high-crime area, it is just a question of time until the car is stolen. Likewise, if the same bank is robbed on numerous occasions, the chances that it will continue to be robbed again and again is easy to predict.

In new situations, however, or in situations in which data have not been or cannot be collected, we have insufficient knowledge on which to base our projections. An example would be the kidnapping of a high-risk-profile businessperson in the absence of any prior threats or other indications that he had been targeted for kidnapping. In such instances, quantification of risk tends to be nothing more than educated guessing. It is in cases such as this that the services of an experienced security professional are needed to reduce subjectivity to an absolute minimum and to deal with the data available, limited though it may be, in a calm, objective manner. This is also true of international and domestic terrorism and, to a lesser degree, workplace violence. Amateurs may become emotional—that is to say, less objective—when faced with unfamiliar and often dangerous situations. The services of an outside consultant or a security professional experienced in such matters are essential in cases of this type, to ensure objective and proper analysis from the outset. Another example is in dealing with the threat of terrorism. Terrorists engage in seemingly random acts of violence. It is difficult, if not impossible, to estimate the frequency of occurrence of random acts of violence. By their very nature, they are highly unpredictable. So, we are left with the nagging thought that "risk analysis" is not an exact science, but an art, and we still have many challenges ahead of us, especially in dealing with the activities of terrorists.

4

Quantifying and Prioritizing Loss Potential

There is no such thing as a "cookbook" receipt solution to security problems. The most effective programs are tailor-made.
—**John J. Janson, Jr. PhD, CPP (deceased)**

As with any complex chain of interrelated issues, overall strength is measured by the weakest link. Very strong security in one area will not compensate for very weak security in another. To proceed to correct conclusions and then to recommendations for corrective action, it is necessary to quantify and prioritize all the loss potentials identified. For the professional, here lies one of the most difficult tasks in the survey process—the task of measurement, or quantification, of exposures. Given adequate historical or empirical data, loss expectancy can be projected with a satisfactory degree of confidence. On the other hand, when there are insufficient data for reliable forecasting because the data either have not or cannot be collected, one is left with the nagging suspicion that conclusions will be nothing more than an exercise in educated guessing—not that there is anything wrong in that, especially when there is little or nothing to work with.

Many risks may be classified as things that might happen but that have not yet occurred. Such risks can either be accepted or minimized, using prescribed preventive measures. Acceptance assumes that the risk is not sufficiently serious to justify the cost of reduction, or that recovery measures will ensure survival, or that cessation of operations, if the risk should occur in its most serious magnitude, is an acceptable alternative. Minimizing the risk assumes that the risk exposure is or may be serious enough to justify the cost of eliminating or reducing the possibility of its occurrence, and that recovery measures alone will not always be effective in ensuring survival. Also, it postulates that the remaining alternative—cessation of operations—is unacceptable.

It is at this juncture that quantifying or prioritizing the loss potential becomes the hallmark of the true professional. We have often told clients that it does not take much talent to prescribe an 85 percent solution for a 15 percent problem. Real talent comes into play when one is able to diagnose client problems correctly and recommend necessary countermeasures to solve them without engaging in overkill. Granted, when we err, it must be on the side of prescribing more rather than less security, but not to the level that turns a college campus, hospital, or resort hotel into a prisoner-of-war camp.

There are always several tradeoffs when one considers the implementation of a new or improved security program. Cost is the most obvious. Less obvious and often overlooked are the inconveniences new security systems cause to personnel and the probable impact on employee morale. This is especially true if the employees perceive (rightly or wrongly) that the inconvenience caused them is greater than the threat. This type of "solution" can cause more harm than if management had done nothing at all. As an example, we observed that when excessive access-control systems were installed in a computer department of an airline reservations center, it resulted in employees propping doors open for simplicity of movement during working hours. When queried, employees said, "The inconvenience was a bigger problem than unauthorized access." In their opinion unauthorized access was controlled at the entrance to the facility. Access control to the computer room was "overkill," especially during work hours when the room was occupied by the computer staff!

Assessing Criticality or Severity

Some authors refer to this stage of the survey as assessing criticality or severity of occurrence.[1] Regardless of what one calls the process, it is vital to search for and locate the proper benchmark to adequately approximate dollar values for the loss probabilities previously identified. Once this is done, the task of comparing the cure to the disease becomes self-evident. One can then develop a list of meaningful solutions with priorities based on a common denominator—the dollar. One technique in use is the three-stage approach, involving prevention, control, and recovery. Prevention attempts to stop undesirable incidents before they get started. Control seeks to keep these incidents from affecting assets, or, if impact occurs, to minimize the loss. Recovery restores the operation after assets have been adversely affected.

Many professionals take the approach that prevention is sufficient, and yet they opt for the installation of various control measures. It is one thing to install fire alarms that signal a serious situation; it is another to respond to and control a fire, and then recover from its effects. Similarly, it may behoove corporate management not only to have adequate security in place to prevent kidnapping attempts but also to design a contingency plan to deal with the kidnapping event, should preventive measures fail and the event become an actuality. In addition, nothing mentioned previously—prevention, detection, control, or contingency planning—precludes the necessity of having adequate insurance to help recover from a serious fire or successful kidnapping, extortion, and ransom event.

Another technique for assessing security is to prepare a segmented schedule of overhead, installation, and operating costs for the security project. All costs identified must be directly chargeable to expected benefits. In this process, it is crucial to show that the benefits (risk prevention or reduction) will outweigh the cost. This is referred to as a cost/benefit ratio and is useful for both existing and proposed security programs and projects. (This will be more fully discussed in Chapter 5, Cost/Benefit Analysis.)

[1]Walsh, T.J., Healy, R.J., 1974. *The Protection of Assets Manual*, Santa Monica, CA, Merrit.

The Decision Matrix

Another simple technique for prioritizing loss potential is the use of a frequency and severity loss matrix as an aid in making decisions about handling risk. Table 4-1 uses the adjectives *high*, *medium*, and *low* as factors to measure both frequency and severity of loss.

The quantification and prioritizing of loss potential should take into account the fact that there are both "intuitive" security control concepts, such as the installation of a burglar alarm at a warehouse, and security control concepts based on detailed cost/benefit analysis. An example of the latter is a multiple-stage electronic card-access control system for the research and development laboratory of a computer chip manufacturer. The procedures for both approaches take into full consideration the following:

- Available information resources
- Reliable probability relationships
- Minimum time and resource requirements and availability
- Maximum incentives for management cooperation
- A realistic evaluation of existing or planned security control effectiveness

The type of protection designed must always be tailored to the specific risk in the real day-to-day working environment of the specific entity being studied. The application of controls simply because they are recommended by some vague standard or acceptable practice, without regard to risk in the real-world environment, often results in controls that are inappropriate, ineffective, and costly. Worse, as is so often seen with inappropriately planned closed-circuit television (CCTV), such controls may generate a false sense of security on the part of management. Remember, CCTV only records what is or has happened; it does little to prevent an unfavorable event from happening. The most obvious example is CCTV in banks, which do a great job of recording the robbery but do little to prevent it from happening.

Another example: I was once asked to review the installation of a CCTV security system for a newly constructed newspaper plant in Southern California. The CCTV system

Table 4-1 Decision Matrix: A Risk-Handling Decision Aid

Severity of Loss	Frequency of Loss		
	High	Medium	Low
High	Avoidance	Loss prevention and avoidance	Transfer via insurance
Medium	Avoidance and loss prevention	Loss prevention and transfer via insurance	Assumption and pooling
Low	Loss prevention	Loss prevention and assumption	Assumption

had been designed for the corporation by a building and facilities engineer. The plant was located in a newly developed industrial park. The CCTV system had apparently been planned without regard to environmental considerations. Upon review, it was determined that the CCTV system—complete with zoom, tilt, and pan lenses as well as videocassette tape-recording functions, all very costly to install and maintain—was operating in an area that had heavy fog about 6 months of the year during nighttime and early-morning hours. Further, the fence line, at its nearest point to the building, was 350 yards away from the closest camera lens! To the question, "Why install CCTV at this location?" the answer was, "We have used CCTV successfully at all our other plants and it just seemed the natural thing to do here."

They did have CCTV at their other plants, but in the plants that I inspected, the CCTV systems were more often than not inoperative, in whole or in part, because of inadequate maintenance and repairs needed on cameras, monitors, and videorecording units. The CCTV systems were regarded by operations personnel as expensive toys that added little to the security of their facility. Management, however, was proud of its security program, having been lulled into a false sense of security by the presence of the CCTV cameras.

The assessment of risk, using actuarial methods to handle large numbers of events or situations, will be further examined in the following chapter. This technique has become generally reliable. However, in our experience, the entire exercise of estimating risks for a specific installation or complex is at best imprecise. Defining risks by using highly specific numbers has not always been validated by experience. Several well-known authorities have concluded that order-of-magnitude expressions, such as low, moderate, and high, to indicate relative degrees of risk are more than adequate for most risk analysis surveys.

The terms *low, moderate,* and *high* equate roughly with probability ranges of 1 to 3, 4 to 6, and 7 to 10, respectively. One is cautioned here to remember that even a low risk should be taken seriously if the potential damage (or danger) is assessed as being moderate to high. An example would be the kidnapping for ransom of a high-profile business executive in a foreign country. One may calculate the risk to be nearly nonexistent (low), but the potential danger to the employee and the impact on the company of such an unfavorable event should always be rated as high.

5

Cost/Benefit Analysis

Violence assessment, like all forms of risk assessment, guides the use of limited resources (time, budget, and personnel) to maximize benefit.

—James S. Cawood, CFE, CPP, CPI, *Violence Assessment and Intervention: The Practitioner's Handbook*, CRC Press, 2003

When we use the approach recommended in the earlier chapters to conduct security surveys, problems (risk) are properly identified, analyzed, and quantified in terms of the seriousness of their impact on the operation or the facility being studied. Only solutions specifically responsive to a demonstrated need or requirement are considered. Only those tools or techniques that perform the needed task most effectively at the least possible cost are designed and then introduced into the system. Efficiency versus cost is, then, the first phase of balancing the cost/benefit ratio, which is essential to the proper development and design of effective security countermeasures.

The following techniques are suggested as an effective way to analyze, develop, and design cost-effective security solutions. These include policy, procedures, hardware (electronics), and manpower utilization programs.

System Design Engineering

If a facility has a security program, an experienced security professional should be able to review it and make recommendations to consolidate, coordinate, upgrade, and improve the existing protection program. If a security program does not exist, security professionals should be able to design one that properly marries the best of the following into a comprehensive and cost-effective security operation:

- Written policy, procedures, and guidelines
- Hardware (lights, lock and key controls, card access, anti-intrusion alarms, to name a few)
- Manpower (guard service or proprietary security personnel)

It is important here to refer the reader to the concept of "synergistic effect." Simply put, if we consider the above three items, each one standing alone will provide some security. All three taken together and woven into a strong cable-like system will provide synergism to the security program. What does this mean? If we consider each of the above alone, they add up to three. If we weave them all together into a synergistic program, they will add up to be greater than the sum of the individual parts. Obviously, for this effect to occur the individual parts must complement and not conflict with one another.

This review can be accomplished by using any one of the many comprehensive review techniques available, such as the one set forth in Appendix A, "Security Survey Work Sheets." By asking questions and directly observing critical operations, such as the adequacy of lights at night in an employee parking lot adjacent to a plant, one can reach certain definite conclusions. Elements of the security program (lights, locks, alarms, guard coverage, and so forth) are adequate, inadequate, or nonexistent. Obviously the two former situations are a subjective matter, whereas the latter leaves no room for argument.

Other, more sophisticated programs use advanced electronic techniques. An example is electronic filtering as applied to access control in security programs for highly sensitive environments, such as research and development facilities. Additionally, we are seeing ever-increasing use of computer programs with models to conduct cost/benefit ratio and computer-aided design (CAD) analysis. Whatever technique is used to determine the cost/benefit ratio, three basic criteria should be considered before the procedures or countermeasures are selected: cost, reliability, and delay.

Cost

Initially, one tends to focus primarily on the acquisition cost. We must, however, also take into consideration life cycle and replacement cost factors as well. For example, the initial base cost to re-core and re-key the locks of an entire facility may be $15,000, plus 10 percent per year for inflation, or $22,500 after a lapse of 5 years.

For a key-and-lock system that includes changeable cores and an in-house capability for cutting keys, the initial capital outlay of $15,000 may well be amortized over 15 to 20 years instead of the original 5 years. Here we have considered all three factors: acquisition cost, the life of the system, and replacement cost. The cost of contract versus proprietary security guard service, or a combination of the two, can also be calculated, using the same principles. The same exercise can work for any type of hardware or electronics equipment with only a slight variation in computation.

In some operations we have observed manpower costs (security personnel) comprise about 85% of the total security budget. Also, manpower costs will continue to increase every year; they do not amortize, as does a capital expense. The planner must consider, wherever possible, how to reduce or eliminate manpower and replace it with a more cost-effective solution.

Reliability

Reliability is especially critical with hardware and electronic devices, such as anti-intrusion and computerized card-access control systems. The state of the art in electronics systems is advancing faster than the speed of sound. Consequently, components are being manufactured, sold, and installed before being properly field-tested. This inevitably leads to difficulties—difficulties that, unless corrected immediately, translate into expensive electronics systems designed to solve problems that, in turn, create new problems, which are often difficult and time-consuming to correct. Over the years since this book was first written, electronics design problems in security systems have diminished. Nevertheless, any system that is installed without first being properly field tested can create problems. I know of only one solution to keep this problem from dragging on for months: build two written requirements into the contract of proposal or purchase. First, require the successful bidder to present for your inspection a site demonstration of a like system or unit installed on a property with the same or similar security or access-control needs. This should be one that has been functioning for a minimum of 6 months without problems. Second, include a clause in the purchase contract withholding the last payment (payments are usually made upon signature [one third], upon installation [one third], and at final system acceptance) until the system has been on line for a sufficient test period, say 90 days, to ensure that the product is problem free and all bugs have been located and eliminated. Such a clause will make it in the supplier's best interest to get the program fully operational as quickly as possible.

A word of caution: no legitimate hardware or electronic security systems supplier will object to the inclusion of the above protective clauses. A supplier that balks at either or both, or cannot or will not guarantee satisfactory results for his or her system, should be avoided. This being the case, it would probably be in the best interest of the buyer to look for a supplier who will agree to your terms and conditions. Nothing is sadder to behold than a well-meaning director of security who has finally convinced management of the necessity for an expensive security system that, after acquisition, continually malfunctions. Acquisition cost, like the state of the art, is ever changing. Sophisticated, computerized security systems can be extremely expensive.

Delay

Time is the third factor to consider. How long will it take, in comparison to other countermeasures that could be used, before the recommended system will become fully operational? Here we may well have to consider the possibility of having more than one countermeasure operating at the same time until the primary system renders the secondary system obsolete or less critical. This is called redundancy.

An example is the introduction of a multilevel, electronic anti-intrusion and card-access system in a facility that houses separate functions under one roof. Some of these functions—such as the research and development laboratories, and the storage area in a warehouse, which contains a $6 million inventory of easy-to-conceal and easy-to-sell items—may require differing levels of protection at different times.

It is probably going to be necessary to incur manpower costs for a guard force to secure the premises 24 hours a day, 7 days a week, until the new access system becomes fully operational. This amounts to 168 hours of guard service coverage. If each guard works a 40-hour shift, it will take 4.2 guards (taking into account time for relief, sick time, vacation, and so forth) to accomplish this task. If the cost for this service is $10.81 per hour, the total cost is as follows:

$$168 \text{ hours} \times 10.81 \text{ per hour} = \$1,816.08 \text{ per week}$$

$$(\$259.44 \text{ per day per post})$$

or

$$\$8,042.64 \text{ per month, or } \$94,436.16 \text{ per year (52 weeks)},$$

or

$$\text{average monthly cost } \$7,869.68^{1}$$

Needless to say, a prudent, cost-conscious director of security will make every effort to phasein the anti-intrusion card-access system and phaseout or reduce the manpower requirements as quickly as possible.

Building Redundancy into the System

To achieve very high levels of reliability in security programs, one should consider building redundancy into them. An example is the use of multiple smoke sensors in a hotel or housing areas to warn of the incipient stages of a potentially disastrous fire. The old saying, "One may well survive a burglary, but a good fire can put you out of business forever," has much meaning for a hotel complex, an accounting and billing office, or computer library. If we install multiple-use smoke detector sensors (ones that use all three detection techniques—ionization, infrared, and the photoelectric cell), the chance of all their modes failing at the same time is statistically 10,000 to 1. Yet in terms of cost, these units can be generally obtained and installed for about $50.00 each—a small outlay compared with the cost of replacing even the least expensive computer equipment, not to mention the potential catastrophe of the loss of life or of losing the materials stored in an accounting and billing office or a computer library, as a result of a fire that could have been detected in its early stages.

Redundancy can also be accomplished by designing a proprietary alarm system, which for a few dollars more can be remotely monitored by a central alarm station as an added backup against the possibility of a power or human failure at the facility's proprietary alarm console. The costs—a dedicated lease line and a rental (monitoring) fee—are relatively inexpensive.

In art museums, multiple redundant systems usage is a common technique. The items on display in an art museum must be readily available for the viewing public;

---

[1] The average monthly cost is arrived at by dividing annual dollars by 12 months. For more specific and detailed costing, the daily rate times the number of days in a given month may be used. Thus, $259.44 \times 31 = \$8,042.64$. For accounting purposes, one can use the same formula to compute the cost of any guard service contract. According to the Bureau of Labor Statistics, U.S. Department of Labor (May 2009), contract security guards earn between $8.10 and $19.34 per hour, with $10.81 as the median and $11.76 as the mean.

thus, security must be as unobtrusive as possible. Many of these art objects, however, are one-of-a-kind treasures and are therefore priceless. Security in the daytime, functional though minimal, may include uniformed security guards, closed-circuit television, anti penetration display case alarms (local, audible, and remote), and anti tamper or anti removal switches behind picture frames or on wall mounts. For night-time security, additional multiple anti-intrusion sensors and motion detectors may be employed to ensure that if an unauthorized penetration of the perimeter barriers happens to occur (such as a remain-behind burglar), the movement and the presence of such a person inside the museum will be detected by multiple sensing devices. Any one of them can be circumvented or might not be fully operational; nevertheless, the odds against all of the devices failing at the same time and therefore not detecting the presence of an intruder are statistically in the thousands (perhaps even in the millions), and there-fore the total failure of all systems at once is virtually impossible. It should be noted here that many art museums have systems installed that are designed to capture an intruder inside the museum, thus negating the possibility that a burglar would accomplish the theft and then successfully flee the premises before the security force can react.

The selection of the right countermeasure to control or minimize the identified risks will take into consideration the fact that written policy, procedures, and guidelines are less expensive than hardware, and hardware (including electronics) is less expensive than manpower. So when looking for the proper "fix," it is well to start with the basics and then work one's way up to the more sophisticated and thus more expensive countermeasures.

As an example, there may be no adequate substitute for the use of mechanical equipment, fences, gates, locks, safes, and vaults. In some locations, however, local real estate codes, covenants, and restrictions (CC&Rs) may prohibit the installation of a 7-foot chain-link fence with the usual three-strand barbed-wire top overhang. In these instances, not uncommon in industrial parks, one must retreat to the exterior wall of the building or buildings as the place to begin perimeter security. It may then become nec-essary to use electronics, alarms, a computerized console, closed-circuit television, and exterior and interior security guard patrols, among other systems, to provide the requisite security countermeasures.

A Security Countermeasure

If the previously described problem is encountered in a new industrial park, one possible solution is to prepare the security countermeasure plan in stages or increments, such as stages 1, 2, and 3, or increments A, B, and C. In this technique, we "cost out" the use of each of the required systems in terms of the minimum level of security that one or more countermeasures will provide. Then, we move up to stages 2 and 3, adding more security countermeasures to the complex plan, adding ever-increasing cost to the project.

This type of program is relatively easy to explain and therefore to sell to management. The underlying philosophy is, "We will try stage 1 or increment A first, at $X. Should stage 1 (increment A) fail to provide the needed level of security, we will move to stage 2 (increment B), and so on, until we solve the problem." This technique will prevent the all too commonly encountered security "overkill" situation. A security risk properly assessed to be in the 15 percent range does not need an 85 percent solution, nor can management afford one. Yet given a brand-new environment, such as a recently developed industrial park in a suburban area, who can predict in advance exactly what level of security will be needed? In the absence of empirical knowledge to the contrary, our systems counter-measure design technique is both efficient and cost-effective. What it says to top management is, "We can start out with the basics, which are the most effective for the least money, and then add to the system (at greater cost) as we develop the necessary historical data to justify spending more money." A security professional who can save money while still doing his job can usually count on job security.

Most cost-effective security systems use a combination of procedures, hardware (electronics), and manpower to achieve the proper countermeasures balance. The first edition of *Risk Analysis and the Security Survey*, in 1980, made the following prediction:

In the next decade we will see more use of security systems that integrate many separate functions, systems that are developed, manufactured, sold, installed, and maintained by one company. These systems will integrate security, communications, fire, life safety, building management, and energy control from one central console or command control center. The big users will be high-rise office buildings, retail stores, and shopping malls. Specialized units will be developed for use at airports, oil refineries, and electronics manufacturing plants, to name a few.

This prediction became an actuality within 5 years. With increased attention now being paid to security as a result of the continuing threat of international and national terrorism, it is anyone's guess where we will be 10 years from now. What one can predict, however, is an ever-increasing presence of security in our daily lives. The public will demand nothing less.

In conclusion, even the best-designed countermeasures system must be proved cost-effective before it can be sold to management. Only by reducing or integrating the largest cost factor—manpower—and replacing it where practical with procedures, hardware, and electronics can we achieve more effective security at less cost. This is a proven technique. Finally, whenever dealing with security vendors, whether for manpower, hardware, or electronics, one should obtain a minimum of three bids based on a written specification of requirements. Not only are competitive bids a proven cost-effective technique, but they also tend to keep everyone honest. Bidders should be advised of the competition, but not necessarily the identities of the other competitors. Sole-source procurement is seldom cost-effective and more often than not provides fertile ground for financial manipulation, which is seldom in the client's best interest. In conclusion, one should always keep in mind one of Charlie Hayden's cardinal rules for conducting security surveys: "The optimum reduction of risk occurs at that point at which further reduction would cost more than the benefits to be gained."

6

Other Risk Analysis Methodologies

He kissed me and our young daughter good-bye. I never predicted or imagined
that he was involved in such horrific activities. He had a kind, caring,
and calming presence about him.

—Samantha Lewthwaite describing her husband Germaine Lindsay, a suspected suicide
bomber who killed 26 people in the London King's Cross attack.

Policy on Critical Infrastructure Protection. It identified 8 sectors of the economy considered critical to national security. Included are telecommunications, transportation, water supply, oil and gas production, banking and finance, electrical generation, emergency services, and essential government functions. This Directive, along with the Bioterrorism Act and other implementing policies, assigned oversight of each function to a separate governmental agency. The protection of the water supply is the responsibility of the Environmental Protection Agency (EPA); the protection of the food supply is the responsibility of the Food and Drug Administration (FDA), for example. These agencies are assigned the task of developing risk assessment and security protocols for the protection of the assets under their purview, with many using a different risk assessment methodology. The EPA uses VSAT, an acronym for the Vulnerability Self Assessment Tool, which is both a methodology and a software tool used to develop security systems capable of protecting specific targets from the acts of specific adversaries.[1] The FDA uses Operational Risk Management, an engineering-based risk management system used by the Federal Aviation Administration and the military to examine the safety and risk to existing systems. It is a tool designed to help identify operational risks and benefits to determine the best course of action for any given situation.[2] The U.S. Department of Agriculture (USDA) Food Safety and Inspection Service (FSIS), in conjunction with the Homeland Security Office of Food Security and Emergency Preparedness (OFSEP), adapted the U.S. Department of Defense's military targeting method called CARVER + Shock to reduce or eliminate the potential risk at vulnerable points along the "farm-to-table" continuum.[3]

In spite of the existence of different risk and vulnerability analysis methods, the basic risk analysis method described in earlier chapters has not changed. The terms *risk analysis* and *vulnerability assessment*, although often used interchangeably, are similar

---

[1] See http://water.epa.gov/infrastructure/watersecurity/techtools/

[2] See http://www.faa.gov/library/manuals/aviation/risk_management/ss_handbook/media/chap15_1200. pdf et. al.

[3] See http://www.fda.gov/Food/FoodDefense/CARVER/default.htm

but different. A vulnerability assessment, according to the U.S. Department of Energy, identifies assets and capabilities in a system, assigns a quantifiable value, identifies vulnerabilities or potential threats to each asset, and eliminates or mitigates the most serious vulnerabilities for the most valuable resources.

Quantitative Risk Analysis assigns the probability of the occurrence of identified hazards and determines their impact or consequence, usually resulting in a value such as an Annual Loss Expectancy or Annual Cost. Qualitative Risk Analysis, more akin to risk assessment or vulnerability analysis, concentrates less (or not at all) on probability and looks at threats (similar to hazard identification: things that can go wrong or attack the system, facility, or assets), vulnerability (things, such as assets, that are prone to attack), and consequences or controls (i.e., countermeasures). Risk assessment often involves the evaluation of existing security and controls and rates their adequacy against threats to the organization. Vulnerability analysis, like risk analysis, drives the risk management process. The methodologies mentioned above are more vulnerability analysis than risk analysis.

National Infrastructure Protection Plan

The Homeland Security Presidential Directive-7 (HSPD-7), Critical Infrastructure Identifica-tion, Prioritization, and Protection, established a national policy for federal departments and agencies (18 sectors) to identify and prioritize critical infrastructures and to protect them from terrorist attacks. Each sector-specific agency is responsible for developing and implementing a sector-specific plan that details the application of the National Infrastructure Protection Plan (NIPP) to the specific characteristics of that sector—that is, water, transportation, food, and so on. The 18th sector is identified as "Critical Manufacturing." This risk management framework provides a common structure to support preparedness, deterrence, prevention, mitigation, response, and recovery. It is applied to terrorist threat scenarios as well as to natural disasters. Organizations that are asset intensive use a bottom-up approach, but the methodology allows a top-down progression for other types of organizations. Information sharing, collaboration between government and private organizations, is a major strategy in the implementation of the NIPP. To achieve its goal of producing scenario-based consequence and vulnerability estimates and the likelihood the scenarios will materialize, DHS recommends thinking of "risk as influenced by the nature and magnitude of a threat, the vulnerabilities to that threat, and the consequences that could result."[4]

This framework consists of the following steps (see Figure 6-1):

- Set goals and objectives
- Identify assets, systems, networks, and functions

[4] Ibid.

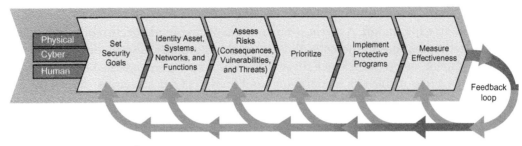

FIGURE 6-1 NIPP framework.[5]

- Assess risk based on consequences, vulnerabilities, and threats
- Establish priorities based on risk assessments
- Implement protection programs and resiliency strategies
- Measure effectiveness

Set Goals and Objectives

The organization begins by deciding what the measurable end product will look like from a risk management perspective. What does the organization wish to achieve by this process is a question that should be answered. The reduction of vulnerabilities, the implementation of risk management strategies, and secured funding are high-level examples.

Identify Assets, Systems, Networks, and Functions

An asset is a person, structure, facility, information, material, or process that has value to the organization. The Department of Homeland Security (DHS) maintains a database of national critical assets, but for the assessment, individual organizations need to identify their own unique or specific assets.

Assess Risk Based on Consequences, Vulnerabilities, and Threats

The NIPP Framework examines the effects of consequence, vulnerability, and threats against the identified assets. Consequence is defined as the public health and safety, direct and indirect economic impact, psychological, and mission impact caused by an incident. Vulnerability is any physical feature or operational attribute that renders an entity, asset, system, network, or geographical area open to exploitation or susceptible to a given hazard, or a characteristic of design, location, security posture, or operation that

[5] National Infrastructure Protection Plan 2009.

renders an entity, asset, system, network, or geographical area susceptible to disruption, destruction, or exploitation. A threat is any natural or manmade occurrence, individual, entity, or action that has or indicates the potential to harm life, information, operations, the environment, or property. Once these three components are assessed for their likelihood against a reasonable risk scenario (or combination of risk scenarios), it uses the following formula to produce a risk estimate: $R = f(C, V, T)$; in other words, risk is a function of consequence, vulnerability, and threat. Scenarios are similar to hazard identification but describe the conditions, constraints, and details of the event at its maximum credible magnitude. An example would include a chemical attack listing the substance, concentration, duration, mode of delivery, and potential attacker, or another example is the effects of a Category 5 hurricane.

Core Criteria

The framework allows sectors to use current and emerging risk analysis and vulnerability methodologies but suggest they follow these guidelines to maintain consistency and allow for a better ranking of risk across all critical assets as long as the analysis is documented (adequately described), and the results are reproducible, defensible (based on sound risk analysis principles), and complete (assesses C, V, and T for each scenario):

CONSEQUENCE
- Document the scenarios assessed, tools used, and any key assumptions made.
- Estimate the number of fatalities, injuries, and illnesses, where applicable and feasible, keeping each separate estimate visible to the user.
- Estimate the economic loss in dollars, stating which costs are included (e.g., property damage losses, lost revenue, loss to the economy) and what duration was considered.
- If monetizing the human health consequences, document the value(s) used and the assumptions made.
- Consider and document any protective or consequence mitigation measures that have their effect after the incident has occurred, such as the rerouting of systems or HAZMAT or fire and rescue response.
- Describe the psychological impacts and mission disruption, where feasible.

VULNERABILITY
- Identify the vulnerabilities associated with physical, cyber, or human factors (openness to both insider and outsider threats); critical dependencies; and physical proximity to hazards.
- Describe all protective measures in place and how they reduce the vulnerability for each scenario.
- In evaluating security vulnerabilities, develop estimates of the likelihood of an adversary's success for each attack scenario.

- For natural hazards, estimate the likelihood that an incident would cause harm to the asset, system, or network, given that the natural hazard event occurs at the location of interest for the risk scenario.

THREAT
- Account for the adversary's ability to recognize the target and the deterrence value of existing security measures.
- Identify attack methods that may be employed.
- Consider the level of capability that an adversary demonstrates with regard to a particular attack method.
- Consider the degree of the adversary's intent to attack the target.
- Estimate threat as the likelihood that the adversary would attempt a given attack method against the target.
- If threat likelihoods cannot be estimated, use conditional risk values (consequence times vulnerability) and conduct sensitivity analyses to determine how likely the scenario would have to be to support the decision.
- Use best-available analytic tools and historical data to estimate the likelihood that natural disasters and accidental hazards will affect the organization.

Establish Priorities Based on Risk Assessments

In an effort to best direct resources, response, and recovery operations, DHS ranks the critical assets of the sectors that face the greatest risk so that mitigation is addressed in a cost-effective manner, based on the highest return on investment or those that would suffer the greatest loss among similarly ranked risks. Requirements based on regulations are also considered.

Implement Protection Programs and Resiliency Strategies

Measures that prevent, deter, mitigate, and reduce vulnerability and that enable response and recovery are implemented in a manner that is comprehensive, coordinated with other sectors and stakeholders, cost-effective, and risk informed (decisions based on risk analysis and in consideration of other pertinent factors).

Measure Effectiveness

DHS examines metrics of the sectors to track performance and progress toward their priorities and strategic goals to check effectiveness of the programs and protective strategies as well as to provide for continuous improvement. Every organization should use good management and quality control initiatives to do the same.

Review

The Department of Homeland Security evaluates risk in terms of $R = T \times V \times C$, where risk (R) is the product of the threat (T), vulnerability (V), and consequence (C). Assets are identified, and the consequences, vulnerability, and threats against these assets are evaluated against credible attack or hazard scenarios. Other forms of risk assessment used by different infrastructure entities adapt their methodologies to conform to the guidelines established under the National Infrastructure Protection Plan Framework.

7

The Security Survey: An Overview

A security survey is a critical, on-site examination...to ascertain the present security status, identify deficiencies or excesses, determine the protection needed, and make recommendations to improve the overall security of the operation.
—**Raymond M. Momboisse, *Industrial Security for Strikes, Riots and Disasters*, Charles C. Thomas Publishers, 1977**

The goal of risk management—to manage risk effectively at the least possible cost—cannot be achieved without eliminating or reducing, through a total management commitment, the incidents that lead to losses.[1] Before any risk can be eliminated or reduced, it must first be identified. One proven method of accomplishing this task is the security survey. Charles A. Sennewald, author and security consultant, has defined the security survey as follows: "The primary vehicle used in a security assessment is the survey. The survey is the process whereby one gathers data that reflects the who, what, how, where, when, and why of the client's existing operation. The survey is the fact-finding process."[2]

Why Are Security Surveys Needed?

There are reports published by the Association of Certified Fraud Examiners (ACFE) estimating that the cost of fraud and financial abuse to American business was in excess of $994 billion per year in 2008 and rising. This figure is believed by most authorities to be very conservative. The sad fact is that no one organization is capable of collecting all the data available concerning fraud. As an example, in America we have an alarming trend in Medicare fraud, costing taxpayers untold millions of dollars. The biggest problem, and the one seen most often by fraud investigators, is that most corporate managers do not know if they have theft problems. Worse, many do not even want to know that they have a problem with employee theft! Some managers seem to prefer to keep things as they are and to regard any suggestion of the need for increased security as a direct or indirect criticism of their ability to manage their operations. We hope that this attitude has changed for the better as a result of the downturn in the global economy. In times of economic difficulties anything that affects the bottom line (profits) is not tolerated. Nevertheless, where fraud exists, most business fraud surveys calculate losses at about 6 percent of annual revenue. Some surveys we have seen reported have concluded that losses attributable to employee theft (internal theft) equal or exceeded profits! This is especially

---

[1] The field of risk management encompasses much more than security and safety. These two subjects, along with insurance, however, are the cornerstones of most effective risk management programs.

[2] Sennewald, C. A., 2004. *CPP, Security Consulting*, third ed. Butterworth-Heinemann, Elsevier, Boston, MA.

true for chain-store operations, in which each individual store is regarded as a separate profit center, and records of inventory shortages are kept for each location as a means to measure management effectiveness.

Crime losses far exceed the losses to business caused by fire and industrial accidents. One professional security organization estimates that the annual loss to business from fraud and employee abuse is twice as great as the total of all business losses due to fires and accidents!

Are you really concerned about how crime may affect your business? Take a few minutes and read the Association of Certified Fraud Examiners' current Report to the Nations on Occupational Fraud and Abuse.[3] Then take a moment to reflect on the following estimates:

- The typical organization loses 5 percent of its annual revenue to fraud. Applied to the estimated 2009 Gross World Product, this figure translates to a potential total fraud loss of more than $2.9 trillion.
- Nearly one-quarter of the frauds involved losses of at least $1 million.
- Occupational fraud most often falls into one of three categories: asset misappropriation, corruption, or fraudulent statements.

In the United States alone, estimates of the cost of fraud vary widely. No recent comprehensive studies could be found that empirically measured the economic effects of fraud and employee abuse. In addition to the direct economic losses to an organization from fraud and abusive behavior, there are indirect costs to be considered: the loss of productivity, legal action, increased unemployment, government intervention, and other hidden costs, all of which affect profitability and the bottom line.

Who Needs Security Surveys?

The same study cited above shows that fraud is distributed fairly evenly across four broad sectors: government agencies, publicly traded companies, privately held companies, and nonprofit organizations. In our own experience, many victims were seen to be growing businesses in which expansion had occurred faster than the development of internal controls to prevent fraud. These were relatively large companies, in which close control over divisions and branches were neglected because of the need to ensure productivity. This fact has been noted by many professional security consultants with whom we have discussed this subject. Every business, no matter how large or small, could profit from an objective review of their internal and external controls. The surveys we have personally conducted, with few exceptions, revealed that the majority of corporate management concern was directed toward external problems, such as theft (burglary and robbery), as the company's most immediate priority. These issues (along with fire protection) are seen as primarily insurance driven. This situation reflects the development and growth of the

---

[3] *2010 Report to the Nations on Occupational Fraud and Abuse*, Association of Certified Fraud Examiners, Austin, TX.

security industry in the United States from a historical perspective. It was once thought that most, if not all, of a company's problems with theft were external in nature.

Up until World War II, before the large-scale expansion of U.S. and Canadian industries and the development of the multinational and international corporations, many companies in North America were what we now regard as small to medium-sized industries and businesses. Most retail enterprises were of the "Mom and Pop" variety. Because of close supervision and the personal identification between management and labor that once existed, internal theft (theft by "trusted" employees) was seldom encountered. Hence, the fire and burglar alarm business was developed and rapidly expanded to protect these enterprises from what were then considered the only two threats to the continued existence and profitability of the business—the possibility of a fire or a burglary.

With the growth of the world economy and the demise of the "Mom and Pop" commercial and variety stores, we witnessed a parallel change in business ethics and standards. These changes affected society as a whole. As an example, in the turbulent 1960s, a new term was coined: "the Establishment." Crimes against the Establishment were then, and still are, perceived by many to be permissible, and in fact not crimes at all. The lack of personal identification with a company by its employees and the dramatic dilution of ethical and moral standards (especially in our financial institutions and on Wall Street) among management, employees, and the general public combined to make internal theft by employees a process driven by rationalization and not logic. After all, who is General Motors, COSTCO, Target, or Toyota? "They" make gigantic profits by "ripping off" the general public (read "us"). The theft by an employee of one small item (or dollar amount) multiplied 1,000 times each year in as many companies can add up to an annual loss that can far exceed the total loss attributed to external theft by a company over the entire period of a firm's existence!

Enlightened students of criminology have come to understand that the most predominant and certainly the most prevalent and most costly form of business crime is employee abuse or internal theft in its various forms. Asset misappropriation accounts for the majority of employee theft. Bribery (corruption) and fraudulent statements, although not as frequent, run second and third. Companies with 100 or fewer employees are said to be the most vulnerable to fraud and abuse.[4] Shoplifting, employee theft, and vandalism all cost American businesses billions of dollars. Table 7-1, based on a percentage of a company's net profits, illustrates how much additionally a company needs to sell to offset losses in stolen merchandise, equipment, and supplies. The theft of one $500 item means the company must sell $25,000 worth of merchandise to break even if it is in the 2 percent net-profit category, and $8,333 if it is in the 6 percent category. Worse yet, it is still without the stolen item, which needs to be replaced! For an example of the impact of loss versus net profits, see Table 7-1.

[4] Association of Certified Fraud Examiners, *Report to the Nations*. Companies earning over a million dollars of revenue per year are particularly hard hit by crime. This is especially true for retail enterprises that operate on a 2 percent profit margin. As serious students of this problem have come to realize, losses due to crime can and do have a dramatic impact on net profits.

Table 7-1   Loss to Sales (Profit) Ratio

	If Company Operates at Net Profit of:				
	2%	3%	4%	5%	6%
Actual Loss of:	These additional sales are required to offset an actual loss:				
$50	$ 2,500	$ 1,666	$ 1,250	$ 1,000	$ 833
100	5,000	3,333	2,500	2,000	1,666
200	10,000	6,666	5,000	4,000	3,333
250	12,500	8,333	6,250	5,000	4,166
300	15,000	10,000	7,500	6,000	5,000
350	17,500	11,666	8,750	7,000	5,833
400	20,000	13,333	10,000	8,000	6,666
450	22,500	15,000	11,250	9,000	7,500
500	25,000	16,666	12,500	10,000	8,333

Attitude of Business Toward Security

In general, we find that most businesses will take the necessary precautions to protect themselves against the entry of burglars and robbers onto their premises. Most will also give protection to high-value areas, such as computer centers, vaults, precious-metal storage areas, and any location where money is the principal product, such as in banks and gambling casinos. Many businesses, however, still do not concern themselves with protection against unauthorized access to their premises; yet 75 to 85 percent of external theft is directly attributable to this source. A prime example is unrestricted access at many warehouses we have surveyed to the shipping and receiving docks by nonemployee truck drivers, making daily pickups and deliveries.

Likewise, if the cost of security protection is regarded as a capital expense or a yearly expenditure, and as reducing the profit line, we can expect a person not sufficiently educated in risk management who is in charge of a facility to authorize only the minimum and least expensive security protection available. If, on the other hand, security is mandated by senior corporate management or required by clients or by various governmental agencies overseeing industries engaged in sensitive government contracts, we find plant managers installing security protection with little regard to costs (nuclear power generating plants, for example).

This latter proposition presupposes another problem. Most plant managers with whom we have worked do not have the foggiest idea what kind of security they need for adequate protection. For sensitive government contract work, security guidelines are often mandated. Nevertheless, many are the managers who have fallen prey to the sales pitches of security manpower and hardware salespeople who proposed package deals that are guaranteed to solve all problems. As a result, we constantly encounter security programs that overemphasize manpower (guards) or hardware (alarms, closed-circuit television [CCTV], or electronics) when the proper solution to the problem is an effective marriage of manpower, hardware, and electronics, with adequately written procedures to deal with the most common occurrences or eventualities encountered in the real, work-a-day world. Life becomes complicated only if we allow it to. Most security solutions are relatively simple in concept and application.

What Can a Security Survey Accomplish?

One approach to determining whether there is a need for a security survey is to find out what services an experienced security expert can provide and then to seek information on security-related (crime-related) losses being incurred by the particular business or company involved. The security expert can be in place (that is, an employee or member of the staff), an outside consultant, or a combination of both.

Security-related issues might reflect any of the problems mentioned in the early chapters of this book. Surveys generally, however, show that most problems encountered in the real world of business are not uncommon ones.

If the company or facility being considered has a security plan, the survey can establish whether the plan is up-to-date and adequate in every respect. Experience has shown that many security plans were established as the need of the moment dictated; most fixes were developed without regard to centralization and coordination. Upon review, many such plans are patched-up, crazy-quilt affairs. More often than not, the policies, procedures, and safeguards need to be brought together for review and consolidated so that the component parts of the plan complement, not contradict, one another.

If the facility has no security plan, the survey can establish the need for one and develop proposals for some or all of the security services commonly found in industrial settings. By conducting a comprehensive survey of the entire facility—its policies, procedures, and operations—one can identify critical factors affecting the security of the premises or operation as a whole. The next step is to analyze the vulnerabilities and recommend cost-effective protection specifically tailored to the problems encountered. The survey should also recommend, as the first order of business, the establishment of sound policies and procedures, as a minimum to define the security mission and to accomplish the following:

- Protect against internal and external theft, including embezzlement, fraud, burglary, robbery, industrial espionage, and the theft of trade secrets and proprietary information.
- Develop access-control procedures to protect the facility perimeter as well as computer facilities and other sensitive areas, such as executive offices.
- Establish lock-and-key control procedures.
- Design, supervise, and review installation of anti-intrusion and detection devices and systems.
- Establish a workplace violence program to help corporate personnel assess internal and external threats.

- Provide control over the movement and identification of employees, customers, and visitors on company property.
- Review the selection, training, and deployment of security personnel, proprietary or contract.
- Assist in the establishment of emergency and disaster recovery plans and guidelines.
- Identify the internal resources available and needed for the establishment of an effective security program.
- Develop and present instructional seminars for management and operations personnel in all the above areas.

This list is by no means all-inclusive. It does, however, set forth some of the most fundamental security programs and systems recommended for development by security surveys that we have conducted.

Why the Need for a Security Professional?

Losses due to all causes continue to represent a problem of major proportions for business and industry. To the extent that the services of security professionals can help in eliminating, preventing, or reducing a company's losses, they are needed. At a security management seminar, Charles A. Sennewald once explained, "Crime prevention, the very essence of a security professional's existence, is another spoke in the wheel of total loss control. It is the orderly and predictive identification, abatement, and response to criminal opportunity. It is a managed process which fosters the elimination of the emotional crisis response to criminal losses and promotes the timely identification of exposures to criminality before these exposures mature to a confrontation process." The proper application of protection techniques to minimize loss opportunity promises the capability not only to improve the net profits of business but also to eliminate or reduce to acceptable levels the frequency of the most disruptive acts, the consequences of which often exceed the fruits of the crime.

How Do You Sell Security?

As mentioned earlier in this chapter, some managers, for a variety of reasons, are reluctant even to discuss the subject of security. In the aftermath of an unsuccessful attempt to burglarize a bank vault, a bank operations manager learned that the bank's anti-intrusion system was 10 years old and somewhat antiquated. "Why didn't the alarm company keep me advised of the necessity to upgrade my system as the state of the art improved?" he asked. The answer he received was, "The alarm system functioned adequately for 10 years with no problem. Would you or the bank have authorized an expenditure of several thousand dollars to upgrade the alarm system before you had this attempted burglary?" The bank representative reluctantly admitted that he would not have. This case is typical of what we call the "knee-jerk reaction" to security. One security consultant with whom the author is acquainted describes it as locking the barn after the horse has been stolen. We all know that the only thing this protects is what the horse left behind.

It is not uncommon to find that management's attention is obtained only after a serious problem, one pointing to the lack of adequate protection, is brought to its attention. The first reaction seen by management is often one of overkill; the response pendulum swings from complacency to paranoia, when the facts indicate that a proper response should be somewhere in between. Given this situation, an unscrupulous alarm salesperson may be presented with an opportunity to prescribe an alarm system worthy of consideration by the manager of the bullion vault at Fort Knox. The true professional will, as the first order of business, try to bring the situation into perspective by reducing the fears of his clients.

There are some things a security director or consultant can do to convince top corporate management that proper security is worth spending some money to obtain. Some methods that have proved successful are the following:

1. Establish a meaningful dialogue with the decision makers in the management hierarchy. First, try to ascertain their feelings about security. What do they really want a security program to accomplish, if in fact they want anything? Do not be surprised to learn that some management personnel regard security as a necessary evil and thus worthy of little attention (that is, time, money, or other resources). Marshal the facts. Research the history of security losses experienced by the company. Use this information to develop trends and projections.

2. When collecting data to support your position, deal in principles, not personalities. Use the technique of nonattribution for all unpublished sources of information. With published sources, such as interoffice memos, extract the pertinent data if possible. Avoid internecine power struggles at all costs. Maintain a position of objective neutrality.

3. Be as professional about security as you can. The better you are at your job, the greater attention you will command from your supervisors. There are many avenues you can explore to develop the information you need, such as developing contacts with other security professionals who share similar problems. Don't reinvent the wheel: attend security seminars, purchase relevant books, study, and do research. Stay current.

4. When making a proposal to management, hit the highlights, and make your proposal as brief as possible. Save the details for later. In any proposal that will cost money, make certain you have developed the cost figures as accurately as possible. If the figures are estimated, label them as such, and err on the high side.

5. It is a wise man or woman who knows his or her own limitations. If you need outside help (and who doesn't from time to time?), do not be reluctant to admit it. Such areas as electronics, advanced CCTV, and sophisticated anti-intrusion alarm systems are usually beyond the capabilities of the security generalist. Do some studying. Know where to go to get the help you need.

6. Suggest that management hire an outside consultant. Competent security professionals have nothing to fear from a second opinion. Often, the "expert from afar" has greater persuasiveness over management than do members of their own

staff. More often than not, the consultant will reinforce your position by reaching the same conclusions you did and making the same or similar recommendations.

7. Present your position at the right time. Recognize that management's priorities are first and foremost the generation of profits. To capture management's attention, wait for the right circumstances. It is difficult to predict when these may occur; therefore, have your facts developed and be ready at a moment's notice to make your presentation. It will be too late to do the research when you are called before the board of directors without notice to explain how the breakdown in security that just happened could have occurred and what you propose to do to solve the problem for the future.

8. Develop a program of public relations. Security represents inconvenience, even under the best of circumstances. Once you have management thinking favorably about your proposal, you will need to sell it to everyone in the organization in order for your ideas to be successfully implemented. Most employees enjoy working in a safe and secure environment. Use this technique to convince employees that the program was designed as much for their safety and security as for the protection of the assets of the corporation.

For a comprehensive treatment of the role a professional security consultant can play and how the consultant can help security professionals properly define their security exposures, refer to Charles A. Sennewald's 2005 text, *Security Consulting*, 4th edition, published by Elsevier Butterworth-Heinemann. Do your homework in a thorough manner, and you cannot help but impress management with your capabilities as a security professional. Remember, be patient. Few have been able to sell 100 percent of their security programs to management the first time out of the starting block. And, if you don't remember anything else, remember, as in comedy, timing is everything!

8

Management Audit Techniques and the Preliminary Survey

The objective of a preliminary survey is to quickly and economically determine if a [security] survey (audit) appears to be desirable and is technically and economically feasible.
—Government Accounting Office Audit Guide

Audit Guide and Procedures

Audit: Aids to Surveys

In the past, security professionals were people with prior experience in law enforcement, military police, or intelligence work. Today, security professionals come into the field with academic training, many with graduate degrees in Security Management. This is seen as another hallmark in the growth of the security profession in the past 30–40 years. Notwithstanding this positive growth in the educational level of present-day security professionals, few have had experience or received training in management audit techniques.

There are many similarities between auditing and investigating, but there is one important difference: Investigators are trained to collect evidence, report the facts obtained objectively, and scrupulously avoid drawing conclusions or making recommendations. These they leave for whoever reads the investigative report. The late J. Edgar Hoover, Director of the Federal Bureau of Investigation, was quoted as saying, "An investigator must never be wrong in reporting the facts of an investigation, and opinions have no place in an investigative report."

This is not to suggest that investigators do not have opinions or draw conclusions from the results of their efforts. They have and they do, but one will look in vain for these thoughts in the investigator's reports. Investigators are taught not to editorialize and to leave opinions for the decision makers for whom the report is prepared. Such training, of course, has a real purpose: It is a proven technique for ensuring objectivity and avoiding the trap of partiality in the difficult search for the truth.

Auditors are trained to appraise the truth or falsity of a proposition—not to take things for granted, jump to conclusions, or accept plausible appearance for hard fact. Audit training postulates that accepting appearance for substance is the surest way to arrive at an improper conclusion. To be able to differentiate between appearance and substance and draw the proper conclusions is the heart and marrow of the auditor's task.

Both disciplines, to be sure, pursue fact—a fact being something that has actual existence, something that can be inferred with certainty, a proposition that is verifiable. Conjecture, on the other hand, involves propositions carrying insufficient evidence to be regarded as fact. The auditor is trained not only to adduce facts but also to appraise, draw conclusions, analyze, and make recommendations from facts—the very techniques the investigator is trained to avoid.

In conducting security surveys, one should borrow heavily from the techniques used by internal auditors. It can be said that most security surveys are, in the truest sense, specialized management audits. In this context, we define internal (management) auditing as a comprehensive review, verification, analysis, and appraisal of the various functions (operations) of an organization, as a service to management.

The auditor works with people first and things second, as does the investigator. Both auditing and investigating presuppose a degree of cooperation from the people one encounters along the path to the objective, that path usually being the obtaining of complete information and accurate data and placing both in the proper perspective. Successful professionals often obtain cooperation from the people they encounter on this journey through the force of their personalities and an empathetic attitude that invariably draws other people out. These individuals, by training or by instinct, understand the art and science of communication—what opens channels and what closes them. But let there be no misunderstanding: Auditors and investigators are often feared. Both represent a force and authority that can be a threat to one's liberty and security, an unknown entity that can adversely affect one's status and well-being, even one's job and livelihood.

The first task of the professional, therefore, is to allay that fear, because fear is a very real impediment to open communication, and open communication is the heart of both a successful investigation and a successful audit. There are many techniques that can be used to allay fear, but perhaps the best one is the use of candor, letting people know, from the very beginning, what to expect. It has been said that we often fear most that about which we know the least. Therefore, in the absence of sound justification to the contrary, there should be complete candor in an effort to gain rapport with everyone with whom one comes in contact, from the outset of the survey or audit until its conclusion. This brings us to some of the methods used by auditors in conducting successful surveys.

Fieldwork

Perhaps as much as 50 percent of survey work is done in the field. The other 50 percent is usually divided between planning the survey (logistics), analyzing the data collected, and then writing the report. However, some security consultants report that they often spend as much or more time compiling data and analyzing it and then writing the report, as they did conducting fieldwork. This is typically the case when conducting a review for a large corporation.

Fieldwork, as considered here, consists of collecting data, records, and written policies and procedures wherever they are located. And oftentimes locating these documents is a difficult task. Complete and through fieldwork will have a significant effect on the operation and thus on the results of the survey. Reduced to its simplest terms, fieldwork is largely measurement and evaluation of the effectiveness of the security program (or the ineffectiveness of one) under review. To be meaningful, measurement must have as its basis an objective standard. By standard, we mean a level of acceptability against which things measured can be compared. Each part of the survey must be approached with the thought that it can be effective if it determines quality in terms that can be objectively measured and compared with an acceptable practice or, better yet, a standard, if one exists.

Only by using these criteria can one measure intelligently and with objectivity. When surveyors cannot measure, they must use extreme caution because in such a case they can produce only subjective evaluations, not objective conclusions based on recognized standards or accepted practices. When no standards or acceptable practices exist, obviously the surveyor must try to develop or construct them. Likewise, when only technical standards exist, measurements obtained must be validated by one who is technically qualified to render such a judgment. As surveyors apply recognized standards or acceptable practices, they must not hesitate to evaluate them to determine whether they are valid and not obsolete. It is no understatement to say that without standards or acceptable practices, there can be no meaningful measurement; without measurement, in turn, fieldwork becomes conjecture and not fact. Remember the saying, "If it can be measured (validated), it's a fact; otherwise it's an opinion."

There are many methods and approaches to fieldwork, and the one selected often depends on the individual approach of the person conducting the survey. Nevertheless, fieldwork usually takes the form of observing, questioning, analyzing, verifying, investigating, and evaluating, though not necessarily in that order.

Measurements usually encompass three aspects of a typical security operation: quality, reliability, and cost. (Obviously more than three aspects can be programmed into the survey, depending on the results desired.) Regardless of how many aspects of the operation are being reviewed, the primary questions become: Is the procedure, technique, hardware, or electronic device being used effective? Does it properly address and solve the problem, with a known degree of reliability? Does it perform at less cost than other measures that might be just as (or more) effective with respect to the problems faced and the overall objective or results desired?

The objective of measurement is to assess the adequacy, effectiveness, and efficiency of an existing or proposed system. This is usually accomplished by applying one or more of the six basic methods of fieldwork that follow.

Observing

Observing is seeing, not just noticing or looking. As Arthur Conan Doyle's Sherlock Holmes was fond of saying, "You see, Watson, but you do not observe." Observing implies a careful, knowledgeable look at people and things and how they relate one to the other. It is a visual examination with a purpose, a mental comparison with practices and standards.

The ability to observe and evaluate comes from training and experience. The broader one's experience, the better one observes, and the more alert one is to recognizing deviations from the norm.

Questioning

Questioning during a survey occurs at every stage of the proceeding. It may be in oral or written form. Oral questions, of course, are the most common and at times the most difficult to pose. To get the truth without upsetting people during the course of a survey is not an easy task. If subjects detect an attitude of cross-examination or an inquisitorial tone, they may promptly raise their defenses, which becomes a barrier to effective communication. The results may be wrong or incomplete answers or, worse, no answers at all. Most successful practitioners have developed the interview technique into an art. They use this, perhaps the most common tool of our trade, with a high degree of effectiveness. Remember, there is a big difference between an interview and an interrogation. An interview presupposes voluntary participation. Interrogation is usually conducted when a participant in a conversation is uncooperative or reluctant to participate in the inquiry.

Analyzing

Analyzing is nothing more than a detailed examination of a complex matter or problem to determine the true nature of its individual parts. It presupposes intent to discover hidden qualities, causes, effects, motives, possibilities, and probabilities. In contrast, if one examines an operation as a whole, one cannot perceive the intricate relationships of the diverse and varied elements that make up a complex function or an unusually large activity. Any organization or operation, no matter how large or complex, can be analyzed by division, by breaking it down into its separate elements and then observing trends, making comparisons, and isolating aberrant transactions and conditions. Frankly, the job of analysis can be done no other way.

Verifying

One verifies by attesting to the truth, accuracy, genuineness, or validity of the matter under scrutiny or inquiry. This implies a deliberate effort to establish the accuracy or truth of some affirmation, by putting it to the test. An example is a comparison with other ascertainable facts, with an original, or perhaps with another acceptable practice or standard. Verification may also include corroboration—the statement of another person or a validation by objective practices or standards, usually found elsewhere.

Investigating

An investigation is an inquiry that has as its aim the uncovering of facts and the obtaining of evidence to establish the truth. An investigation may occur as a part, or as the result, of a survey, but it is not restricted to some impropriety. If in fact an impropriety is discovered during the survey, it must be divorced immediately from the audit and referred to the proper authorities for appropriate handling.

It is not unusual in the course of a management audit to uncover suspected fraud. We did just that in the course of a security survey at a large hospital and medical complex. The receiving (intake) office had an employee who had been stealing money and jewelry from patients. The discovery necessitated a theft investigation, which proved successful and resulted in the conviction of the responsible employee and the recovery of some of the stolen property. When uncovered, fraud and other forms of employee malfeasance must be dealt with immediately and outside of the scope of the audit or review. (See Appendix B, Danger Signs of Fraud, Embezzlement, and Theft.)

Evaluating

To evaluate is to estimate worth by arriving at a judgment. It is to weigh what has been analyzed and determine its adequacy, effectiveness, and efficiency. It is one step beyond opinion, in that it represents the conclusions drawn from accumulated facts. Evaluation, by necessity, implies professional judgment. Professional judgment is a thread that runs through the entire fabric of the security survey.

Evaluation in a survey occurs constantly throughout the duration of the project. In the beginning, one must determine which programs and procedures will be reviewed, which processes and operations are to be tested, and how big a sample must be obtained for the test to achieve the degree of reliability needed. Finally, as the results of the survey accumulate, one must evaluate what the results imply—fact finding without evaluation is a clerical, not a professional, function.

Evaluation obviously calls for judgment, and good judgment is perhaps the one single characteristic which separates the amateur from the real professional. Questions and imponderables are bound to arise. If you find none, it may be time to leave well enough alone and get out of the game. The true art is in recognizing where the questions lie. Once we identify the problem, we are well along the way toward the solution. Like a physician making a diagnosis, only by properly identifying the problem can the security professional comfortably arrive at the proper solution.

Mature professionals can evaluate results almost intuitively, and more often than not they are correct. This talent comes from years of experience. We of lesser experience can benefit from a more structured, formal, and organized approach to the evaluation of the results obtained. For example, in evaluating a failure to meet specific standards or acceptable practices, one might ask the following questions:

- How significant are the observed deviations?
- Have they or will they prevent the operation from achieving its objective or mission?
- Who or what can be hurt or injured?
- Could the injury perhaps be fatal to the enterprise?
- If corrective action is not taken, would the deviation be likely to recur?
- How did the problem surface in the first place?
- What were the causes? What event or combination of events caused the failure to occur?
- Will the event or combination of events cause the observed results or failure every time?

Obviously, to recommend corrective action, we must first answer some of the above basic questions. In following these procedures, surveyors find themselves in a constant state of evaluating. One must constantly query everything under review, with questions such as the following:

- What is the real problem? (Not necessarily what management thinks it is—what is it really?)
- What are the relevant facts? What are the processes, systems, procedures, policies, and organizational structures? What were they in the past? What should they be in the future?
- What is presently being done about this problem at other locations within the company? At other locations or companies?
- What are the causes? The number and variety of causes? The root cause as well as the surface cause? When and how do these causes affect the overall problem?
- What are the possible solutions, the alternatives, the cost, and the answers to the problem? What are the possible side effects, advantages, or disadvantages to the proposed solution?

Only by constant probing, questioning, analyzing, and evaluating do surveyors uncover system and performance defects, establish reliability, and develop cost-effective solutions to their clients' problems.

The Preliminary Survey

Definition and Purpose

The basic purpose of the preliminary survey is familiarization, based on more than a mere discussion or brief observation of the tasks to be reviewed. It presupposes an ability to perceive the true objectives of the operation and to locate and evaluate the key control points, if any exist. Also, one must understand the management concepts being used and the qualifications and abilities of the employees responsible for the success of the operation. (See Box 8.1 for a sample statement of purpose.) A properly planned and implemented preliminary survey allows one to develop a well-thought-out audit program, to employ one's efforts efficiently and economically, and to form a firm foundation for the detailed examination that will follow.

Although a poor preliminary survey (or none at all) can easily result in a poor audit, a good preliminary survey can ensure an intelligent examination and may also substitute for many parts of the final examination. It is the simplest way to cut through the mass of detail that often obscures the objective and to get the job started quickly and on the right track. Another advantage of the preliminary survey is to test the client's sincerity and avoid misunderstandings regarding the client's expectations from the project at the outset.

Many security and safety operations in large companies are extremely complex. The immediate task becomes one not only of identifying and understanding these operations but also of

---

BOX 8.1   PRELIMINARY SURVEY—STATEMENT OF PURPOSE (EXAMPLE)

Attention: Name of Client

Reference: University of _____

   The preliminary survey will be a primary overview of the university to identify major problem areas within the system affecting security. This initial review will become the basis for defining the parameters of the final survey. It will include the following:

- Interviews of officials concerned with administration of revenue-producing and accounting programs to identify loss exposure.
- A physical tour of selected campus locations to familiarize consultant with the revenue-generating operations of the university.
- Preliminary interviews with selected on-site operations personnel for basic orientation with cash-handling methods and control procedures.

   The agreed upon scope of the final survey will emphasize internal control procedures, with physical security and emergency planning secondary. A complete evaluation will include specific recommendations for appropriate administrative controls, hardware application, and personnel to complement the system already in use to achieve effective cost-control security.

   We will also provide downstream inspection to ensure that the agreed upon recommendations are being properly implemented.

Sincerely,

James F. Broder, CFE, CPP, FACFE, Security Consultant

PMG

---

analyzing and evaluating them and recommending improvements designed to accomplish the job with greater efficiency and at less cost. Admittedly, this is no easy task in a large and complex industrial, research and development, or military environment. This is especially so when one adds to the already identified internal complexity of the operation such factors as the external environment (community), legal restrictions, ecological and environmental protection procedures, public relations, employee relations, union activities, government regulations and restrictions (especially the Occupational Safety and Health Administration, or OSHA), and stockholder interest. Additionally, since the events of 9/11, we must also consider the prospect of international terrorism, as it may (or may not) affect some of these organizations.

   As complicated as the assignment may appear initially, it is well to remember that there is no mass of data so large, and no operation so complex, that it cannot be given a semblance of order and then be arrayed, evaluated, and scrutinized in a logical, organized, methodical manner.

   A preliminary survey should as a minimum answer the following questions:

- What is the operation?
- Who or what is responsible for the operation?
- Why is it accomplished (where and when)?
- How is the operation accomplished?

In short, get what is important, and get it with a minimum of delay. Focus on the highlights and forget, for the moment, the details; they will come later. However, don't let haste interfere with order and methodology. To go into the program's initial operation without a well-prepared agenda may well leave the client with the impression of disorganization, the exact opposite of what one would hope to instill at this (or any other) stage of the management review.

We were contacted some years ago and asked to do a physical security survey of three large petrochemical plants near Bombay, India. The client asked for a brief proposal outlining our plan to complete this complex project. We recommended first doing a preliminary survey and provided an estimated cost (travel, time, and expenses) to do the survey work and to write the preliminary survey report. Our planned approach was well thought out. In our opinion (then and now) a preliminary survey was essential as the first order of business. We estimated the job to take about 3 days (a total of 24 hours) of consulting time and prepared a written proposal setting forth suggestions. Upon submission, however, our proposal was rejected. The client had expected us to absorb the cost associated with the preliminary survey, in return for the possibility of being selected (maybe) to do the principal project, which would have taken about 6 months to complete. We, of course, never did the project.

We later learned that this was the customary way some people in that country did business. Notwithstanding, the proposal we sent included a comprehensive outline for performing a preliminary survey. The project was so large and decentralized that in our opinion, it would have been impossible to accomplish the principal task without first performing a preliminary survey. Obviously, it is best to ensure that if such misunderstandings occur, they occur at the outset and not during the course of the project.

The Initial Interview

At the initial interview (or "opening conference" in audit parlance), the nature of the questions asked will vary depending on whether the survey is organizational, functional, or operational. If it is organizational, people-oriented questions will be the general rule. Functional surveys are more concerned with the actual work flow, which more often than not crosses organizational lines. In operational surveys, one is primarily interested in hardware and, equally important, in the software and related procedures concerned with the effective use of the hardware. Regardless of the type of survey, the initial interview should elicit answers to the following questions: What does management perceive to be the major problem areas? What does management hope the survey will accomplish in regard to solving these problems? Remember two important factors: one, we live in a world where people want instant results and generally have unreasonable expectations, but to do an adequate job takes time; and two, time is money. Everyone should leave the table with a complete understanding of what is going to be accomplished and what it will cost to do the job.

At the initial interview, it is wise to have prepared in final form a document called the management memorandum. This memorandum, to be published on company letterhead and signed by the highest authority possible, introduces the survey team. It describes the objective of the survey, solicits the assistance and cooperation of all company employees, and authorizes access to all documents and information that may be requested in the course of the survey. Copies of this signed memorandum, when possible, should be sent to all affected department heads in advance of the arrival of the survey team. In this regard, it may be helpful for the people doing the survey to be introduced at a meeting of all department heads.

At this meeting, the management memorandum can be read and distributed, and any questions regarding operational authority can be answered. At a minimum, someone from the company must be assigned to escort and introduce members of the survey team to department heads and employees with whom they will be working. We recommend the department head meeting because such a meeting can go a long way toward establishing rapport and cooperation. It also gives the department heads directly involved a chance to ask questions and receive answers that ideally will allay any fears or apprehensions they brought with them to the meeting. At the very least, it serves as a vehicle to get them involved in the survey program at the outset.

After the preliminary survey has been completed, it is useful to have a second meeting with the client. The purpose of this second meeting is fourfold:

1. To give the client a brief report of initial impressions obtained
2. To explain how the surveyor perceives the objectives, activities, or functions under review
3. To establish a meeting of the minds as to just what will be accomplished
4. To briefly outline the general plan of attack

It is essential that both parties to the survey be in total agreement with regard to the objectives of the project. In this way, any misunderstandings can be resolved so that the work can proceed without delay.

Obtaining Information

What Information to Obtain

The preliminary survey, and in fact the primary survey as well, will move along rapidly and systematically only if one has a clear idea of what data are needed and where to find relevant information. Some, but by no means all, of the basic sources one should acquire are as follows:

The Charter for the Organization. Copies of policy statements, directives, statements of functions, responsibilities, goals, and delegations of authority will be needed for review. In this regard we have found the company's annual statement to shareholders to be a good source of basic information. In addition, job descriptions of all security personnel directly involved in the activity will need to be reviewed.

Beyond the written word, it is essential for one to focus on the objectives of the organization and operation—what is its real mission (not necessarily what the official statements say it is). It is not uncommon to find that official job descriptions are mere window dressing or that they have not kept pace with changing times and aims. The documentation of the operation may include the following:

- Organizational charts
- Position descriptions of the operation in the overall company structure
- The nature, size, and location of ancillary or satellite activities
- Interfacing operations and their relationship to the activity under review (safety, for instance, when the primary review is security)

Financial Information. One will want to obtain for review all financial data that have a bearing on the subject under scrutiny, either directly or indirectly. Obtain copies of the company's annual budget for security, if available.

Operating Instructions. It is essential to obtain an accurate picture of the flow of records and other data. One of the simplest ways to review this is by flowcharting the activity. Flowcharts (discussed later in this chapter) can provide a useful picture of the operation and at the same time highlight gaps and duplications in procedures, as well as pinpoint risk areas for later scrutiny.

Problem Areas. During the entire survey, one should keep in mind the problems outlined by management at the preliminary (opening) conference, as well as any deficiencies found in prior surveys, audits, or reviews. Also, focus attention on the procedures or controls that have supposedly been designed to alleviate the difficulties and problems.

Matters of Special Interest. One will be especially interested in exploring any new areas mentioned during discussions with management that were of concern to them during the preliminary (opening) conference.

Sources of Information

Some, but certainly not all, of the principal sources of information available during the survey are as follows:

1. Discussions with supervisors and employees directly engaged in the activity under review. One cannot overemphasize the importance of these people, because:
   a. Not only are they intimately aware of the problems, but they often have thought out the solutions as well.
   b. It is essential to obtain their cooperation because in the final analysis, these are the people who will be responsible for implementing many of the recommendations made as a result of the survey.
   c. If they feel they have played a real part in the development of solutions to the problem, they will be more inclined to work for the success of the recommendations. The reverse is, unfortunately, also true.
2. Discussions with supervisors downstream and upstream of the operations under review
3. Correspondence files

4. Previous survey, audit, or inspection reports
5. Incident and crime reports
6. Budget data
7. Mission or objective statements or reports
8. Procedural (operational) manuals
9. Reports by or to government agencies, both state and federal (OSHA, for example)

Physical Observation

Observations or inspections should be conducted in two phases. The first is a familiarization tour of the entire facility to obtain "the big picture." At this time, the various departments are identified, and the managers and supervisors are seen in their normal work environments.

Notes are made, but few questions are asked at this point. In a small operation, one tour may be sufficient to accomplish the desired objective. In a large, complex operation, it may be necessary to make a second or even a third tour before feeling comfortable with the facility and the environment. The second (or subsequent) tour may be made in connection with flowcharting various parts of the activity or operation.

Flowcharting

Flowcharting is an art that with proper practice can become an invaluable survey tool. Making a flowchart is the easiest way to obtain a visual grasp of a system or procedure, and it is a ready means of analyzing complex operations that cannot easily be reduced to a simple narrative. Figure 8.1 shows some standard flowchart symbols and a legend describing each symbol used. Sometimes a simple sketch may suffice; at other times, it may be necessary to use plastic overlays to describe detailed and complex operations. Figure 8.2 is an example of a formal flowchart describing a complex operation. Flowcharts, however, need not be formal or greatly detailed to accomplish the task. Table 8.1 is an example of a simple, informal flowchart.

Summary

Fieldwork is measurement—it is measuring "what is" against "what should be." This requires both the methods of measurement and the existence of acceptable practices and standards for comparison. Security surveys are usually concerned with measuring at least three basic factors: quality, reliability, and cost.

To measure, one must go out into the field and perform surveys or, if a program already exists, to review and to test it for reliability. Fieldwork is performed using the techniques of observing, questioning, analyzing, verifying, investigating, and evaluating. The largest portion of fieldwork is the gathering of data and accumulating evidence, which must then be analyzed and evaluated before recommendations can be developed. It is in the proper evaluation of the data obtained that true professionals meet their test.

Symbol	*Explanation*

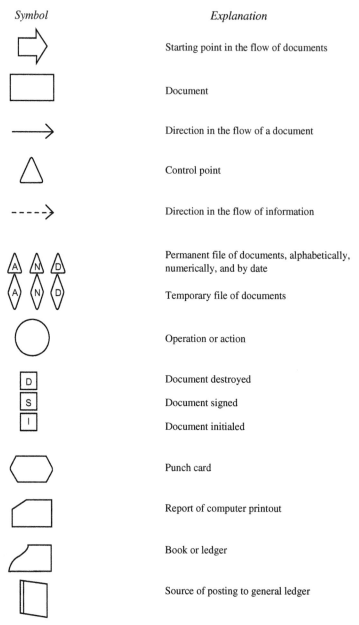

Starting point in the flow of documents

Document

Direction in the flow of a document

Control point

Direction in the flow of information

Permanent file of documents, alphabetically, numerically, and by date

Temporary file of documents

Operation or action

Document destroyed

Document signed

Document initialed

Punch card

Report of computer printout

Book or ledger

Source of posting to general ledger

FIGURE 8-1 Standard flowchart symbols.

The preliminary survey charts the course for the main voyage. It often provides a clear enough view of certain activities to eliminate the need for further review of these operations. The time spent on the preliminary survey is well spent. Done properly, it will ensure a more efficient and economical final review.

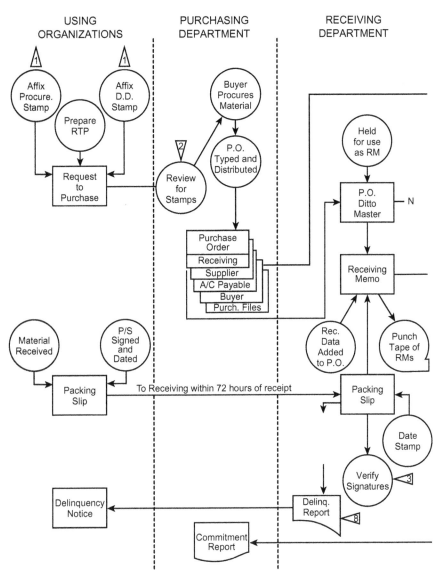

FIGURE 8-2 Formal flowchart.

In the preliminary survey, one gets to know people, understand operations, and focus on objectives, controls, and risks. One is then in a better position to perform the main survey in an intelligent, effective, and efficient manner. The preliminary survey is a road map, which we find essential for a long and difficult journey.

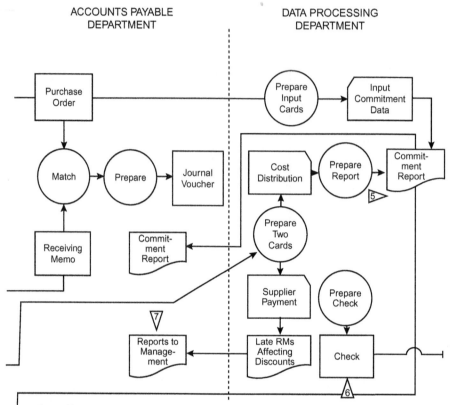

Controls:
Stamps controlled by registers in Purchasing Department. Only authorized
departments and personnel may validate RTPs or use D.D. system.
Unvalidated requests are returned to requesters.
Improperly approved P/S's challenged.
Payments not made without proof of receipt.
Invoices not required.
Record of commitments helps establish cash needs.
Check signed in Finance Department (not on chart) where it is mailed to supplier.

    Reports to cognizant managers on unearned discounts monitors
timeliness of processing P.O.'s, P/S's, RMs, and payments to supplier.

    Reports warn that D.D. privileges may be withdrawn.

Abbreviations:
D.D. – Direct Delivery
P.O. – Purchase Order
P/S – Packing Slip
RM – Receiving Memo
RTD – Request to Purchase

FIGURE 8-2 (*Continued*)

Table 8-1 Informal Flowchart: Procurement of Materials

Purchasing Department	Receiving Department			A/C Payable	Inspection	Stores
	Office	Dock	Hold Area			
Purchase orders and changes are prepared and sent to:	The master P.O. is held in temporary files awaiting receipt of materials and shipping notice.	Materials and shipping notice are received.	Materials are held until the Receiving Memo is prepared.	Evidence of receipt is matched with copy of P.O. No invoice is required.	Material is inspected.	Materials are stored awaiting requisitions from using departments.
1. Supplier 2. Accounts Payable 3. Receiving Department 4. Buyer 5. Purchasing files		Upon receipt of shipping notice, the receiving information is added to the ditto master of the P.O. to create the Receiving Memo.	The S/N is sent to Receiving Office. The materials are sent to the hold area.	Thereupon the materials are sent to Inspection.	If match is satisfactory, payment to supplier is approved.	Unsatisfactory material is sent to hold area. Satisfactory material is sent to stores.

9 The Survey Report

Did you ever stop to think how odd it is that you have to learn how to write your own language? Why? What is there to learn?
—**Rudolf Flesch and A. H. Lass, *The Way to Write*, Harper & Row, 1963**

There is little likelihood that the average security professional was born with the writing mastery of, say, Ernest Hemingway. Writing is an art, and like most arts, it must be constantly practiced if it is to become a useful and natural talent. Although few of us can learn to write with the flow and style of a great master, we can, with effort and practice, greatly improve our ability to communicate by use of the written word.

Many of us find fieldwork the most exciting and challenging part of our daily assignments. We all know many professionals who are extremely competent investigators and auditors but who, when it comes to writing reports, leave much to be desired. Good writing requires good thinking. If one's concepts are confused and tangled, one's reports will reflect the same problems. If one's thoughts are muddy and don't seem to establish a bridge between cause and effect, the resulting written report is bound to reflect this confusion. Unfortunately, we are likely to be judged more often by our ability to write a good report than by our ability to do good fieldwork. Our professional ability and efficiency will be demonstrated before the eyes of many by clear, concise, complete, and accurate reporting.

I need cite but one example to make my point. We were contracted to do an investigation/audit regarding a possible kickback scheme between a vendor and a staff employee of a public agency in a large city of southern California. When we submitted our final written report to the client, we expected it to be reviewed by no more than five senior management officials. Instead, the report was reproduced, and copies were presented to the entire board of directors. Also, copies were furnished to the district attorney's office for review and possible submission to the grand jury. And, eventually, a copy of our report was leaked to the press! Many people, most of whom the writer never met, were thus in a position to judge the quality of our work by merely reading the report. The message here is, "Every report you write bears your name. Write one bad or inaccurate report, and it will haunt you forever."

"I Must Write, Therefore I Shall"

What makes good writing? The answer seems to lie in good fieldwork, a well-structured outline, copious notes or working papers, and dogged persistence. If one has done an adequate job in the field, one's notes and working papers will be full of facts and figures.

Only then can one be sure that adequate material exists for a good report. At this point, it is well to remember the Chinese proverb, "A journey of a thousand miles begins with the first step." Sit down and start writing. The first order of business will be to prepare an outline. An example of an effective reporting outline can be found in Box 9.1.

BOX 9.1   SECURITY SURVEY REPORT

   I. Purpose. State the reason for conducting the survey. The purpose could encompass all of the subject matter in this outline or be restricted to a specific portion or spot problem; it could encompass one location or cover the entire corporation.

  II. Scope. Describe briefly the scope of the survey effort. What was actually done, or not done, in some cases? For example:

    A. People interviewed

    B. Premises visited, times, and so forth

    C. Categories of documents reviewed

 III. Findings. This section includes findings appropriate to the purpose of the survey. For example, a survey report on physical security only would not contain findings related to purchasing, inventories, or conflicts of interest unless they have a direct impact on the physical security situation.

    A. General: Provide brief descriptions of facilities, environment, operations, products or services, and schedules.

    B. Organization: Provide brief description of the organizational structure and number of employees. Include details regarding the number and categories of personnel who perform security-related duties.

    C. Physical security features: Describe lock-and-key controls, lock hardware on exit and entrance doors, doors, fences, gates, and other structural and natural barriers to access and entry. Describe access controls, including identification systems and control of identification media; the security effect of lighting; the location, type, class, installation, and monitoring of intrusion alarm and surveillance systems; guard operations; and so forth.

    D. Internal controls: Describe methods and procedures governing inventory control, shrinkage, and adjustments; identification and control of capital assets; receiving and shipping accounts receivable; purchasing and accounts payable; personnel selection; payroll; cash control and protection; sales to employees; frequency and scope of audits; division of responsibilities involving fiduciary actions; policy and practices regarding conflicts of interest; and so forth.

    E. Data systems and records: Describe physical features and procedures for protecting the data center, access to electronic data processing (EDP) equipment, the media, and the data; types of application systems in use; dependency; backup media and computer capacity; and capability for auditing through the computer. Identify essential records and how they are protected and stored.

(Continued)

> **BOX 9.1   (Continued)**
>
> F. Emergency planning: Describe status and extent of planning, organization, and training to react to accidents and emergencies, such as emergencies arising from natural disasters, bombs and bomb threats, kidnapping and hostage situations, and disorders or riots. Consider these in three phases: preemergency preparations, actions during emergencies, and postemergency or recovery actions. Discuss coordination with and utilization of external resources—for example, other firms or government agencies.
>
> G. Proprietary information and trade secrets: Discuss the extent to which these categories of information are recognized and how they are classified; means of protection—for example, clearance, accountability, storage, declassification, disposal; and secrecy arrangements with employees, suppliers, and so forth.
>
> IV. Conclusions. Evaluate the protective measures discussed in part III, Findings. Identify specific vulnerabilities and rate them as to seriousness—for example, slight, moderate, or serious, or similar terms of comparison.
>
> V. Recommendations. Using the systems design technique, the writer develops specific recommendations for appropriate applications of hardware, administrative controls, and "person power" that will complement the protective measures already in use to provide effective controls of the vulnerabilities identified in part IV, Conclusions. Cost/benefit ratio considerations are applied to each recommendation and to the structure as a whole.
>
> ---
>
> *Note*: The client's experience is always reflected in the appropriate sections of the report.
> *Source*: Compliments of Charles E. Hayden, PE, CPP, Senior Security Consultant (Retired)

If you sit around and wait for divine inspiration to give you the perfect beginning, you may well wait forever. You have to be able to develop a discipline that says, "I must write, therefore I shall," and immediately take pen to paper. It will matter little that first passages of the draft are poorly phrased or could be worded better. Doubtless, the final report will have been reshaped and revised a number of times. The rule to remember is: "There is no such thing as good writing; there is only good rewriting." Only after painstakingly preparing dozens and dozens of reports does one's writing skills improve. Only then does the job become less painful and the product more professional. Finally, when writing truly becomes a joy, professional reports will follow. Few of us, however, will reach this exalted plateau, and it is not the purpose of this text to chart a course to becoming a professional writer. It is our purpose, however, to describe some of the time-tested methods for writing better reports, thus increasing one's ability to better serve clients and management.

Survey reports generally have two functions: first, to communicate, and second, to persuade. The findings, conclusions, and recommendations in the report are very important to management. For the report to communicate effectively, the channels must be clear, the medium must be incisive, and the details must be easily understood. Also, the story must be worthy of the material. Too often the skill and effort expended during the fieldwork portion of the survey are lost in the murky waters of a poorly written report. A dull or poorly written report will not penetrate the upper circles where the story needs to be told and where the persuasion needs to occur. Write your report expecting it to be read by the top decision maker in the organization, and you will always be safe.

Five Criteria of Good Reporting

The ability to write good reports can be developed. With the proper desire, the right amount of effort, the right standards, and the proper techniques, much can be accomplished. We stated earlier that fieldwork is largely measurement, which implies the existence of acceptable practices and standards (criteria). Reports can also be measured against standards. A good report must meet the following criteria: accuracy, clarity, conciseness, timeliness, and slant (or pitch). To assist the reader in improving one's ability to write reports, we will consider each of these criteria separately.

Accuracy

A reader should be able to rely on the survey report, because of its documented fact and inescapable logic. Additionally, the report must be completely and scrupulously factual, based entirely on the evidence at hand. Likewise, the report should speak with authority. It should be written and documented so as to command belief and convey reliability. Facts and figures, as well as statements and recommendations, must be supported by the evidence. Statements of fact must carry the assurance that the person doing the survey personally observed or otherwise validated them. When it is necessary to report matters not personally observed, the report should clearly identify, if practicable, the true source. The writer must be careful to avoid personal attribution. Internal politics are such that in some organizations a supervisor or manager who reports candidly may become the subject of later retribution. Like a newspaper reporter it may be necessary to protect one's sources of the information. Another important facet of accuracy—and one often overlooked—is perspective, the reporting of facts in the proper light. As an example, it would be reporting out of perspective to show how one of a dozen related activities may be deficient without showing how all of the activities relate, one to another, in order of importance. This is sometimes referred to as "balance."

For a report to be accurate, then, it must encompass truth, relevancy, and perspective. It must also have balance.

Clarity

One cannot write clearly about subject matter one does not understand. Clarity implies communicating from one mind to another, with few obstacles in between. To accomplish this, the report writer must have a firm grasp of the material at hand. Until reporters reach this level of understanding, they cannot take pen in hand and start to write a report. They would be better off returning to the field and continuing their activities until they develop the comfort level necessary for them to continue.

Poor structure is an impediment to clarity. An orderly progression of ideas lends to clarity and thus enhances understanding. Therefore, as most professional writers have found, one of the best aids to effective written communication is the prepared outline, as shown in Box 9.1.

Acronyms, abbreviations, and technical jargon should be avoided when possible. If it is not possible to avoid their use, then at least give the reader at the outset an explanation of the term or initials. For example, at first mention: "Department of Defense (DOD) Bulletin #12 states ..." not "DOD Bulletin #12 states ..." Do not assume that everyone reading your report will be familiar with a term, acronym (initials), or technical jargon or slang terms common to the environment being studied. Quite to the contrary, good writers assume that their audiences are not familiar with these oddities and take the time to explain them in order to enhance clarity and understanding.

Likewise, reporting a finding without properly setting the stage can lead to misunderstanding. Only by reporting relevant information and background data can the author expect the reader to understand the process or condition and thus appreciate the significance of the finding. If one is recommending a new procedure, one should first state what procedure, if any, now exists and why it isn't working. This makes the reader fully cognizant of the procedure in question and much better positioned to consider the proposed change favorably.

Long discussions of technical procedures will bore the reader and muddy the waters. Here, the liberal use of flowcharts, schedules, exhibits, and graphs can be a real aid to clarity and ease of understanding. Be ever mindful of the old Chinese proverb, "One picture is worth a thousand words." Such aids benefit both the reader and the writer of the report. If it is axiomatic that one of the purposes of the survey report is to stimulate action, along the road to change, then the report must be an effective stimulant. To be effective, the writing in the report must be clear!

Conciseness

To be concise means to eliminate that which is unessential. Conciseness, however, does not necessarily mean brevity. The subject matter may dictate expanded coverage. Conciseness does mean elimination of that which is superfluous, redundant, or unnecessary. In short, it is the deletion of all words, sentences, and paragraphs that do not directly relate to the subject matter in question. This is not to suggest an arbitrary, telegraphic style of writing, though in some instances this style may be appropriate. It is, however, to suggest that easy flow and continuity of thought do not depend on excessive verbiage. Short, simple, easy-to-understand sentences are the general rule. Whenever one comes head-on into a long, complicated sentence, continuity of thought on the part of the reader generally is replaced by confusion. Long sentences should be dissected and rewritten to obtain a comfortable, happy medium.

Do not, however, confuse conciseness with lack of data. There must be sufficient detail in the report to make it meaningful to all levels of the audience. Those involved in the intimate day-to-day operations, as well as the president and chairman of the board of directors, must have sufficient information to understand the problems before they can make informed decision.

Timeliness

Often it takes a while to get management to approve having a security survey or audit initiated. Once the decision is made and the project begins, management waits impatiently for the final result of the survey, the written report. The report should answer all of management's needs for current information. Therefore, the report must be submitted in a timely fashion. One must be careful, though, not to be in such a hurry to communicate one's findings that the report is otherwise unacceptable. The survey report must meet all the other criteria for good writing, as well as timeliness. Remember, we live in the age of instant gratification!

How do we then meet objectives, timeliness, and thoroughness? One solution may be the interim, or progress report. When, during the course of a survey, we come across a finding of such importance or magnitude that the details must be reported to management without delay, we can prepare an interim report or memorandum of special findings. An example may be a serious safety deficiency that if not corrected immediately could lead to injury or death. Another example is if one uncovers the unmistakable danger signs of fraud, theft, embezzlement, or industrial espionage (see Appendix B). Interim reports communicate the need for immediate attention and, one would hope, immediate corrective action. In this regard, one would probably first communicate the deficiency to top management orally, then follow-up immediately with the written interim report or memorandum.

Interim reports are designed to be short and to the point. They should address one subject, perhaps two at most. These reports should be clearly identified as "interim," and they must contain, usually at the bottom of the page, written disclaimers against accepting them as the final word on the subject being reported—"in the absence of a full inquiry," or words to that effect. The interim report's purpose is to give management an opportunity to focus immediately on the reported condition and get corrective action started without waiting for the final report to be published. Interim reports notwithstanding, the final survey report should always be submitted in a timely manner.

It is helpful to establish, at the outset of the project, goals and time frames, which include the issuance of the final report. Schedules are necessary to keep survey projects under control. The larger and more complex the project, the more important scheduling becomes. This is especially true for those parts of the survey that are least desirable; report writing unfortunately falls into this category for many people. We have found that once the fieldwork is completed, there is a tendency to delay the writing of the report.

Nevertheless, the end result of the task is the written report. Management has every right to expect that reports be submitted expeditiously.

Another thought on this subject: We have seen otherwise outstanding fieldwork totally destroyed because a report was not submitted to the client on or before an agreed-upon deadline. A security survey can be an expensive proposition. Clients often become upset, and rightfully so, if they feel they are not getting what they are paying for, in a timely manner.

Slant or Pitch

Finally, the writer should consider the slant or pitch of the report. Obviously, the tone of the report should be courteous. We should also consider the effect it may have on operations personnel. It should strive for impersonality and thus not identify or highlight the mistakes of easily identified individuals, small units, or departments. The report must not be overly concerned with minor details or trivialities. It must avoid sounding narrow-minded, concentrating instead on that which has real meaning and substance. The report should clearly be identified with the needs, desires, and goals of sound management principles. Pettiness must be avoided at all costs. Remember, careers may be at stake, so caution should be the watchword.

Format

A report is generally divided into two parts. The first concerns itself with substance—the heart and soul of the project. The second is form, and form usually is dictated by the type of report being prepared, whether it is formal or informal, final or interim, written or oral.

The form of the report can also depend on those to whom the report is being directed. What do prospective readers expect? How much time will they be able to devote to reading the report? The answers to these questions often become the deciding factors regarding the report format used.

Also, different writers employ different formats. There is no universal style or format that can be recommended to satisfy all needs. Novices will usually begin by experimenting until they find a format that best suits their style of delivery. When they find an effective format, they can modify it to meet the specific requirements and tasks at hand. The following elements are usually found in survey reports.

Cover Letter

The cover letter can serve many purposes, not the least of which is to be a transmittal document for the report. The writer may wish to include in the cover letter a brief synopsis of the findings. This would be especially helpful to members of management who are concerned only with the broad overview of the project. The writer must make certain that the details contained in the synopsis are factual extracts from the body of the report. The tendency to take poetic license and "summarize the details" must be carefully avoided, or meaning may become distorted. Whenever possible, cover letters should be limited to one page, for the benefit of the busy executive; also, anything beyond a one-page transmittal may invade the province of the attached report.

Body of the Report

Title

The title should fully identify the name and address, with zip code, of the entity being surveyed. It should also include the dates between which the survey was conducted, but not the date the final report is being submitted to management—that goes on the cover letter. Also, the title should clearly state that this is a "SECURITY SURVEY (AUDIT)," in the top center of the first page, in capital letters.

Introduction or Foreword

The introduction should be brief and clear and should invite further attention from the reader. Its purpose is to provide all the data necessary to acquaint the reader with the subject under review. At this point, the writer may wish to identify the sites or departments toured, the people interviewed, and the documents examined. One point of information that is essential here is the authority for the survey; this is usually contained in the opening sentence—for example, "This survey was initiated under the direction of the XYZ corporate management, specifically to identify and evaluate. ..."

Purpose

The purpose section is a brief description of the objectives of the survey. It must be in sufficient detail to give the reader an understanding of what to expect as a result of the survey. The purpose should be spelled out precisely at this point so that when the findings are reported, they can be seen to conform to the statements contained in this section. This technique makes it easier for readers to find the substance of any particular item without reviewing the entire report. As an example, in the statement of purpose, both the objectives and findings may be listed in numerical or alphabetical order, for ease in comparison between the two.

Scope

The statement of scope should be a clear delineation of exactly what areas are being reviewed and to what extent they are being examined. It sometimes helps to clarify the issue by specifically spelling out areas that are not covered in a particular survey. This is especially helpful when the title of the report is so broad or general that it may lead the reader to expect much more than the report actually delivers. In a brief report, it may be advantageous to combine the scope and purpose for the sake of brevity.

An example of a statement of scope is, "We limited our review of the purchasing department to those activities that directly relate to sole-source acquisitions. We did not review competitive bid activities in the course of this examination."

Findings

Findings are the product of fieldwork. They are the facts produced by the interviews, examinations, observations, analyses, and investigations. They are the heart and marrow of the survey project. Findings may be positive (favorable) or negative (unfavorable). Findings can depict a satisfactory condition that needs no attention or an unsatisfactory condition deserving of immediate correction.

Positive findings require less space in a report than negative findings. Because they do, however, represent a survey determination, it is essential that they be included. Some writers do not report positive findings because they see no reason to burden management with matters requiring no corrective action. Others report favorable findings to show objectivity and to give balance to the report. We favor reporting both positive and negative findings so that management can see the operation under review in total perspective.

The statement of reported findings usually includes a summary of the findings and the criteria or standards of measurement used. Also included are the conditions found, and, for deficiencies, their significance, causes and effects, and recommendations for corrective action.

In preparing to present a finding, one should be able to answer the following questions:

- What is the problem?
- What are people (procedures) supposed to be doing about this problem?
- What are they actually doing, if anything?
- How was this situation allowed to happen?
- What should be done about it?
- Who is responsible to ensure that corrective action will be taken?
- What corrective action is necessary to remove the deficiency?

Using these questions and other obvious questions as criteria, the writer can be satisfied that he or she has done an adequate job, in a fair and impartial manner. Only then can the report writer set forth the findings in a way that will satisfy the client that the recommendations have merit and meaning. It is well to remember, however, that the sole responsibility to initiate action, based on the recommendations, is the duty of the person who receives the survey report. The security consultant can also assist the client in the implementation phase of the recommendations, to the extent desired by the client.

Statement of Opinion (Conclusions)

Not all reports contain the opinions of or conclusions reached by the members of the survey team. Those that do usually provide a capsule comment that reflects the professional opinion and judgment of the surveyor regarding those activities that have been reviewed. When they are reported, comments can be both positive and negative.

Opinions are nothing more than the professional judgments of the people making the review. Some professionals believe that management is entitled to receive and review their opinions and conclusions. Some believe otherwise, eliminating conclusions entirely from the report and moving instead directly to the findings and recommendations.

Regardless of the report-writing technique used, when one sets forth a conclusion or an opinion, it must be supported by fact and fully justified. Also, the opinion must be responsive to the purpose of the review as set forth in an appropriate part of the report, or it may well be superfluous.

Although it goes without saying that opinion more often than not finds fault, one should not hesitate to express positive opinions when compliments are deserved. This should be done not only in the interest of fairness but also to lend balance to the report. An example of an opinion (conclusion) is, "In our opinion, the procedures designed to ensure an orderly evacuation of bedridden patients in the event of fire or natural disaster were found to be adequate given the obvious limitations of a high-rise environment at this hospital." Or, "In our opinion, the transit system was determined to be safe and secure based on our analysis of the Reports of Serious Incidents made available for our review."

A word of caution! Due to the ever-increasing problem of civil lawsuits regarding third-party liability (adequate vs. inadequate security), we recommend that reporters studiously avoid putting conclusions in writing. This is especially true for negative conclusions. Remember, your report may well be used in future litigation. Caution is again the watchword in this regard.

Summary

Reports are evaluated according to their accuracy, clarity, conciseness, timeliness, and slant (or pitch). Whatever format is used (and formats come in many styles), it should include one or more of these common elements: a cover letter and the body of the report, which includes a title; a foreword or introduction; the purpose; the scope; the findings; the opinions; the conclusions; and the recommendations. Good writing takes constant practice and effort, but any writing can be improved by using certain techniques and adhering to established criteria or standards. Begin by preparing a logical outline as the skeleton to which the muscle can be added later. A liberal use of flowcharts, schedules, exhibits, photographs, and graphs can be a real aid to clarity and understanding. Finally, remember the golden rule of report writing: "There is no such thing as good report writing; there is only good rewriting."

10

Crime Prediction

What will amount to adequate security varies in each individual case. No case has determined how many guards might be required or provided any other specifics as to what constitutes adequate/reasonable security.
—Lopez v. McDonald's, 193 Cal. App. 3d 495 (1987)

This chapter examines crime as an element of hazard identification and provides some guidelines to predicting its probability and estimating its criticality. With this information, the security manager will be better equipped to justify the allocation of resources (budget) to mitigate the effects of these hazards. "By analyzing statistics and methods of operation, specific crime-conducive conditions will be obvious."[1] The risk for crime can come from within the organization or from third parties on the outside. Because business owners owe a duty of care to employees, patrons, and guests, this chapter also outlines methods to establish whether the business or landowner is "on notice" that future crime is foreseeable and therefore required to take action to prevent the crime, to mitigate its effects, or to warn patrons and employees of the danger.

Unfortunately, there is no exact formula the security manager can use to make these determinations. This is due in part to the following:

- Differing laws among states
- Differences in the duty owed and the interpretation and applicability of case law for various types of property. Restaurants, shopping malls, and theaters, where guests are "invited," may be treated differently by the courts than an industrial site, college campus, or doctor's office.
- Changes in case law and decisions
- Different interpretations regarding the assignment of responsibility. Some courts believe that if the perpetrator of the crime is at least 50 percent responsible, third-party liability is dismissed. Other juries have awarded victims large sums even though they found the perpetrator of a crime only 35 percent responsible, assigning the majority of the responsibility to the property owner.

The prediction of crime is, like risk analysis, an inexact science and is often based on the professional's best educated guess. The methodologies used to predict crime are subject to more debate than those of risk analysis. Many factors influence criminal behavior.

---

[1] D'Addario, F.J., 1989. *Loss Prevention through Crime Analysis*. National Crime Prevention Institute, Butterworth-Heinemann, Boston, MA.

Precise prediction is difficult, if not impossible. Statistical models to predict crime, such as the Burgess method, configurational analysis, multiple regression, multidiscriminant analysis, and log-linear analysis, are usually not required to meet the goals of the security manager. Simple methods exist that can give the security manager a systematic indication of future crime risk. Relying on the criminological theory that recidivism is the best predictor of future crime (though some theories hold that age, demographics, or causation are the major predictors), we can apply this rate to individual categories of crime to estimate the potential that crime in the surrounding community will "spill over" and affect the safety of employees at a target location.

Analysis of Internal Crime

As in risk analysis, the prediction of internal crime relies on historical data. We can expect past or current crime rates to continue or increase into the future if conditions responsible for the criminal activity (opportunity, for example) do not change. The probability of hazards and events, to a great extent, depends on consistent conditions over a time period sufficient to draw statistical inferences from past data. For example, if a company averages a certain number of worker terminations per year, we can predict that the company will probably experience close to the same number in the future if no factors are introduced to mitigate this risk.

Crime analysis focuses managers on the past, not the present. State or federal Uniform Crime Reports (UCRs) can be used as data sources, but their information is already a year old by the time they are published. By the time countermeasures are devised, capital budgets approved, and service or equipment proposals submitted, we implement countermeasures based on data that are 2 years or more out of date. Projecting crime trends will give management a better understanding of present and future crime exposure.

Given that a small percentage of employees cause most crime, the arrest or departure of a single employee could radically affect future rates. Many security professionals believe internal crime, especially theft, is the result of drug- and alcohol-dependent employees. The introduction of preemployment and random drug testing, education, and treatment programs will surely affect the future projection for drug use, injuries, and theft.

If the business has already compiled adequate historical data, it becomes a simple matter to project the trends into the future. Software programs are available that sort incidents by building, site, and time. They generate incident reports and produce trend analyses[2] by location, day, time of day, modus operandi, and other identifiers to pinpoint expected losses in specific areas or divisions, or by types of crime. If accurate records have not been maintained within the security department, check with the accounting, human resources, internal audit, and risk management departments for information.

---

[2]The trending of lagging and leading indicators can help to verify the effectiveness of security surveys and controls. Analysis of the trends may indicate the need to refresh the security survey.

When gathering data, especially if it is anecdotal, use caution that it is not misreported or misclassified; many nonprofessionals, for instance, confuse burglary with robbery.

Adjustments must be made for factors that can influence the future occurrence of the crimes or incidents being projected. Staffing increases and changes in workforce demographics and employee attitudes (morale, job satisfaction) must be considered. How will changing rates of domestic violence against women affect the workplace if the workforce is predominantly female? Do labor contracts expire soon?

Projections can be represented as a rate; UCRs prior to 2004 listed the number of crimes per 100,000 population or as the number of crimes expected per year. A retail organization can express its rate as the number of crimes (such as robbery) for the total number of stores. The use of rates will maintain consistency in comparisons if the base (for instance, the number of stores) increases or decreases, but when deriving rates, be very careful that numerical biases are avoided. Look for cyclical trends, such as rising theft during the holiday season or after new-product releases. Use this information to pinpoint areas of concern and to utilize limited resources in the best way. Concentrate patrols in high-crime or high-potential-crime areas of a campus, building complex, or parking lot. Focus on certain crimes, such as rape, that the projections identify as tending to occur, for example, more often during certain months of the year.

Analysis of External Crime

The intent of this section is to guide policy and decision makers to realistic evaluation of the risk associated with criminal behavior from the community that may affect the health and safety of employees and assets. Its assumptions and generalities are not sufficiently focused to allow predicting the future criminal behavior of specific members of that community, or of a specific applicant, employee, contractor, or former employee. The prediction of criminal behavior is an inexact science, open to many debatable issues; errors in prediction are therefore inevitable. The security professional should evaluate the results of these predictive methods in light of the totality of the circumstances, of currently accepted criminological theory, and of the professional's experience. If adjustments seem necessary, look to the costs and personal losses that a false negative may create. Is it in the best interest of the firm to err on the side of extra protective measures against a rape, even though the analysis may indicate the probability is very low (say, one in 25 years), rather than devote more resources to a higher-cost, higher-probability event, such as shoplifting?

External crime can be analyzed for three purposes. One is to determine, as accurately as possible, the true potential that crime in the surrounding community will affect the health, safety, and assets of the organization. The second purpose is to determine whether the business is legally "on notice" that injury to employees, patrons, or guests, by third parties, is foreseeable. The third is to learn how a business should go about gathering data to support or defend a claim of negligent security after an injury has occurred—that is, how does one determine whether the injury was foreseeable?

Many security managers simply looked at the crime rates for their locations and benchmarked the data with that of a similar organization or city with comparable demographics and population. When using or deriving crime rates or survey data from valid research, always compare apples with apples—measure parking lot crime against parking lot crime from other areas, and compare it to the norm for the study areas. For example, a survey found that 35 percent of restaurant industry workers admitted to taking company supplies for personal use. Unfortunately, we cannot expect all restaurant employees to follow the same pattern because the study focused only on fast-food establishments.

Prediction of the "spillover" of crime from the surrounding community is based on many assumptions. In an industrial setting, we need to assume that the level of security protection is no different from the local standard. It makes little sense to compare burglary statistics for homes or businesses that have a minimal level of protection with those of a location that has the best locks, the highest fences, the brightest glare lighting, closed-circuit television (CCTV), and an armed security force with attack-trained dogs.

The following method is an attempt to estimate the chance of external crime affecting the workplace, absent any security measures beyond the local standard. This method is based on criminological theories and established trends, but it has not been validated in a court of law or by a professional review board. We have found it both a useful and accurate tool to justify manpower and budget decisions based on projected needs.

1. Select the crimes to measure or predict. Include Category I ("Indexed") crimes:
   - Murder
   - Non negligent manslaughter
   - Forcible rape
   - Robbery
   - Aggravated assault
   - Burglary
   - Grand theft
   - Motor vehicle theft
   - Arson

   Include other crime exposures that would have a negative impact on the company, especially those that may result in litigation. These may include simple assault and battery, vandalism, trespassing, drug sales, and the like. Be sure to consider any special risks, such as the presence of local extremist groups.
2. Research crime data for the offenses selected above. Data on crime rates are obtained from local, state, and federal sources. Understand that "there are concerns that criminal justice data collection mechanisms are woefully inadequate and not standardized across the state and federal systems."[3]

---

[3] The Joint Emergency Preparedness Program (Jepp), *Domestic and Sexual Violence Data Collection: A Report to Congress under the Violence against Women Act* (Washington, D.C.: National Institute of Justice, 1996). Although improved, this is still true today.

Crime statistics are inaccurate because of underreporting and differing interpretations of the definitions of crimes. Increases or decreases in the rates are also affected by the level and focus of enforcement (that is, targeted enforcement) and by the efficiency of the enforcement agencies. The Federal Bureau of Investigation (FBI) UCR summarizes crimes under the "hierarchy rule," recording only the most serious crime within an incident. The UCR is the nation's primary source of information about crime and arrest activities of local law enforcement agencies. It is relied on by the general public as an indicator of community safety. A different, complimentary system, designed to alleviate many of the current problems with the UCR, "moves beyond aggregate statistics and raw counts of crimes and arrests that comprise the summary UCR program, to individual records for each reported crime incident and its associated arrest."[4]

This program, the National Incident-Based Reporting System (NIBRS), eliminates the "hierarchy rule" and looks at detailed offense, offender, victim, property, and arrest data in 22 crime categories and for 46 offenses. Although much improved, implementation of this system has been slow since its inception.

Determine the number of crimes committed for a ½-mile, 1-mile, or 3-mile radius, as appropriate, possibly adjusting for any natural boundaries that may skew the results, if the local jurisdictions maintain data at this level of detail. Crime-tracking software used by many police departments has the ability to list, tally, and plot graphically the various types of crimes around a specific location. If this is not available, use crime district information or, as a last resort, use citywide statistics. Extract the raw numbers from the rate, if a rate is used, by multiplying the basis by the rate. Record these numbers over time: the prior 2, 3, 5, or 10 years. Record the percentage increase or decrease in crime rates over each year for the target areas. Use this percentage for the calculation in step 5 below.

3. Adjust the data in each category by the ratio of reported to unreported crime. Victimization and criminological studies conclude that 63 to 67 percent of violent crimes are unreported. Only 16 percent of rapes are reported to the police. By adjusting the raw figures by these percentages, you should arrive at the "true" (at least a more accurate) rate of crime for each category. Compare your adjusted figures with those derived from victimization studies. This may take some research because these numbers are also subject to change.

4. Reduce the numbers by the recidivism rate (rate of reoffenders). There is discussion about the validity of this weighting because it is not clear whether the 20 percent who do not return to prison have been rehabilitated or are simply evading recapture. How this adjustment will miss new offenders is unknown. Other factors that can affect recidivism include overestimation of the number of reoffenders because of political agendas and racism.

[4] Chaiken, J.M., PhD., July 1997. *Implementing the National Incident-Based Reporting System: A Project Status Report*. U.S. Department of Justice, Washington, D.C.

Some believe the numbers should be increased or decreased by changes in the area's overall crime. Simply adjusting overall crime rates up or down according to the recent trends does not account for the variations in the individual classes of crimes. Generally speaking, property and violent crimes are committed by younger age groups, and white-collar crime is committed by older age groups. Granted, the particular trend for each individual crime can be applied to the data, but this does not consider variations in recidivism for individual crimes. Crimes of violence tend to be impulsive and therefore have lower recidivism (most homicides are one-time affairs), whereas the perpetrators of property crimes tend to continue the practice. The best result is obtained by an examination of the areas' demographics and by adjusting the data for sex and age factors.

5. Calculate the expected recurrence of crime for your location in terms of how many will occur in a certain period—for example, "We can expect one rape in the next 15 years, assuming no additional mitigation is introduced" or in terms of full percentages.

When analyzing rates, remember that certain crimes can be cyclical. For example, cases of rape increase in the summer months; vandalism and sabotage occur more often before and during labor disputes.

6. If possible, set probable loss figures for the predicted events. If your analysis predicts that the truck transporting finished goods from manufacturing to the warehouse will be hijacked once every 5 years, and the average or maximum shipment is worth $4.2 million, you can calculate the impact of this loss. Remember to subtract insurance reimbursements and to add contractual penalties for nondelivery of product, potential loss of market share, extra costs for remanufacturing, the cost of posttraumatic stress counseling, and other indirect costs. Financial loss can also result from civil litigation, loss of morale, restricted access during crime scene investigation, cleanup costs (blood and glass), recruitment costs for a new security manager or a replacement for the injured worker, and increased worker compensation premiums.

Consider short-term or crime-specific items, such as workplace violence or domestic violence spillover. The above method is not very useful for understanding the potential for domestic violence spillover because the comparative "population" is different. In this case, the demographics of the workforce, compared with the rate of victimization, should provide a more accurate projection. As with comparisons discussed previously, the actual projections are affected by the number and type of formal and informal controls in place.

Methodologies that rely exclusively on historical data or other static factors may not best protect the organization against losses from external or internal influences. Past behavior and historical data are not in themselves predictive of future behavior. The analyst must have the ability to project fundamental dynamic relationships into the future. Thus, "clinical" criteria developed from the most recent trends, conditions, and the analyst's experience are needed to add practical value to the results. Reliance on purely historical data may not identify a relatively sudden rise in high-technology invasion-style robberies until the problem becomes widespread. The analyst must use great caution that subjective criteria do not add ambiguity or bias the results by unintentionally or subconsciously weighting factors to justify an agenda and that the added variables are not redundant.

Inadequate Security

The occurrence of crime on property controlled by a business may place its owners "on notice" that the recurrence of a similar crime is "foreseeable." If a person is then injured by a third party, the victim may have a cause of action against the business or property owner for inadequate security. Management ignorance often is responsible for liability in injuries caused by third-party crime. The wise business owner will conduct a foreseeability study to estimate the level of risk and from its results determine what degree of security protection is reasonable. Although there are few facts and valid studies to support expert opinion, attorneys will argue whether the presence of additional security personnel, increased lighting, attention to procedures, or other measures could have prevented an attack.

Many business owners and corporate managers are motivated to conduct a foreseeability study by fear of litigation. Absent litigation, a violent act on the property will almost certainly create an unwanted cost to the organization through reduced employee morale and customer confidence. Additional justifications for a foreseeability study are that its results may accomplish the following:

- Lead to a summary judgment of a case
- Pinpoint areas of concentration for guard staff and patrols
- Lead to better utilization of resources and equipment
- Improve overall security and financial planning
- Help to justify security budgets and programs
- Aid in site selection for new facilities
- Lead to better understanding of crime risk

Third-party liability cases are expensive and often difficult to defend. Judgments for security-related negligence often exceed insurance coverage. Firms have found themselves underinsured, despite high premiums, or not insured at all. Policies that cover the loss of customers, reputation, or future business do not exist. For plaintiffs to prevail, they must generally demonstrate the following:[5]

- The business or property owner had a duty to protect.
- The business or property owner breached the duty.
- The breach of the duty (crime) was the legal (proximate) cause of the injury.

---

[5] Exact legal requirements vary among states.

Crime analysis completed for security-planning purposes is usually not sufficient for presentation to a court subsequent to a negligence (inadequate security) claim. The necessary scope of the analysis, including the type of data analyzed, will change. Property (as opposed to violent) crime, concentration on the specific cause of the injury, and the sources of the data become more important. You must match or exceed the sources of information the opposing parties intend to use in their attempt to establish notice. If the opposition bases its analysis on data that include arrest and incident numbers, you need to do the same. (Arrest information, however, is illegal to obtain in certain states. You may be at a disadvantage if the opposition has access to police contacts, but this information may be subpoenaed or discovered by court order.) This is one of many reasons the security manager or investigator must work closely with legal counsel.

In California, the duty to protect ends at a public area, such as a sidewalk or grass strip, unless it is shown that the business or property owner took control over this public area. There are, however, notable exceptions to this rule. Control could become an issue if employees or patrons must walk across an "uncontrolled" (that is, unowned) property to get to the controlled property. The business or property owner does not always need to own, possess, and control the property in order to be held liable; some courts consider that control alone is sufficient, but others hold that no liability is established in this instance. In one court decision, the actual or "apparent" control over immediately adjacent property and the foreseeability of injury created a duty on the part of the property owner to protect the victim from the danger (or to warn the victim of the danger). A duty may exist if the design of the building or passageways forces employees or patrons to pass through dangerous areas.

As a security manager, use a liberal approach in the analysis. Although the courts may rule in favor of the business in a given case, the time, expense, and adverse publicity of such litigation is to be avoided. The control of an area is usually a triable fact. Additionally, courts view the issue of derived benefit to the injured person differently. At least one court takes the position that "liability does not depend upon whether the defendant derived a commercial benefit from the property" (*Princess Hotels International, Inc., v. Superior Court*, 33 Cal. App. 4th 645 [1995]). It often comes down to what testimony the judge will include or exclude in establishing control and foreseeability for a reasonable distance from the property. For planning purposes, consider crime at adjacent properties even if no legal control over the property exists. Although you may not be responsible in court, an injury could have a damaging effect on your employees.

Once duty and notice are established, it becomes a matter of causation—what level of security protection is reasonable, and would it have deterred, mitigated, or prevented the offense?

There must be a causal connection (proximate cause) between the negligence to protect and the injury or attack. Proximate cause is often difficult to prove—for instance, did the lack of additional security officers or the lack of increased lighting contribute to the attack? This may be difficult to prove or disprove in court. Standards for the amount and type of security the court will find reasonable vary from one type of property to another. Whereas the court may find a shopping mall liable for not providing uniformed security officers, they may rule differently (find no requirement for security officers) in the case of a small business. "Standards or minimums for security can never be ironclad. Security procedures must be adapted to local conditions and changes. That is why the trial attorney requires a security expert to define security negligence."[6] The exact character of the injury is not the correct standard; rather, the question of foreseeability must be decided by the type of harm likely to be sustained.

How to Establish Notice

Notice is based on prior similar incidents. A high rate of embezzlement or other type of white-collar crime does not place the business or property owner on notice for a rape. This does not mean that a high incidence of robbery does not put the owner on notice for rape. Courts have concluded it is not necessary to decide whether particular criminal conduct establishes notice, but rather that it is necessary "to evaluate more generally whether the category of negligent conduct at issue is sufficiently likely to result in the kind of harm experienced that liability may appropriately be imposed on the negligent party" (*Ballard v. Uribe*, 41 Cal. 3d 564, 573, fn.6 [1986]). Also, "It is possible that some other circumstances such as immediate proximity to a substantially similar business establishment that has experienced violent crime on its premises could provide the requisite degree of foreseeability" (*Ann M. v. Pacific Plaza Shopping Center*, supra. 6 Cal. 4th 679, fn.7, 1993). The dissent to this opinion, shared by other courts, holds that "similar" means "identical" at a very specific location (for instance, a rape next door does not put the establishment on notice for rape, only a rape that actually occurs on the premises). Other decisions support the majority decision. *The California Appellate Court in Lisa P. v. J. Gordon Bingham* (43 Cal. App. 4th 376 [1996]) held that prior armed robberies were not similar in nature to the rape of a clerk but still had the effect of putting the defendants on notice. Subsequent cases have modified that decision (*Sharon P. v. Arman Ltd.*, 21 Cal. 4th 1181 [1999]), but the point here is the manager must be aware of the range of possibilities they may encounter.

A business or landowner who knows, or reasonably should know, that criminal behavior is occurring on or close to his or her premises must investigate to determine whether the criminal behavior is likely to pose a risk to those who enter the property in the future, and whether some aspect or feature on the property (lack of lighting or inadequate locks) encourages the criminal conduct, or at least makes it easy to perpetrate. This seemingly negates the concept that business and property owners in some jurisdictions receive

---

[6]Bottom, Jr., N.R., PhD., 1985. *Security Loss Control Negligence*. Harrow Press, Maryland.

"one free crime," one for which notice is not established until the crime actually occurs on the controlled property.

As discussed above, the crime that puts the business on notice usually must be relevant—examine violent crimes in comparison with prior violent crime. However, be prepared to discuss crimes, such as burglary, that, although not violent in themselves, may have a violent outcome.

To establish notice, analyze crime data over varying time periods and distances from the premises. No consistent standard or formula exists to tell the analyst what geographical area and time period to examine. The area and period are best defined by counsel. The analyst will combine the parameters provided by counsel and utilize the various sources of data available to identify the relevant incidents within these boundaries. The analyst will then examine the details of the individual cases to confirm further their relevance to the circumstances that would or would not place the business or property owners on notice. Although anecdotal evidence supplied by employees or others may not be statistically useful, it may be introduced into evidence by the opposing party and would therefore become relevant to the case. The results of the analysis are presented in a report or memo.

After-the-fact analysis of notice is a straightforward process. The challenge arises when the security manager conducts this type of study before any litigation. He or she must anticipate the types of incidents that would cause negligence. The analyst must examine a broader range of scenarios and draw inferences on issues of control, proximate cause, and other negligent security issues. The analyst must consider a wide range of injury, category I crimes, as well as property crimes, for a focused location or for multiple locations, but must also compare these incidents for a more generalized area. A college campus with many differing locations may have differing crime rates, security exposures, and conditions that invite crime.

Previous courts have rejected crime in the neighborhood as inadequate evidence for foreseeability inside a shopping mall. It is important to make comparisons as similar as possible—shopping malls with shopping malls, manufacturing with manufacturing. But in these cases, more is better. Include crime in the neighborhood as part of your analysis; it then becomes a matter for the courts to decide what data are relevant. Normally the incidence of crime is examined at the adjacent property or from a 2,000- to 3,000-foot radius. Include a comparison to other cities, counties, states, and regions. If possible, compare individual districts within the city or similarly ranked districts in other cities with the same population.

The number of police calls for assistance in the area during the past 5 years can be used to show notice. Determine the ratio of calls that did or could result in violence to property, such as burglary and vandalism. The opposition will use this to prove, or try to prove, that the business knew, or reasonably should have known, of prior incidents near the business. Courts usually look at a 3-year review of rates of crimes against people (category I). Crimes against people account for 99 percent of the inadequate-security litigation exposure. The courts, however, may consider a 5-year period as reasonable, and plaintiffs have used 10-year periods when it is to their advantage. It then becomes a matter of convincing the judge which study is the most reasonable.

Review

The prediction of crime both internally and externally is an inexact subject that will help the security manager or business owner to better understand the risk of crime so that resources can be efficiently assigned, preventative measures implemented, or patrons warned of potential danger. Prior similar crimes on or near a property may place those in charge "on notice" that future crime is foreseeable. This is determined in part through the use of crime and other comparative data. If the business or property owner breaches a duty to protect, and the breach is the proximate cause of an injury, liability may be established.

11

Determining Insurance Requirements

Insurance has become part of the overall loss prevention plan supporting proper security. . . . The more comprehensive the security plan, the lower the cost of the insurance.
—Robert J. Fisher and Gion Green, *Introduction to Security*, Seventh Edition, Butterworth-Heinemann, 2004

Having once been employed by what the *Wall Street Journal* described as "the world's largest insurance brokerage firm," this author is aware of just how little I know about the complex business of insurance. Nevertheless, working with some of the industry's most outstanding brokers and risk managers from a number of Fortune 500 companies does give one an appreciation for, and some insight into, the vital role insurance plays in risk analysis and management. The reader is cautioned, however, that this chapter is only a brief introduction to some aspects of insurance of which the security professional should be aware. The advice of competent insurance professionals should always be obtained before deciding on, recommending, or considering insurance matters.

Risk Management Defined

Risk management can be defined as the process by which an entity identifies its potential losses and then decides what is the best way to manage these potential losses. Once a risk is identified, analyzed, and evaluated, the optimum method of treating the risk can be chosen and put into effect. Security-related losses can occur as a result of a variety of factors, such as internal and external theft and manmade or natural disasters. Also, losses through fire, safety problems, and product or third-party liability are some of the immediate concerns of corporate risk managers.

A properly performed risk analysis can be used for many things, but its end result is a definition of the effect risks have on a particular company, in terms of the company's potential for loss. The analysis should tell where, when, and how the risk is likely to be incurred. It should also indicate the extent of loss or liability if the risk does in fact occur and how badly the company would be injured. The risk manager can then design a program to cover the potential losses, exposures, and liabilities. In dealing with most risks, the company is faced with three basic options:

- The risk can be avoided, eliminated, or reduced to manageable proportions.
- The risk can be assumed or retained.

- The risk can be transferred to a third party. (Transfer to a third party generally implies transfer of liability to an insurance carrier.)

Risk Control

The process of eliminating or reducing risk to manageable proportions is somewhat self-explanatory. This is usually done by programming security and safety procedures to do away with problems or to reduce them to acceptable or manageable levels of severity. This is commonly referred to as loss prevention or control.

By assuming the risk, the company makes itself liable for the loss, if any, that may be incurred. If the potential loss is deemed to be within the limits of an expected and otherwise acceptable dollar figure, the risk may be acknowledged and left alone. No effort is made to control, eliminate, or minimize the risk. No action is taken to correct the situation, and no insurance is purchased to cover it. In some cases, a company may develop some form of self-insurance, whereby the exposure or liability is assumed by the company itself. In most instances of risk assumption or retention, the risk is perceived to be small enough that management is willing to assume total responsibility and absorb, out of operating expenses, any losses that may occur—the rationale being that the cure would be worse than the disease. This is especially true in retail stores in which the costs of some losses can be passed on to the consumer. This is sometimes referred to as "the cost of doing business."

When a company transfers a risk, the risk manager, who usually works in conjunction with an insurance broker, endeavors to find the best insurance program available from carriers in the marketplace that provide the needed type of coverage. This is no simple task; it includes, among other things, determining the best deductible and premium payments available, which in the case of large companies often run into the millions of dollars annually.

A risk manager is constantly faced with the problem of the selection of the best method (or if necessary, some combination of methods) of handling each identifiable risk. Regardless of the method or combination of methods used, some basic considerations affect most insurance programs. To be insurable, a risk must substantially meet the following requirements:

- The risk should be worth the cost and effort to insure.
- The risks are calculable, through large numbers of similar risks.
- Losses can be clearly established as to occurrences and amounts. (This is especially important to the security professional who usually investigates the loss in question.)
- Losses must be accidental in nature, unexpected, and unintentional on the part of the insured.

Crime Insurance

Crime insurance is usually obtained to supplement a company's security program. Although the presence or absence of insurance has no deterrent effect on crime, it does reimburse the company in whole or in part for losses sustained in a burglary or robbery or from internal theft. Crime insurance should, like other coverage, be tailored to meet the specific needs of the client. As one insurance broker advises, "If you want to know exactly what your crime insurance policy covers, ask what it doesn't cover."

At a minimum, most crime insurance programs begin with a "3D policy"—comprehensive dishonesty, disappearance, and destruction—a blanket crime or broad-form storekeepers' policy. This coverage will usually reimburse a company for losses due to employee dishonesty or counterfeit currency, as well as for loss of money, securities, or merchandise through robbery, burglary, or mysterious disappearance. These policies also generally cover certain types of check forgery and damage to the premises or equipment resulting from a break-in.

Some forms of specialized crime insurance coverage usually available for consideration include the following:

- *Mercantile safe-burglary policy:* covers loss of money, securities, and valuables from a safe or vault, and pays for damage to the container and any other property damage as a result of a burglary.
- *Mercantile open-stock policy:* mostly used by retail firms as coverage against burglary or theft of merchandise, furniture, fixtures, and equipment on premises, and pays for damage to property resulting from a burglary.
- *Fidelity bonds:* reimburse the employer for loss due to embezzlement and employee theft of money, securities, and other property. (Bonds cover certain positions; employees who directly handle money, cash receipts, and merchandise are often bonded.)
- *Forgery bonds:* reimburse merchants and banks for any loss sustained from the forgery of business checks.

Insurance premiums vary for these programs according to the type of business, store location, number of employees, maximum cash values, amount of security equipment (such as alarms) installed on the premises, and prior losses. Merchants operating in some high-risk crime areas and thus needing insurance are often the least able to afford the coverage due to high premiums. Further, it is difficult to find insurance companies willing to underwrite crime coverage in high-risk crime areas. Companies that experience a number of robberies or burglaries usually face escalating premiums or, worse, canceled policies.

For many small businesses and commercial enterprises, insurance and an antiquated burglar alarm may be the only form of protection affordable. For large companies and corporations, things are much different.

It is generally believed by knowledgeable corporate management, especially risk managers, that insurance should be used for protection only against risk that cannot be avoided or controlled through the effective use of property, casualty, and security techniques. (In the insurance industry, property protection is synonymous with fire, and casualty is synonymous with safety.) This change in philosophy has come into vogue because more and more managers are getting the message that loss prevention, through risk avoidance, elimination, or control, is the best approach for the preservation of corporate assets. It is also recognized that most insurance programs do not fully compensate a firm for a loss, regardless of the type of coverage.

A vice president of a large California construction company complained bitterly because the police were unable to send a detective to one of the firm's construction sites— a newly developed industrial park—to take a report and conduct an investigation into the theft of a $15,000 compressor that had been delivered to the site the day before. The police officer explained that he would have to take the report over the telephone. Further conversation with the officer revealed that little active investigation would be conducted in a case of this nature: it was regarded, the officer said, as "a problem between you and your insurance company." What the officer was not aware of, and what the vice president soon found out, was that the construction company's insurance policy had a $100,000 deductible clause—the theft of the $15,000 compressor would not be covered! The big problem, the vice president lamented, was not so much the cash outlay for a new compressor as the time it would take for delivery, which was going to cause him major difficulties in meeting his contracted deadline.

What this example illustrates is that management must become more interested in avoiding loss, and not rest with the comfortable thought that the company is insured against "any and all eventualities." There is no insurance company of which we are aware that would insure this vice president against the mental aggravation he went through; he is now, however, a believer in loss-control procedures as a solution to these kinds of problems.

A word about deductibles: these clauses are intended to reduce the cost of insurance premiums by deliberately excluding small, frequent losses while covering large, serious ones. The premiums are less expensive for two reasons: small claims are excluded if they fall under the dollar amount of the deductible, and the carrier's administrative cost of settling claims is also reduced.

It is neither the intent of this text nor the purpose of this chapter to do more than explain some basic insurance considerations that most security professionals need to understand in order to do their job properly. Further information can be obtained by contacting the local chapter of the Risk Insurance Management Society (RIMS). The address and telephone number of the RIMS national headquarters is listed later in this chapter. Another good source is the business section of the public library. There is, however, one highly specialized form of insurance coverage that, because of historical developments during the past 40 years and the increase in international terrorism, should be given special consideration in this text—kidnap, ransom, and extortion insurance, known in the industry as "K & R coverage."

K & R (Kidnap and Ransom) Coverage

K & R insurance has been offered by Lloyds of London Underwriters for more than 100 years. During the 1970s, as the demand for K & R coverage increased, a number of other insurers entered this market. Generally, there is little to be concerned about with respect to the financial security offered by such companies as the American International Group

(AIG), Insurance Company of North America, the Chubb Insurance Company, and Lloyds Underwriters. Premium costs and the scope of coverage afforded under each of the companies' policies are generally the basis for deciding from which carrier one decides to purchase insurance coverage.

Such an analysis is normally done at the time the risk manager or broker chooses to explore insuring this risk. Because there is competition in the insurance business—not only from a premium-cost standpoint but also with regard to breadth of form—any coverage comparison here would serve little purpose. Generally, the basic coverage provides reimbursement for loss of monies surrendered as a ransom payment for actual or alleged kidnapping, or following receipt of a threat to injure or kidnap an insured person. In addition, some unique features, such as the following, are usually incorporated into the policy contract:

- *A business premises extension* reimburses for any monies that must be brought in from outside for any kidnap situation if the money is lost while it is on the premises.
- *A transit extension* reimburses for any monies that are stolen between leaving the premises and reaching the kidnappers.
- *A reward extension* includes coverage for monies paid to informants whose information leads to the arrest and conviction of the individuals responsible for the kidnapping or extortion.
- *A personal assets extension* reimburses the insured for their personal assets that are used as a ransom payment if the demand is made on the insured person and not the corporation.
- *Negotiations, fees, and expenses* reimburse for reasonable fees and expenses incurred to secure the release of a hostage, including interest on a bank loan to pay a ransom payment.
- *A property damage coverage* extension provides coverage against threats that cause physical damage to property.
- *Defense costs, fees, and judgments* cover costs resulting from any suit for damages brought by an insured person. The importance of this extension is underlined by an occurrence wherein a kidnapped executive later sued his employer for $185 million in damages, claiming that the employer had not exerted sufficient efforts to free him and had not taken steps to protect him from such an occurrence after being warned that the executive might be a target of abduction.

There are some general requirements, relating to secrecy concerning the fact that the company has K & R insurance coverage, of which the security professional should be aware. Here we caution that the specific details of each policy must be studied and adhered to in the event of a kidnap; otherwise, the incident may be noninsurable. Some of these considerations include the following:

- The ransom or extortion demand must be specifically made against the named insured.
- The extortionate demand must be made during the time frame of coverage as set forth in the policy.

- The company has taken every reasonable precaution to ensure that the existence of the coverage is not disclosed to anyone except senior officials of the corporation.
- If a kidnap occurs, every reasonable effort is made to determine:
  1. That an insured person has been abducted (note: not all policies cover all employees)
  2. That the police or Federal Bureau of Investigation (FBI) has been notified before payment and that instructions and recommendations of the police and FBI in the best interest of the victim are accomplished to the extent possible
  3. That the insurance company is notified at the earliest practical time
  4. That the serial numbers of the ransom payment are recorded

Some underwriters (insurance carriers) require that immediately upon obtaining coverage, written policy and procedural guidelines be established to eliminate the possibility of confusion with regard to the handling of these matters. Box 11.1 lists a number of topics that must be considered in establishing procedural guidelines. As will be discussed more thoroughly in Chapter 18, Crisis Management Planning for Kidnap, Ransom, and Extortion, a more prudent course of action is to develop a crisis management program specifically tailored to the requirements based on the insured corporation's requirements. One of the benefits of such planning is to eliminate confusion before the incident occurs. Once a kidnapping occurs, confusion usually reigns supreme if there is no well-thought-out and rehearsed plan. Even under the best of circumstances, there is high drama and many emotional issues involved when a kidnapping is first reported.

BOX 11.1   EXECUTIVE PROTECTION PROGRAM OUTLINE

I. Home and Family
  A. General information
  B. Telephone numbers, including cell phone and vacation numbers
  C. Biographical data and full descriptions (include DNA samples)
  D. The executive
  E. The executive's spouse
  F. The executive's children
    1. General information
    2. Babysitters
    3. Schools
  G. Residence in general—home checklist
  H. Training for security awareness
  I. Doors and locks
  J. Alarm systems
  K. Lighting
    1. Exterior
    2. Interior
  L. Fence and barriers

BOX 11.1 (Continued)

    M. Window grilles
    N. Dogs
    O. Safe room
 II. Office and Work
    A. Premises—general
    B. Access control
    C. The executive
    D. Executive profile
    E. Employees and associates
        1. Office rules and work procedures
        2. Meetings
    F. Security guards
    G. Bombs
        1. Surveys
        2. Target hardening
        3. The bomb incident
        4. Letter and package bombs
        5. Antibomb curtaining
    H. Threats
        1. Telephone
        2. Written
    I. Hostage
    J. Hostage calls
III. Travel
    A. Automobile
    B. Chauffeurs
    C. Defensive driving
    D. Walking
        1. Jogging, golfing, and tennis
    E. Elevators
    F. Taxicabs
    G. Aircraft
        1. Company
        2. Commercial
    H. Overseas or long-distance travel
    I. Reservations
 IV. Personal Protection
    A. Firearms—defense
        1. Laws
    B. Choosing a defense weapon
    C. Firearms proficiency
    D. Weapon—method of carrying

(*Continued*)

BOX 11.1 (Continued)

    E. Bodyguard
      1. Selection
      2. Personal qualifications
      3. Professional qualifications
      4. Guarding—locally
      5. Guarding—away from home
      6. Visitor protection
      7. Motorcades
      8. Public appearances
    F. Protective clothing
V. Crisis Management Team (CMT)
    A. Defined
      1. Purpose
      2. Composition
      3. Scope
    B. Organization and planning
      1. Readiness plan (prevention)
      2. Contingency plan
        a. Worst possible case scenario
      3. Training the team
    C. Intelligence training
    D. Law enforcement liaison
    E. Public relations considerations
    F. Ransom
      1. Policy and procedures
      2. Limitations
      3. Negotiations
    G. Civil liability considerations
      1. Injury of employee(s)
      2. Wrongful death claim
      3. Stockholder suits

These are some subjects that may need attention. The list is by no means exhaustive. It is our experience that a comprehensive K & R plan should be tailored to the organization, its executives, and the personnel insured. We have found no other way to ensure that all the bases are covered. And, once the plan is developed, it must be constantly reviewed and updated as events and personnel change. An outdated plan is worse than no plan at all.

For serious students of risk management and security professionals who want to understand better how insurance relates to their jobs, we recommend contacting Risk Insurance Management Society (RIMS) Publishing, Inc., 1065 Avenue of the Americas, 13th Floor, New York City, NY 10018 at (212) 286-0202 and subscribing to *Risk Management*, a monthly magazine published by the RIMS.

Emergency Management and Business Continuity Planning

12 ::::

Emergency Management – A Brief Introduction

Stress makes you stupid.
—Unknown

The prospect of planning for emergencies and catastrophic events, including the likelihood our fellow citizens or employees will create untold grief simply to satisfy the needs of their hatred or to impose their religious or political beliefs on others, can be overwhelming. Added to the enormous responsibility placed on the shoulders of emergency managers, security practitioners, and business continuity professionals to effectively protect life, the environment, property, and our economic livelihoods, many perceive their duties daunting at best. Much of the "old school" thinking and the mental paralysis of a "head-in-the-sand" attitude displayed by some who are tasked at lower levels within an organization with the development of emergency management programs (including business continuity) are also based on a misunderstanding of human behavior and systematic organizational response to adversity, and on an unrealistic expectation of the actual capabilities of the government and other response and supporting resources. Managers, in decades past, attempted to list all possible (and sometimes impossible) hazards, assets, individual processes, and threats to their jurisdictions or organizations, and then devised a response plan for each permutation, resulting in volumes of plans that were unmanageable and duplicative. Some of the new Emergency Management and Business Continuity standards resulting from post–September 11 call for, in part, just this sort of planning! Utilizing an "All Hazards Planning" or a Multi-Hazard Functional Planning (see upcoming chapters) approach to emergency management, where commonalities are addressed instead of the focus on plans for every adversity imaginable, is a more effective way to overcome many of the above roadblocks.

Comprehensive Emergency Management

The concept of Comprehensive Emergency Management (CEM), introduced in the late 1970s, took us out of the "Civil Defense" mindset of emergency management with the introduction of an all-hazards approach to the management of emergencies and disasters. Its four phases—Mitigation, Preparedness, Response, and Recovery[1]—represent an integrated approach to the management of emergency programs and activities for all types of emergencies and disasters (natural, manmade, and attack), and for all levels of government and the private sector. It provides a framework for a complete planning process that avoids the tendency to plan for only one element of an emergency. Very often, a plan or organizational responsibility will focus on a single element such as business continuity (recovery), and not incorporate the necessary elements of mitigation or response, for example. When one person is responsible for emergency response and another for business continuity, the ineffective use of resources, overlaps in effort, and disconnects in the transition from one phase to another are typically found. A truly effective planning process will include serious attention to each component in a coordinated fashion.

The concept of CEM is also a risk management method increasingly applied to security planning. The various ways to treat risk—risk avoidance, risk assumption, risk transfer, and risk control—are in one way or another addressed by the CEM model. It forces managers to think in a less myopic way when devising a complete risk-based loss prevention and protection of assets or counterterrorism program.

The United States National Response Framework (which replaced the National Response Plan) and other similar documents look at this model as prevention, protection, response, and recovery. Emerging standards also are morphing this concept, although not stated as such, to include prevention (and deterrence), preparedness (readiness), mitigation, response, continuity, and recovery. In reference to its required programs, the ASIS SPC.1-2009, Organizational Resilience Standard defines these terms as:

- Prevention and deterrence: Avoid, eliminate, deter, or prevent the likelihood of a disruptive incident and its consequences, including removal of human or physical assets at risk.
- Preparedness (Readiness): Activities, programs, and systems developed and implemented prior to an incident that may be used to support and enhance mitigation of, response to, and recovery from disruptions, disasters, or emergencies.
- Mitigation: Minimize the impact of a disruptive incident.
- Emergency response: The initial response to a disruptive incident involving the protection of people and property from immediate harm.
- Continuity: Processes, controls, and resources are made available to ensure that the organization continues to meet its critical operational objectives.

---

[1] Because the term has its origin in governmental emergency management, "Recovery" in this sense means the reestablishment of infrastructure, the local postdisaster economy, and other associated issues; however, it can also include the terms *disaster recovery* and *business continuity*.

- Recovery: Processes, resources, and capabilities of the organization are reestablished to meet ongoing operational requirements within the time period specified in the objectives.

Standards

The ASIS International's SPC.1-2009, Organizational Resilience Standard is accepted as a standard by the American National Standards Institute (ANSI) and is either adopted or under consideration for adoption in many locations internationally. It is one of several that are emerging or changing to adapt to the post–September 11 mindset. Based on the British Standards BS 25999 1-2 Business Continuity Management, they, along with the National Fire Prevention Association's (NFPA) 1600 Standard on Disaster/Emergency Management and Business Continuity Programs, each with a slightly different focus, have taken the lead to define the development and management of Emergency Response and Business Continuity Programs.

The standards are aligned with International Organization for Standardization (ISO) management systems to allow for integration into other business systems and standards, and impose a process management framework using the Deming Model of Plan, Do, Check, Act (PDCA). Many of the standards (ASIS and NFPA in particular) contain appendices that help implement their requirements and, in most cases, can be used to form the basis of a Business Continuity Program.

A standard is a formal document and set of rules or specifications that establishes a norm or a voluntary requirement that can be used as a basis for comparison. They attempt to codify existing and minimal practices, if not best practices, and attempt to ensure the viability, quality, and reliability of programs. They force commonality and consistency, and help ensure programs are in place. The use of these standards in some locations (NFPA 1600 in Canada and certain organizations within or funded by the United States government, for example) are mandatory. Even in jurisdictions that have not codified or adopted the standards into law, they establish a level of care that organizations can be legally (civil and criminally) responsible to follow (detractors of standards, and of these standards in particular, point out that their contents establish too many built-in legal pitfalls). The wording of the standards and the willingness of post-9/11 courts to find terrorist and disaster events foreseeable, may force private employers to comply with these standards. Other incentives to comply include:

- Insurance. The 9/11 Commission encouraged the insurance and credit rating industries to include compliance with these standards (as Standard and Poor's has now implemented) in their ratings.
- Internal/external audit requirement.
- Third-party requirement. A customer may require as a prerequisite for engaging in business.
- Competitive advantage.

Private Sector Preparedness Accreditation and Certification Program

The Private Sector Preparedness Accreditation and Certification Program (PS-Prep) is a voluntary certification mechanism for private organizations using the standards adopted by the US Department of Homeland Security (DHS—see above for a short discussion about just how voluntary the program is). DHS has adopted all three standards for PS-Prep: ASIS, BS25999, and NFPA 1600. The program was initiated by Title IX of the Implementing Recommendations of the 9/11 Commission Act of 2007 and seeks to accredit and certify an organization's efforts that enable preparedness, disaster and emergency management, and business continuity using these standards. Certification is established by an accredited third-party organization. A separate accreditation, the Emergency Management Accreditation Program (EMAP), is used by governmental agencies to assess jurisdictions against NFPA 1600. EMAP begins with self-assessment and peer review. Certification in these programs will help to satisfy the requirements and advantages listed above (EMAP is not part of PS-Prep).

National Incident Management System (NIMS)

The United States established the National Response Plan (NRP), which specified how the resources of the federal government will work in concert with state, local, and tribal governments, and with the private sector to respond to "incidents of national significance." The National Response Framework outlines an all-hazards organized approach to domestic incident response by identifying the major roles, structures, and response methods used by these organizations. It relies heavily on the National Incident Management System, a command and control structure that provides a template for working together to prevent or respond to threats and incidents regardless of cause, size, or complexity and consists of the following components:

- Preparedness
- Communications and Information Management
- Resource Management
- Command and Management
- Ongoing Management and Maintenance

NIMS relies heavily on the use of the Incident Command System.

The Incident Command System (ICS)

The ICS is a hierarchical management system used by governmental agencies, fire, and police to respond to an emergency. It is primarily a field response system but is adapted for use in the Emergency Operations Center (EOC). Devised by the fire service in 1971, it provides guidelines for common multiagency operating procedures, terminology, communications, and management. Its modular structure allows for a consistent and coordinated response to incidents of all types and complexity. Because of the growing interdependency among the response organizations of industry, business, and governmental agencies, the use of the ICS by business and industry is becoming commonplace, especially in light of the above standards. Emergency response teams (ERTs) may already be required by law or industry standard to utilize this system. If the organization's response requirements don't warrant the use of an ERT, business owners and responsible managers (including recovery planners) should still be aware of the methods and protocols, such as ICS, that are used by the local jurisdictions to manage emergencies at their site.

ICS is a tool that relies heavily on the concept of Management By Objectives (MBO). Response objectives are set by the senior responder and delegated to the subordinate positions after agreement that the objectives can be met. The senior responder is referred to as the incident commander. By using this approach, the incident commander can coordinate the response to complex and technical incidents without unreasonable expectations. ICS is also sensitive to the basic management principle of span of control that limits the ratio of subordinates. If the incident is small and the response is relatively simple, the ratio is eight subordinates to one manager. If the crisis expands and becomes more complex, the span of control is reduced to provide the most effective leadership. Some believe the span of control in an emergency situation should be five subordinates or less.

ICS is divided into five major functional units. The fire service version is expandable to 36 positions, but most are not relevant to business response. The five units (called sections) are as follows:

- Incident command
- Operations
- Planning and intelligence
- Logistics
- Finance and administration

With small incidents, it is not necessary to establish all sections of the ICS. In this case, the incident commander (see below) will directly manage or assume the duties of each of the sections or activate the sections as additional personnel arrive. Operational need is the primary factor in determining what is activated.[2] Each unit is headed by a section chief and may be further divided into subsections as required by the complexity of the incident or need to maintain the proper span of control.[3]

---

[2] ICS for executives: *Standard Emergency Management System Executive Course, Student Reference Manual*, State of California, August 1995.

[3] The Law Enforcement version of ICS uses the term "Officer-in-Charge" or "OIC."

Incident Commander

The incident commander has overall responsibility at the incident or event. A distinctive vest that contains the words "Incident Commander" is worn as identification. The incident commander determines objectives and establishes priorities based on the nature of the incident, available resources, and agency (or company) policy.[4] The role of the incident commander is usually filled by the first responder to arrive at the scene, who is relieved of this duty when a more senior responder or a designated incident commander arrives. A *command* post is set up at a safe distance near the location of the emergency where the incident commander will manage the response. Once established, the command post should not be moved unless the conditions of the emergency pose a threat. It can be located in the field, at a vehicle, inside an office, or where reliable communications (electronic and verbal) and security (access control) can be maintained. When appropriate, it should be within view of the incident but away from noise or activity that may interfere with the command efforts.

Management must delegate (ahead of time) to the incident commander the authority to make the tactical decisions necessary to stabilize or end the emergency without interference by those who would normally possess some degree of authority. Management's role is in the Emergency Operations Center (EOC) to make strategic decisions based on the events or to allocate resources among multiple incidents, generally not at the scene of an incident. Reliable communications between the EOC and incident commander are essential. The incident commander follows preexisting policy set by management and will use standard forms and checklists to ensure that all tasks are completed. Software programs are available to aid in the management of the emergency in the field, but for these tools to work effectively, as we will point our later, they must be practiced and reside on systems that can withstand field conditions and the possibility of limited resources such as electrical power or extra batteries.

Some of the specific duties of the incident commander are as follows:

- Overall field management and responsibility of the emergency
- Coordination with the EOC or other incident commanders. The incident commander of the firm's ERT should co-locate with the fire or police department incident commander (an element of Unified Command).
- Ultimate responsibility for the safety of responders
- Approval of all Incident Action Plans and resources
- Situational analysis
- Setting objectives and priorities
- Delegating authority as necessary
- Primary responder until others arrive

If the size of the emergency warrants the establishment of the following positions, assistants to the incident commander include an information officer, safety officer, and liaison officer.

---

[4] Ibid.

Information Officer

The information officer, or public information officer (PIO), is the news media or community contact for the event. In a business environment, the public relations representative will fill this role and should be less (or not at all) subordinate to the incident commander, as he or she would be under the government's version.

Safety Officer

The safety officer ensures that regulatory compliance is maintained and develops measures to ensure the safety of all assigned personnel. The safety officer is often responsible for evaluating changing conditions and should have the authority to withdraw responders or to suspend an operation without clearance from the incident commander.

Liaison Officer

The liaison officer assists the incident commander on larger incidents to which representatives from other agencies may respond by coordinating their involvement and providing them with information on conditions, objectives, and resources.

Operations

The operations section implements the action plans and objectives issued by the incident commander. These are the "doers" of the response. They participate in the selection and reality-checking of goals and direct all resources necessary to carry out the response. A constant flow of situational information and milestone achievement is communicated back to the incident commander. Operations can be subdivided into functional or geographical divisions as needed. Examples include first aid, search and rescue, and hazmat cleanup.

Planning and Intelligence

The planning and intelligence section develops the Incident Action Plans (IAP) to implement the goals and objectives of the incident commander. As part of their plans, this section also determines what resources are needed to accomplish each task. Members of this section must gather information about the incident before they can devise a meaningful plan. In a large-scale incident, this section will accomplish the following:

- Collect intelligence (analyze conditions and the scope of the incident)
- Project or predict changing conditions
- Prepare action plans
- Prepare contingency plans if conditions, events, or resources change
- Track resources available, in service, and used

Technical advisors are included in the planning section to provide expert advice when needed. Chemists, safety engineers, toxicologists, industrial hygienists, meteorologists, radiological technicians, and structural engineers are examples of the types of experts that might be included in the response.

Logistics

The logistics section obtains all resources and services needed to manage the incident. This section delivers personnel, equipment, food and supplies, restroom and shower facilities, and so forth. The logistics section simply supplies resources. The planning section is responsible for resource management and use.

Finance and Administration

The finance and administration section maintains records and documents the history of the response. It projects, tracks, and approves expenditures by the logistics section, and completes a final cost analysis of the response. Documentation of times, events, and actions is important to the postincident analysis, insurance reimbursement, criminal prosecution, and defense of a civil action.

Example

This is how it might work—the incident commander might issue the command (goal) to extinguish a fire. The planning section determines that the fire is small in origin, involves general combustibles, and will require dousing with water by one hose team (Incident Action Plan). If no hose is available, logistics will find one. The operations section will then grab the hose and put out the fire. This is noted by the incident commander and the planning section personnel, who are also responsible for tracking any resources used. The cost of the hose, the damage caused by the fire, and the time and other expenses used to put out the fire are tracked and reported by finance and administration.

ASIS and NFPA 1600

Although previous versions of NFPA 1600 directly suggested the use of the Incident Command System, the 2010 revision and the ASIS Organization Resilience standard don't directly refer to ICS. ASIS does briefly mention ICS and NFPA, under its specification for Incident Management and its Explanatory Material (Appendix A), and describes some of its elements such as Management by Objectives (MBO). Given NFPA 1600's historical intent, one can assume this is what they expect. We can think of no other response system that calls for the use of MBO. Popular in the 1970s and 1980s, MBO is a management principle that is seldom used today in business organizations. The Incident Command System is very effective when managing the response to an emergency because, in part, it fits the command and control environment familiar to police, fire, and military agencies. Most business organizations are not organized in this manner, and one of the basic tenets of business continuity planning is to make as few changes to routine (i.e., duties and reporting structures) as possible, especially under stressful or unfamiliar situations. To effectively apply this system to a business environment often requires a change in the organization's normal structure that must be constantly practiced to be effective when it counts. This represents a time commitment that many organizations are reluctant to make.

British Incident Management System

The British Incident Management System, as described in their Major Incident Procedure Manual, was created by the Metropolitan Police in 1985 to focus on the Strategic, Tactical, and Operational aspects of emergency and organizational response. With parallels to NIMS and ICS, these response functions are labeled Gold (Strategic), Silver (Tactical), and Bronze (Operational).

Located at a distant command center, the Gold Commander, in coordination with the Gold Commanders from other related incidents, develop incident response strategies, and control the resources of their jurisdiction. Also detached from direct, at-the-scene operations, the Silver Commander will translate the strategies from the Gold Commander into a set of tactical actions that are implemented by the Bronze Commander. One or more Bronze Commanders directly control the jurisdiction's resources at the incident. This system relies on a functional, not a rank-based hierarchy.

Unified Command

Unified Command is used when multiple agencies or jurisdictions are involved in the response to an incident. Its structure is that of the ICS, but unified command planning combines the objectives of all incident commanders into one Incident Action Plan so that the overall response can operate as if it were a single-agency incident, so that working at cross-purposes to other agencies is avoided. Resources are shared, but participating agencies do not have the authority to approve or disapprove the objectives of others. In the response to certain types of events, such as a terrorist bombing, certain agencies will take a lead role and therefore act as incident commanders.

Despite the necessity of Unified Command to coordinate the response to a large incident, it is often difficult to implement because of the territorial or jurisdictional battles that still exist in today's response community. While the use of ICS is becoming the norm, frequently exercises that incorporate the use of Unified Command, especially those that involve public–private interface, are not completed as often as necessary to be effective. Public agencies and the business community must continue to recognize the need to work together.

Multi-Agency Coordination System

The Multi-Agency Coordination System (MACS) is just as the name suggests; it is the coordination between agencies responding to multiple incidents managed from the Emergency Operations Center, not from the tactical level in the field. Its primary functions are to ensure that the direction of the event or events is communicated and conducted under consistent objectives, assign or resolve any issues of competing resources, and coordinate interoperability of technical communications as well as to act as a source of public information.

Emergency Operations Center

The Emergency Operations Center (EOC) is a location where the management team members or emergency managers and their staff meet to direct or coordinate the response to a large-scale incident or to begin the direction of a business's recovery. To the business continuity planner, the EOC, or command center, as it is sometimes called, is where management coordinates the actions of the individual recovery teams, monitors the progress of the recovery, and passes requests and information up, down, and across the structure of the recovery organization. Although its primary function is strategic, the EOC can make such tactical decisions as the allocation of resources between competing teams, incidents, or plans, specifically if ICS or the above constructs are not used.

The organization must anticipate the complexity of its response and recovery and design an EOC that accommodates its operational requirements. From the business perspective, the physical and operational layout of the EOC can be as simple as a conference or hotel room. Companies with numerous sites in different geographical areas need or at least should design theater-style complexes that have adjoining rooms and integrated audiovisual, computer network, communications systems, and support personnel. Large EOCs can utilize an ICS structure internally and configure the seating or work space along ICS functional lines. The EOC should be located in a secure, structurally safe location that is centrally located and easily accessible to the responders even when transportation systems are likely disrupted. An alternate EOC, such as a mobile trailer, located a distance from the main site, should always be available. In today's technological and highly mobile age, virtual EOCs are possible to accommodate those executives who are traveling out of the area or for those who encounter transportation difficulties. The potential failures of a virtual EOC due to technological inadequacies caused by the incident, the lack of continued training, and practice by the EOC in the operation of the supporting software should be considered when planning for such an arrangement.

The EOC centralizes control of disaster response among individual recovery teams or among multiple governmental agencies where response to issues have not or cannot be reliably predicted or planned for in advance (as we will discuss in subsequent chapters, continuity planning, especially in a business environment, should decentralize decision making and task implementation responsibility to the individual recovery teams as much as possible, negating the need for the EOC to assume a more tactical or planning role).

The primary functions of a governmental EOC are as follows:

- Coordinate the response to large or multiple events
- Create or refine policy
- Allocate resources
- Collect and manage information about the incident, responses, and decisions
- Release information to the public
- Maintain appropriate records

For most businesses, the EOC is a large conference room that has a sufficient number of phone jacks, room for status boards, and separate work space for the management team leader and recovery coordinator. The phone jacks should be wired to allow the firm's phone switch to be bypassed in case of power failure or the destruction of communications equipment. Extra phone sets and fax machines, as well as all supplies needed to operate the EOC (radios, extra batteries, overhead projectors, whiteboards, forms, and office supplies), should be stored in the room or close by. Duplicate supplies and equipment should be stored in the alternate EOC if possible. The need for supplies can be minimized by the use of EOC software residing on a dedicated server located in the EOC that is connected to laptop computers with backup power available. Be aware, however, that the use of technology carries the roadblocks described above (the Operations Chief in a recent incident took 45 minutes to log onto his system because he forgot his password, and the stress and workload thrown upon him in the initial moments contributed to his difficulties).

The flow of information, both from outside and within the EOC, is critical to the decisions made during a crisis, especially in a large-scale operation. Technology, such as computer networks and multimedia displays with colorful graphics, will speed the delivery of accurate information to decision makers and will greatly reduce fatigue. External information must flow between the incident commander (or recovery team leaders) and the EOC. Situational or conditional reports must be available. Cable television, local news feeds, broadcast radio, as well as access to governmental information systems add to this need. The EOC must have the ability to communicate with the outside world even when power and normal modes of communication are disrupted. Backup (redundant) communications, such as satellite telephones and amateur radio, should be utilized.

Other considerations for the EOC include the following:

- Don't overwork the EOC staff. Decisions made under stressful conditions are often less effective. Add long hours to the stress, and the quality of the decisions can deteriorate even more. Shifts should not exceed 12 hours. Arrange for rest periods, breaks, professional massages, and plenty of food and water.
- Access control (security) is important, especially in a large EOC. Unauthorized visitors, media personnel, and managers not directly involved in the EOC operations must be denied access to the center. Consider a badging system to help control access to the EOC.
- A videoconferencing capability is useful, as is an ability to monitor ATV (amateur television) broadcasts. During a major flood, the local ATV club rented a helicopter and sent live video back to an EOC.
- All EOC operations, rooms, and equipment should be connected to backup power generators.
- The EOC must be "user friendly" with respect to both comfort and functionality. Poor lighting, high noise levels, difficult-to-read visuals, poor ergonomics, and other negative "human factors" will tend to fatigue the staff sooner and adversely affect their ability to make intelligent decisions.
- The EOC must be designed to support multishift staffing and to operate continuously for extended periods.

- Keep operations as quiet as possible. Establish separate meeting rooms and a soundproof radio room; deliver television audio through headphones, and use telephones that light instead of ring.

Summary

The four phases of Comprehensive Emergency Management (CEM): Mitigation, Preparedness, Response and Recovery enable effective planning for emergencies and disastrous events of all types and scale. Recent planning models use prevention, protection, response, and recovery, and some standards have redefined the elements to include prevention (and deterrence), preparedness (readiness), mitigation, response, continuity, and recovery. Risk avoidance, risk assumption, risk transfer, and risk control are in one way or another addressed by the CEM model. A Business Continuity Planning Process that does not include all elements of CEM will be less than successful.

Several standards—NFPA 1600, BS25999, and ASIS Organizational Resilience—are adopted by the U.S. Department of Homeland Security for the voluntary Private Sector Preparedness Accreditation and Certification Program (PS-Prep). These standards impose a consistent set of planning and response principles on the emergency management and business continuity community. They rely on a process management framework using the Deming quality control model of Plan, Do, Check, Act (PDCA). In certain instances, the use of NFPA 1600 is mandatory.

The National Incident Management System (NIMS) is a command and control structure envisioned under the U.S. National Response Framework that provides a template for working together to prevent or respond to threats and incidents regardless of cause, size, or complexity. It relies heavily on the Incident Command System (ICS), a hierarchical command and control system used to respond to emergencies and disasters. Its principles are increasingly applied to the management of large events and to business continuity planning. It solves many problems once encountered in response operations by defining common terms and reporting structures, maintaining a manageable span of control, resource management, and is goal oriented. Its principles can also be used to direct an Emergency Operations Center (EOC) where the strategic coordination of an incident, or of multiple incidents, will occur.

13

Mitigation and Preparedness

When I said my business was preventing disasters, a woman from Hong Kong asked me, "How can you stop a typhoon?" Of course, I can't, and neither can anyone else on this planet.

—John Laye, FBCI Contingency Management Consultants, Moraga, CA. Rest in peace, John

Mitigation

Mitigation is sustained action that reduces or eliminates long-term risk to people and property from natural hazards and their effects. According to the Federal Emergency Management Agency (FEMA), mitigation refers to specific actions that can be taken to reduce loss of life and property from manmade hazards by modifying the built environment to reduce the risk and potential consequences of these hazards. It is vulnerability reduction; it reduces the potential for future losses. It is a strategy to eliminate or to reduce the impact of any impediment toward reaching a goal. Its definition and usefulness go beyond natural hazards to include manmade hazards—a definition that is now expanded to include technological hazards and terrorism, such as the effects of hazardous materials accidents and the use of weapons of mass destruction. The concepts of crime prevention and prevention in general fall within this category of Comprehensive Emergency Management.

Mitigation is often a concept underutilized by the business continuity planner, but its importance in many aspects far outshadows their focus on simply producing a continuity plan.[1] A good mitigation program that eliminates or reduces the risks can prevent the need to implement a continuity plan in its entirety, if at all. The investment in a good mitigation program can return literally millions of dollars in damages avoided and help to ensure the survival of the firm after a disaster. It makes communities and businesses disaster resistant and can reduce damage to property and other assets. A good example described in the FEMA literature[2] is the case of the Kingsford Manufacturing Company's charcoal plant, which sustained $11 million in flood damage and did not return to full production for 6 months after a 2-month shutdown. Eleven years later, the plant suffered another $4 million in damage, again the result of flooding (it was shut down twice during that year).

---

[1] A mitigation program and plan are required by both the NFPA 1600 Standard on Disaster/Emergency Management and Business Continuity Programs (2010 edition) and the ANSI/ASIS SPC.1-2009 Organizational Resilience Standard.

[2] Protecting Business Operations, FEMA Publication 331, August 1998. (Washington, D.C.: US Government Printing Office).

The company invested $2.85 million in mitigation (the construction of a levee and other measures) and has avoided not only additional major flood damage but also relocation, helping to sustain the economy of the area. Because the company was able to remain in production in subsequent years, it avoided the loss of retail shelf space (the desirable location and amount of space in a supermarket shelf that draws the most attention to the product), which can take years to regain.

Mitigation can reduce the occurrence of a hazard or a loss. Stronger building codes may not prevent an earthquake but can drastically reduce the damage caused by one. Mitigation can reduce exposure to civil or criminal liability in the event of a terrorist attack or technological accident. Mitigation actions may help reduce insurance premiums. Mitigation allows for a smoother and therefore faster recovery, reduces the "oops" factor as unforeseen incidents and consequences are reduced, and as previously mentioned, helps to avoid the need to restore operations.

Mitigation is cost-effective. According to FEMA, mitigation measures increase construction costs for new facilities between 1 and 5 percent, but when considering earthquake remediation, the cost can be five times higher than the original cost. For every dollar spent on mitigation, three are saved on damages avoided. Warner Brothers estimates they saved $1 million in losses during the Northridge, California, earthquake because of their mitigation efforts. A company that manufactures equipment for the paper and feed industry spent $40,000 in flood mitigation and saved $3 million after Hurricane Eloise, which included $2 million in lost revenue avoidance. Seafirst Bank spends $17.00 to prevent the loss of $3,000 computer systems, estimating a 4 to 5 percent mitigation cost versus replacement, saving $30 million in replacement costs.

The cost/benefit ratio of mitigation strategies is more difficult to determine when dealing with terrorist acts because, absent very specific and credible intelligence, the recurrence rate is more of a guess. Quantification of the damage may not be known, unless you can use maximum probable loss or engineering modeling, and the time period of exposure may vary. Deterrence has always been difficult to quantify.

Mitigation is generally considered the first phase of Comprehensive Emergency Management, but it is not a linear process. Many believe it is important to integrate the recovery and mitigation phases but acknowledge that mitigation takes place during the other three phases of emergency management. They point out that after a disaster, the availability of funds and interest in taking action is at its highest. The need to make repairs also signals a good time to build in mitigation measures, the cost of which may be partially absorbed by assistance programs. After the effects of the disaster are stabilized, one can analyze the damage and design strategies to prevent a repeat of the disaster's consequences. In a business environment (and in the general community for that matter), there is sufficient history, knowledge, and experience to identify beforehand the types of hazards that may affect your location and to then predict their likely effects. You don't need to first break your hand before you learn to wear boxing gloves in the ring. Similarly, you don't need to watch your facility burn down before you can consider installing an automatic fire sprinkler system. Mitigation should start the process and end the process. If you identify what to mitigate, go through recovery, and then find that something was missed, you must go back to mitigation. After action analysis can be considered a form of mitigation if you identify a better method or hazards that were not anticipated. Mitigation planning is often a part of the business impact analysis phase of the business continuity planning process (see Chapter 15), because the methods used to identify and quantify critical functions, the people who should be involved in the process, and the need to conduct inspections are similar and usually the most expedient. The mitigation plan is often combined with the business impact analysis report, which outlines the cost of recovery strategies that are often a form of mitigation.

The FEMA mitigation methodology and much of the FEMA literature are geared toward mitigation planning on a regional basis, not for individual business enterprises. This approach encompasses five major steps: organize resources, assess risks, develop a mitigation plan, implement the plan, and monitor progress. Although the emphasis is on community and governmental protocols, such as memorandums of understanding, and includes steps to satisfy the requirements of the Disaster Mitigation Act of 2000 (in which the U.S. government will fund state and local mitigation projects), many of the steps can be used to develop information that is beneficial to corporate mitigation efforts. For example, the "assess risks" phase includes the following steps:

- Identify all hazards that may affect the community.
- Narrow the list to hazards most likely to cause an impact.
- Develop a hazard profile; that is, determine how bad it can get. This information is used to determine the assets in the hazard area you need to inventory. Areas that can be affected by the hazard are mapped. This tells where the impacts are likely to occur.
- Inventory assets. List assets that will be affected by the event. Governmental planning considers assets to include hospitals, schools, infrastructure and utilities, and the like. Assets affected may vary by the different types of hazards and are inventoried for each. The value of the assets is estimated in this step.
- Estimate the losses. This answers the question: how will assets be affected by the different hazards? The loss of structures, contents, use, and function is determined and summed to arrive at a total loss for each hazard event based on the total or percentage of damage.

Similarly, in a corporate planning process, we identify the hazards; devise strategies to reduce or eliminate the impact of the hazards; select the most practical, cost-effective solution; and gain approval and funding to implement the solution. Once these measures are in place, they must be monitored and maintained to ensure they are functional when needed the most.

Hazard Identification

If a hazard is not identified, it cannot be prevented, prepared for, or mitigated. Many methods exist to assist in the identification of hazards, but few give one a precise formula that applies to all environments or that will help to reveal conditions that are not previously known.[3] This is due in part to the perceptions, values, knowledge, and methodology used by the planner. Unless your processes are very complex, hazard identification in a corporate environment looks outside the organization as well as internally at hazards and risks to processes, materials, equipment, and human resources. As discussed later, interviews with the process owners should reveal hazards that may require mitigation. The ability to identify hazards is really a state of mind, based on a healthy degree of paranoia, knowledge of history, and information about cause and effect.

History

It is important to understand the hazards in your community. Even if your facilities are located on high ground not subject to flooding, if the city's sewage treatment plant located in a flood zone is subject to damage in the next flood, you may be shut down for health reasons. An examination of the city's past disasters will reveal this concern. Most natural hazards are easy to identify because they often recur on a roughly periodic basis or typically occur in the same region under study. If your manufacturing site is located at the base of a mountain that has produced landslides in the past, you may be affected by land movement in the future. Earthquakes recur in the same region according to very broad, often predictable time periods. Major tornados and storms are more common in certain locations. Look at the history of the region and of the specific location and list the type of natural hazards and the expected magnitude of their effects. Predict the future impact of past events, taking into consideration recent mitigation efforts or conditions that contribute to a greater impact, such as recent construction near coastal waters prone to tsunamis. Research newspapers and other historical records. Libraries are a good source; locally or federally declared disaster reports are another. Review existing plans and reports that can be found on the Internet, land use plans, and geological reports. Interview experts in the community, including Office of Emergency Services (OES) directors, universities, architects, and fire and police personnel.

Speak with the "old timers" from the area, or seek the judgment of experts, such as a geologist, to offer their opinion of the stability of a hillside, for example. In the United States, FEMA, the U.S. Geological Survey (USGS), the National Oceanic and Atmospheric Administration (NOAA), the U.S. Department of Justice, and Homeland Security maintain websites that contain information useful to hazard identification. Insurance loss control

---

[3] NFPA 1600 and some security literature recommend classifying hazards according to the broad categories of Natural Hazards (Geological, Meteorological, Biological), Technological (Power Outages, Equipment Failure, Data Corruption), and Human Caused (Accidents, Mistakes, Terrorism).

histories, experiences of other businesses in the area or of those in the same industry can answer many of these questions.

Inspections

A basic component of hazard identification begins with a complete physical inspection of your facilities and their surroundings. Look at your facility from both a macro and a micro view. What hazards exist in the community that will affect your site? Is it located in a high crime area? Can nearby businesses or their processes negatively affect your site? If your plant is located next to a munitions or fireworks factory, or if the tax revenue office is located on the next floor, you may find this of some concern. Look at conditions, equipment, processes, and the environment and visualize ways they may negatively affect your concern (see the section on Cause and Effect, later). If there are trees on the site, do they pose a fire hazard? Can they blow over and injure someone or damage power lines? Can someone climb up to a less protected second-story window? Look for items that could cause collateral damage. For example, if someone placed an explosive device next to an ammonia tank, would the resulting release of gas cause a greater problem? A physical inspection should identify nonstructural hazards, such as file cabinets not bolted to the floor or wall in seismically active areas. Is the building's air vent unprotected and accessible to a terrorist? Failure analysis, knowledge of history, and understanding of cause and effect are important skills needed to identify hazards and risks during inspections. The experience of the inspector cannot be understated.

Checklists

Checklists can help with the inspection process by giving the inspector clues about what to look for. Checklists, however, are not the only tool used for a complete identification of hazards. Checklists can never be complete because the number of environments and hazards are varied and numerous. They should be used as reminders and as a springboard for further thought and should include items that answer the following questions:

- How can employees be injured?
- How can critical systems be damaged?
- What single points of failure exist?
- What hazards can disrupt operations?
- How will hazards affect the environment?
- What hazards can have a public relations impact?
- What hazards might generate regulatory issues?

The checklist should view perils under both normal and disaster conditions.

HAZUS

HAZUS (Hazards United States) is one of many risk assessment software programs that analyzes potential losses from earthquakes, floods, and hurricanes. It is used with

Geographical Information System (GIS) software to map the effects of disasters and can estimate damage from these hazards before and after the event. Depending on the skill of the user, HAZUS can estimate physical damage to residential and commercial buildings, infrastructure, and other critical facilities; calculate the economic loss from business interruptions, repair and reconstruction costs, and lost jobs; and determine the social impacts of the disaster. It can be used to model technological hazards (nuclear and conventional blast, radiological, chemical, and biological) that supplement the natural hazard loss estimation capability. HAZUS is available free of charge from FEMA.

Process Analysis

To excel at this important task, one does not generally need to use each formal methodology, such as Hazard and Operability (HAZOP), Failure Mode and Effects Analysis (FMEA), Preliminary Hazard Analysis (PrHA), or others used in the engineering and safety fields, but if your processes are very complex, you may want to seek outside assistance.

The concept of a HAZOP study involves investigating how a process might deviate from its design intent. It uses a multidisciplinary team methodically that brainstorms the process design, following a structure provided by "guide words" and the team leader's experience. Guide words are simple words used to qualify or to quantify the design criteria and to stimulate the team's thinking toward the discovery of deviations. Guide words include "no," "more," "as well as," and "other than." They are used to ensure that the design is explored in every conceivable way. The causes of meaningful deviations are identified and their significant consequences mapped.

FMEA is a structured technique used to analyze a design or a process to determine shortcomings and opportunities for improvement. It is a tool to identify the relative risks that are designed into a product or process and to initiate actions that reduce the risks with the highest potential. Risks are rated relative to each other using a Risk Priority Number (RPN) for each failure mode and its resulting effects. The RPN is calculated by the product of the severity rating, the occurrence rating, and the detection rating. A 10-step process is typically used to arrive at a mitigation plan (control plan) and a recalculation of the RPN to determine its effectiveness.

PrHA is a line-item inventory system of hazards and their risks. It is an approach that identifies known hazards and their potential consequences to develop an expected loss rate (probability stated as a loss event or unit of time multiplied by the potential loss). As with the above methods, it results in a hazard mitigation plan.

Experts

Consultants can provide a fresh outsider's view and identify potential hazards that you may miss. Terrorism, information, or telecommunications experts can make up for any technological shortcomings of the hazard identification effort. Security and business continuity planning consultants and engineering firms abound, with varying degrees of ability, but your insurance carrier's loss control department may have the ability to provide assistance at a reduced cost. Engineering firms can evaluate the structural performance of your facilities during a natural event. Equipment manufacturers may provide ideas or engineering solutions for hardening their products or may provide ideas that other customers have implemented. In some industries in which it may take up to 6 months to replace a piece of equipment or to recreate its environment, and in situations in which the firm is using equipment that is not replaceable, mitigation may be the only continuity strategy available. Department heads or process owners are probably the first source of effective mitigation ideas.

Cause and Effect

A primary goal of security, business continuity planning, and hazard identification is to anticipate the unexpected. Take what you have learned about history, methodology, and your professional experience and use this information to project possible future scenarios. Use the techniques of scenario planning. "Scenarios form a method for articulating the different pathways that might exist for your tomorrow, and finding your appropriate movements down each of those possible paths. Using scenarios is rehearsing the future. You run through the simulated events as if you were already living them. You train yourself to recognize which drama is unfolding. That helps you to avoid unpleasant surprises and know how to act."[4] Scenario planning is a discipline that all security managers and continuity planners should master. It is a tool to identify and devise strategies based on future variables or changing conditions. Use it to uncover the sources of a crisis, whether a physical, technological, or human-caused hazard. If you are responsible for a petroleum refinery and you are faced with a fire and explosion, you may also need to deal with a toxic vapor cloud release, injuries and damage in the surrounding community from the cloud, injuries and medical response in the refinery caused by the explosion that could overtax emergency services in the community affected by the incident, hazardous material cleanup, control of contaminated water used to put out the fire, and so on. Scenario planning will help you identify all of these concerns. It may be frustrating, however, that management may not agree to fund mitigation projects based simply on your good imagination.

Unlike natural hazards that follow the laws of nature, occur more often, and are therefore more predictable, the acts of a terrorist are more difficult to anticipate. Because targets are often mobile and the terrorist can select those most vulnerable and that return the highest "yield" to their objective, many argue that the threats they pose cannot be identified or predicted. Others argue that it is possible, at least to the extent that we can minimize the damage. Again, scenario planning is useful to make these predictions.

Terrorist acts that are easier to predict involve those who think rationally, as opposed to those whose thought processes are on the fringe or delusional. Organized groups such as Al Qaeda, the Irish Republican Army, and others tend to follow the same methods and use the same destructive tactics throughout their existence. Kidnapping may be the

[4] Schwartz, P., 1991. *The Art of the Long View*. Doubleday Dell Publishing Group, New York.

preferred method of one, whereas bombings may be used by another group, often with little variation. As with any type of crime, we can try to put ourselves in the place of the criminal or terrorist and devise not only targets and methods of attack but also ways to prevent their occurrence or to mitigate their effects. To do this, especially when dealing with terrorist, hate, antigovernment, or apocalyptic groups, we must place their frame of reference—that is, their way of thinking—into ours. Their culture, belief systems, and values could be far different from ours.

In devising these scenarios, we must also determine what is and is not technically possible. Keep in mind that Aum Shinriko, a Japanese apocalyptic group, employed top scientists and spent $6 million in their failed attempt to kill thousands in the Tokyo subway (only 12 people died, because their devices would not work as intended). Much information is available on the technical difficulties of the delivery of weapons of mass destruction. Beware, however, that the terrorist can probably figure out ways around technical difficulties or needs to get lucky only once.

Methodology

As described above, humans are creatures of habit, and many organized terrorist groups or criminal syndicates tend to operate in the same manner throughout their history. Certain animal rights factions tend to release animals; others use arson, and still others use explosives to make their point. The Abu Sayyaf Group in the Philippines specialized in kidnapping and murder for ransom before their involvement with Al Qaeda, when they modified their tactics to include bombings. When identifying hazards and risks from criminal groups and terrorists, understanding their methodologies will direct you toward mitigation specific to their attacks.

Mitigation Strategies

To help find solutions, investigate what others, especially those in your industry, have done, if anything, to solve similar problems, risks, and hazards. The Internet is a rich source (www.fema.gov and www.dhs.gov) of information. Federal, state, and local emergency services organizations often have information and publications that describe specific mitigation strategies.

Mitigation strategies are both general in nature and specific to the hazard. Some specific mitigation considerations are included in Chapter 14, Response Planning, listed under the hazards discussed. Use these as a guide to developing your mitigation program and as a foundation to expand on the ones already presented.

General mitigation strategies can be loosely classified under the following headings. These categories are presented only to give the reader an idea of types of strategies available:

- Risk management
- Engineering controls
- Regulatory controls

- Administrative controls
- Service agreements
- Redundancies and divergence
- Separation of processes

Risk Management

The principles of risk management can be used to identify effective mitigation strategies. The hierarchy of control holds that the elimination of a hazard (risk avoidance) is the first and most effective method to control a hazard. If the hazard no longer exists, you don't need to worry about it. Relocating your facility on higher ground further from a river and moving operations to a lower risk area are examples. The hierarchy lists the preferred order of controls, from the most effective to the least effective: elimination, substitution, engineering, and administrative controls are generally the steps followed in the hierarchy.

Substitution involves replacing a hazard with a process that is less hazardous or nonhazardous. A high-technology manufacturer used a chemical that was making its workers sick. The company found a way to produce the product with a less toxic substance and saved many dollars in health costs.

Engineering Controls

Absent the complete removal of a hazard (which is usually impossible or impractical), engineering controls are probably the next best form of mitigation, especially when compared with codes, standards, and administrative controls, because they do not generally rely on human intervention or action. The flood diversion dam discussed earlier is a good example of an engineering control. Bomb blast resistant designs, earthquake bracing, and "crime prevention through environmental design" are other examples.

Regulatory Controls

Codes and standards are an example of regulatory controls. They serve to eliminate or reduce hazards to the built environment through land use permits (don't build in a floodplain or on unstable ground) and building codes that require minimum safety, construction, and engineering practices. Life safety codes define the required number of exits and specifications of evacuation stairwells. Automatic fire sprinklers are an example. Use caution when designing mitigation strategies based on these standards because they often represent minimum practices. Codes and standards are often revised after a disaster when engineers learn what went wrong with the previous version of the code.

Administrative Controls

Administrative controls include policy and procedure. A requirement that consistent data platforms are used throughout the organization and a requirement that business continuity planning is included in all new project designs are examples. This is considered one of the least effective controls because it is dependent on human intervention.

One company's policy failed when the roof of a building partially collapsed because the person assigned to measure and remove the snow accumulation was late because of the storm.

Service Agreements

Service agreements include contractual obligations for repair personnel to respond to an issue within a certain amount of time, the overnight replacement of damaged equipment, or priority service in times of increased demand, such as backup electrical generator fuel replenishment during a disaster or extended power outage. A common example is the subscription of a service to deliver a number of workstations preconfigured with your software image (that is, the standard programs you use) to your alternate work area within 24 hours of your need.

Redundancies/Divergence

In business continuity planning, the decentralization of facilities, systems, equipment, and processes goes a long way to help ensure the survival of the organization in times of a disaster. In today's business environment, many organizations believe it is in their financial interest to centralize as many of their processes as possible. Despite this, there are many opportunities to build redundant or diverse functionality into equipment and processes. Redundant array of inexpensive drives (RAID drives), dual power supplies and other critical system components, and uninterruptible power supplies (UPS) are examples of basic mitigation at an equipment level. A second, redundant data center and a data center configured on a load-balanced arrangement with diverse telecommunications routing are examples on a larger basis.

Separation of Hazards

Another form of mitigation involves keeping critical equipment, personnel, and processes away from hazards that can affect their functionality. Large water pipes running directly over the server room or mainframe can be a hazard. Similarly, locating your critical processes next to the boiler room could become a problem if there is a fire or explosion. One company at risk for mail bombs located their chief executive's office just above the room where the mail was sorted. To mitigate these risks, one needs to be separated from the other.

Specific Mitigation

Some examples of specific mitigation include the following:

- Alternate power sources
- Alternate communications
- Policies and procedures
- Data backup
- Records management
- Facilities salvage and restoration

Alternate Power Sources

Power surges, spikes, and drops account for the more common data "disasters." These utility problems damage or destroy sensitive computer systems, research, and production equipment. Important data files can be corrupted. If power is lost, work in progress can be lost. Power losses over wide areas are expected to increase over the years as more demand is placed on aging power grids.

The most common and least expensive mitigation to this risk is to install an individual UPS on each piece of critical equipment. A UPS is basically a device that delivers "conditioned" power (current protected from significant spikes or drops) to the equipment. It also contains a battery that, in case of a complete loss of power, will allow the unit to continue operation until a backup generator takes the load or the equipment can save its data and execute its shutdown routine. Some high-technology equipment can be damaged if it is simply shut off without going through this routine.

If the loss of electrical power will have a serious impact, consider bringing in redundant power from a different grid. This will prevent power losses caused by local conditions, such as lightning strikes or downed lines. Feed the power from a different direction and to a different part of the site.

Another common mitigation strategy is to install a backup generator capable of running critical (or emergency) systems as long as a fuel supply is available. If possible, have extra fuel on hand (if electrical power is out, the pumps at the local gas station will not work). Generators are powered by diesel, natural gas, or gasoline. Those supplied by natural gas might also be fed by separate sources or routes. Smaller generators mounted on trailers can be rented and brought on site. If this is the strategy selected, consider the installation of a "quick fit" device (transfer switch) outside the building. This device would be hardwired to the electrical distribution panel; the generator is simply plugged into the building, saving many hours of connection time. Power supplies and generators should be tested regularly under load conditions.

Alternate Communications

Most organizations are highly dependent on communications for voice and data transmission. The loss of voice and data transmission can quickly have a severe impact on the organization. Equipment failure, software glitches, cable cuts, hackers, and fires in cable vaults or central stations can cause the loss of this function for days. Phone companies devote a tremendous amount of resources to ensure the reliability of their networks, but after a disaster, communications become the most quickly affected utility. Because of increased demand, the telephone network can become overloaded and cease to work, even if the equipment is undamaged. This can happen internally, for instance, if a well-publicized event causes a sudden influx of calls that can overload the switchboard's ability to handle the traffic.

Strategies to mitigate damage to the communications system and to recover its function include the following:

- Service and replacement agreements
- Bypass circuits and fax lines

- Divergent routing
- Cellular backup
- Satellite systems
- Hot/cold sites
- Third-party call centers

When equipment fails or is damaged or destroyed, it will need replacement. However, many organizations cannot afford to be without communications for the time required to reorder, deliver, and install new systems. Most communications vendors offer 24-hour equipment replacement agreements for an up-front additional cost, with annual renewals of the agreement. If the impact of the loss of communications is severe or time dependent (such as in a catalogue sales operation), the organization may use a telecommunications hot site. A call to the telephone company will transfer the company's lines to a hot site or third-party vendor where compatible equipment is installed and waiting to go so that the move will be transparent to the customers. Like hot or cold sites for computer systems, there are subscription, setup, declaration, and user fees involved. Modern call centers have operators on standby to answer questions or take messages until the firm is set up in the hot site. This initial switchover can take less than 15 minutes if the company is faced with a local (not regional) disaster.

Many systems are equipped with a number of "power failure" circuits that bypass the phone switch (your main on-site phone-switching equipment) and directly access an outside line. This capability, and the location of these circuits, should be confirmed and used when necessary. If a fax server is not in use, phone lines for facsimile machines can be "borrowed" for voice or data communication. Check to see if your handsets are compatible (digital versus analog) with these circuits.

Voice and data are transmitted around the world through a variety of modes—overhead cable and fiber, underground cable and fiber, microwave, and satellite, to name a few. One of the more common causes of communications failure is a cable cut by a contractor digging a trench. Landslides and bridge collapses also disrupt communication cables. Diverse routing is one method used to protect against these dangers. With diverse routing, your main circuits may pass through southern states, while your secondary or diverse circuits use cables located in northern states. Unfortunately, diverse routing can in reality mean only that your circuits use separate pipes—buried right next to each other. This can be true even if your primary and secondary cables are carried by different communications providers.

After the Loma Prieta earthquake in California, interest in cellular communications for emergency and recovery operations increased. Cellular systems minimize the use of ground-based cable, and the abundance of cell antennas adds redundancy to the system. Many believe they would be more survivable during and after a disaster. The switch to digital cellular makes data transmission and portable Internet access over the cellular system an acceptable strategy for a portion of the organization's recovery needs. The planner should use this strategy with caution, however; as in land-based systems, the increased number of users will congest the network even during nondisaster times. We see this occurring now in large metropolitan areas. After 9/11, newer forms of communication worked better than cellular, but these will suffer the same problems in the future. In fact, the U.S. government may approve the reallocation of cellular channels in an emergency to give governmental services more reliable use of the cellular network. In business continuity, in planning communications and telecommunications, diversity and redundancy are key to success. To add a degree of reliability, consider subscribing to two different providers, if this option is available in your area. Because texting requires less bandwidth, consider this mode over voice communication when appropriate.

Microwave is a form of high-frequency radio transmission beamed from point to point. Microwave transmissions can be used to reestablish communications between buildings across a campus, city, or wider area. They can provide diversity in both voice and data communications, and they are not very susceptible to cable cuts. The use of microwave for diverse routing can be expensive, but it easily spans difficult terrain and provides large bandwidth capabilities. Transmission towers, however, are susceptible to destruction or misalignment by high winds and earthquakes.

Satellite transmission is used for primary or diverse routing, or as a backup communications channel. It is especially useful for continuity planning in that mobile transmitters can be connected to the site and used to reestablish communications very quickly. Their failure rate is very low, and they are affected only by the difficulty with transportation after a disaster. Bandwidth and security are very good. Cellular phone companies, in some locations, are now using satellites as their cell antennas.

Third-party call centers and answering services, used for overflow customer support or order entry, can also become primary call centers after a disaster. Investigate the ability of their equipment and staff to handle the increased call volume. Consider sending some of your staff members who are familiar with the product and company to assist or train the center's staff.

Policies and Procedures

Most "data" disasters are caused by people, not by equipment failure or natural catastrophes. Policies and procedures regarding data systems must be identified, implemented, and enforced to ensure a smooth and effective recovery. A data backup policy is an obvious starting point, but backup policies cannot be completely effective if the client does not store its data on the server. Most client/server architectures allow users to access two areas to store data—locally on the user's C-drive and also on the server. Although there are programs that back up the user's local drive (if the user's computer is running), this is a time-consuming operation and may back up unnecessary files. To avoid this, all users should store their data files on the server. Policies that restrict unauthorized installation of personal software or downloading material from the Internet will help to avoid disasters resulting from computer viruses.

Data Backup

Most organizations can afford to lose a day or two of data or can tolerate the time required to reconstruct a small amount of lost work-in-progress. The primary strategy used in these organizations involves the nightly backup of each day's transactions onto magnetic tape. This is often done despite a lack of a policy establishing its need. Each organization must develop a data backup policy that requires the following:

- Nightly incremental backup of the server
- Weekly full backup
- Monthly archiving
- Yearly archiving

Backup tapes should ideally be taken each day off-site to a storage facility that specializes in the safe and secure storage of computer backup media. Offsite storage facilities should be audited annually to ensure the following:

- Good physical security
- That authorization lists are up-to-date and enforced
- That good fire prevention measures are in place
- That the building is structurally sound
- That it can find and deliver tapes and manuals in a reasonable time

If this is not possible, try to store tapes in a fire-resistant cabinet located in a separate building. During an evacuation, take the tapes (or whatever media used) with you, if this can be done safely. At the very least, backup tapes should be sent off-site weekly, and the previous week's tapes returned and recycled.

Records Management

Many businesses that have not recovered after disasters have considered the loss of their business records the primary cause. Most regulations affecting business continuity planning refer to record retention and recovery. It should be apparent that a major focus of business continuity planning is the preservation and recovery of vital records. The insurance industry refers to these records as "important papers." The loss of customer, accounts receivable, and asset lists can severely damage future sales, cash flow, and insurance reimbursements. Also, failure to protect corporate records could in some instances bring criminal sanctions against management.

Records are identified as vital if they are important to the continued and future operation of the company. Their loss will have a severe impact or make it difficult to remain in operation. Some firms distinguish records as either vital or important, depending on how much they would be needed after a disaster. Vital records are those absolutely required to recover and restore operations; important, or key, documents are those that will reduce recovery time. Examples include the following:

- Customer lists
- Securities and stock records
- Corporate minutes
- Deeds

- Articles of incorporation
- Bylaws
- Other corporate financial records
- Leases
- Patents and trademarks
- License agreements
- Accounts receivable
- Banking statements
- Tax documents
- Treasury records
- Payroll and benefits information
- Research and development
- Information and specifications
- Insurance policies
- As-built drawings
- Clinical trial results
- Food and Drug Administration security files
- Food and Drug Administration filings that support regulatory requirements
- Business continuity plan
- Negotiations records
- Asset lists
- Laboratory notebooks

Once records are categorized, there are a number of options for safeguarding against their destruction if they cannot be quickly recreated from their original source. The type of media they are stored on and the time dependency of the information may affect your recovery strategy. Most records are paper documents stored on-site. At the very least, they should be stored in locked, fire-resistant cabinets. The most important documents should be photocopied and stored off-site at a facility that specializes in document storage. As with data storage, always audit the security, fire safety, rapid retrieval, and environmental controls used by the storage company. Copies can also be stored at off-site locations within the organization. Document tracking and rapid retrieval can, however, become a problem when storage is internal.

Many records are copied onto microfiche, a technology that is becoming less popular with the increased use and lower costs of document scanning. Document scanning decreases retrieval time and reduces storage space. The documents are stored electronically. They can be sent directly to electronic vault facilities for storage. Once vital records are identified, it is important to implement policies and procedures that ensure their routine backup. Not all documents need backup; many can be recreated from the originals at vendors, customers, or regulatory agencies. Back up those documents that cannot be recreated without difficulty and those for which the delay required for their recreation cannot be tolerated.

Facilities Salvage and Restoration

A general recovery strategy that leads organizations to identify alternative processing and manufacturing facilities before a disaster and to develop plans that allow for the rapid transfer of operations to these facilities is typically very effective in getting back to some level of service. The goal of restoration is to return the organization to a predisaster state or to a defined strategic position. This cannot be done, however, in a temporary facility. The damaged facility must be rebuilt, relocated, or repaired. The rebuilding or permanent relocation of the organization is an issue to be addressed by the management team subsequent to a disaster. The planner will discuss these possibilities with management ahead of time and arrange for agreements and resources to expedite the search for new facilities—during conditions in which, inevitably, space would be scarce and prices inflated.

Most organizations return to their original facilities after a disaster, especially if the buildings are structurally sound. Before this can happen, the facility must be cleaned and made habitable. In the aftermath of a fire, there will be heat and soot damage to equipment. Smoke from a fire is acrid and corrodes sensitive electronic components, even after the fire is out. Water from sprinklers or firefighting hoses can damage documents and cause dangerous molds to grow. The drying and dehumidification of buildings requires specialized equipment and techniques. These services are provided by "restoration" companies, which either do the work themselves or subcontract to companies that specialize in the following areas:

- Salvage and debris removal
- Electronic component cleaning and repair
- Soot removal
- Dehumidification and drying
- Document drying and recovery
- Reconstruction (painting, plumbing, masonry, and drywall)
- Water extraction and moisture control

The safe recovery of vital records and books and the prevention of disease from water damage caused by flood, fire sprinklers, or roof collapse require special expertise and equipment. Restoration companies that specialize in a particular type of damage have expertise, experience, and equipment not generally available to internal facilities staff or to general contractors. Most general contractors can clean and rebuild a facility after a fire, but they may be unaware of the techniques used to eliminate the odor of smoke.

It can be very expensive to scrap damaged equipment, especially if it is unique or replacement lead times are extensive. Insurance policies may not pay the full replacement value of the damaged equipment. According to companies that specialize in the decontamination of equipment, restoration can save up to 75 percent over replacement costs, and restoration can be completed in a few weeks instead of the months potentially needed for replacement.

Most restoration companies will, for no charge, inventory assets and maintain construction plans off site. The use of these services can expedite the recovery process and allow management to concentrate on other recovery issues. In areas susceptible to earthquakes (that is, virtually every part of the United States), one of the first tasks necessary is the cosmetic repair of cracks and other damage. A large part of the psychological healing after a disaster is the return to normal surroundings. Employees staring at cracks in the wall are reminded of the disaster; restoration companies, if agreements are reached ahead of time, can erase the reminders that hinder a return to normalcy.

Cost-Effectiveness

Mitigation solutions must be cost-effective and technically feasible, and must not create additional hazards or problems. This should be obvious but must be considered. Mitigation is often a "big-ticket item." Be sure you have as much justification as possible before presenting your plan for approval and funding. You may need to prioritize your projects or spread them out over several budget cycles. Try to find solutions that can solve multiple risks at the same time. Identify and cost-out alternative mitigation solutions to problems. This can help you achieve the most cost-effective measures or at least have others ready if management does not approve or fund your original plan.

One way to win approval for your mitigation plan is to demonstrate the amount of the loss your solutions will prevent (losses avoided). Provide a dollar value estimate of the structural, content, and displacement costs that would have occurred if the mitigation action were not taken. Use information from your business impact analysis to identify any potential loss of revenue. Include any maintenance and upkeep costs associated with the solution as well as any nonmonetary considerations, such as increased employee morale, customer satisfaction, or other subjective justifications. The losses avoided are most easily estimated for structural mitigation actions.

Displacement costs represent the dollar amount required to relocate a function, or a building's functions, to another location on a permanent or temporary basis. Consider moving costs, replacement costs if not included above, costs to prepare the new space, lost subleases, and other expenses as required. If moving to a temporary location, include the costs to move back to the original or to a permanent facility.

One possible shortcut to determine the expected future loss based on a past incident for which the previous impact is known is to multiply the impact (dollar loss) by the increase or decrease in exposure. Adjust the resulting figure for inflation.

A building or facility replacement cost is usually expressed in terms of cost per square foot and reflects the present-day replacement value. The cost may be offset by insurance, but be careful because that coverage may include only "present value" (depreciated value) and not total replacement. Include the cost to demolish a damaged facility, especially if there is some type of overriding expense, such as asbestos removal. Anticipating management's decision to replace a facility exactly as previously configured, or to use the opportunity for expansion, is probably best not attempted. Determine if repairs to the current building, if not replaced, will require expensive building code upgrades. Projected repairs may be predicted by multiplying the replacement cost adjusted as necessary by the percentage of damage.

It may be advantageous or useful to understand the value of the functional loss of a building or facility. One method to make this determination is to add the budgets (or appropriate percentage of the budgets) or annual sales of the groups located within.

You may need to determine the content value of an entire facility if it is completely destroyed. Few businesses have complete inventories listed by location. The risk or insurance manager may be the best source of content data. Depending on your intended use (business impact analysis or mitigation analysis), you may need to use depreciated values or replacement values.

Preparedness

Before September 11, 2001, when we spoke of "preparedness," it generally referred to the steps an individual or organization took to place it in a better position or to enable it to respond to and to survive the effects of a disaster. Unfortunately, we now need to include those steps necessary to prepare for a terrorist event.

Preparedness is having your plans and resources in place, keeping them updated, testing both the plan and those required to implement the plans. It gives you the capability to manage and respond to an incident.

In the United States, National Preparedness goals are established as a result of the Homeland Security Presidential Directives to develop capabilities to prevent, respond to, and recover from terrorist attacks, major disasters, and other emergencies. The directives also establish measurable targets, priorities, and methodologies for prevention and response to terrorist and natural threats. They are intended to guide federal, state, and local entities to determine how to devote limited resources to strengthen their preparedness efforts. The implementation of standardized response and management plans, collaboration and information sharing, and the strengthening of response capabilities are some of its goals.

Preparedness is an important step in the Comprehensive Emergency Management cycle because materials needed to respond to an event must be in place and ready for utilization. When there's a fire, you don't call the purchasing department to order an extinguisher. Likewise, after a disaster, supplies and resources, as we have and will continue to mention, may be in short supply. The more you have beforehand, the better your chances of survival. We most often think of preparedness as supplies and action, but it can also consist of knowledge. You need to know what to prepare for. Information and intelligence about the threat and its properties and consequences will enable a more rational and effective response. Preparedness can also include the following steps:

- Development of response procedures
- Design and installation of warning systems

- Travel advisories and employee tracking
- Establishing partnerships with local government (or with the business community if you are in government)
- Exercises
- Training
- Collecting personal information for kidnapping and ransom planning
- Stockpiling supplies and materials
- Entering into mutual aid or service-level agreements

Home and Personal Preparedness

Preparedness at home is important to the business or governmental operations because when workers know their families are safe, they are more apt to stay at work and deal with issues, or they are better able to respond back to work because they have less to deal with at home. Business and government should hold preparedness classes for their workers and provide them with a list of supplies and resources they will need. Basic topics can include the following:

- Emergency contacts and meeting locations
- Emergency supplies (food, water, medical, sanitary)
- Structural and nonstructural mitigation
- Insurance considerations and documentation
- Fire prevention and control
- First aid and cardiopulmonary resuscitation (CPR)
- Shelter-in-place instructions
- Emergency procedures
- Evacuation routes
- Warnings and media sources
- Light search and rescue
- What to do and not to do

Many local fire departments conduct Citizens Emergency Response Team (CERT or NERT) training to prepare residents for emergencies and disasters. Where available, consider sponsoring these programs for employees and their families.

Emergency Supplies

Many businesses also stockpile a 3-day or more[5] cache of disaster supplies stored in containers located in the parking lot. The rationale is that the company may be responsible for those employees stranded at the workplace or the supplies might be used for workers

---

[5] After Hurricane Katrina, many emergency managers recommend a 5- to 7-day supply, and others recommend a 2-week or more supply of provisions at home.

who need to remain to engage in recovery operations. They are also aware that after a disaster or terrorist event, these supplies may become scarce. Although this is a commendable practice, it can be expensive because the supplies need to be replenished every 3 to 5 years; also, locating all the supplies together exposes them to vulnerability to damage, theft, or sabotage, and cannot provide for the needs of individual employees. An alternate method, at least for smaller organizations, is to issue employees a carry bag of supplies (food, water, basic first-aid supplies, light sticks, battery-powered radio, and so forth) when they are first hired. Employees then keep these in their desk or car with instructions to customize the contents to their individual needs (prescription medications, reading glasses, pictures of loved ones, clean underwear, and so on). The bag can contain the company or department logo. Specific guidelines and lists of recommended supplies are found on websites of local government and other disaster or homeland security agencies.

Keeping spare parts or components such as access controller boards, RAID drives, or preconfigured laptop computers is another dimension of emergency supplies.

Public–Private Partnerships

Information sharing is important to business leaders in the midst of a crisis or emergency. By developing close relationships with governmental officials (emergency services, police, and federal agencies), you may gain better access to information and resources. Businesses may also have resources they can make available to government during times of need, not necessarily during a crisis or disaster. These partnerships can be difficult to foster because there can be an element of mistrust, regulation, and bureaucracy in the relationship between business and government. The advantages to both sides can be tremendous. Mutual planning and response, resource sharing, and information collaboration can make the difference in a successful response and recovery. These partnerships must be developed whenever possible; they allow you to make only a single telephone call to get what you need.

Vendor Relations

Vendor relationships are often cited as the most important element that gets organizations back in business quickly after a disaster. A simple phone call to the "right person" can get a replacement delivery expedited. Companies are willing to pay up-front for agreements with vendors such as structural engineers, real estate brokers, contractors, and equipment rental firms to give them priority treatment.

Mutual aid agreements, or any arrangements you can make beforehand, will help to reduce costs and to ensure that things that are needed will have a better chance of presenting themselves for service.

Justification

Unlike mitigation, preparedness measures, although generally far less expensive, may be more difficult to justify or to elicit cooperation and participation. The threat of the disaster or attack must be seen as high in the short term, and the source predicting or advising of the threat must be seen as credible. Preparedness information is most effective if it is presented repeatedly through different media and in a form that is easy to use or to recall. Although many acknowledge the need, procrastination may delay implementation until it is too late. An interesting belief held by some is that once they have experienced a disaster and have survived, nothing worse can happen or that they don't need further preparedness. Organizational preparedness may be more motivated by regulatory requirement or by the fear of litigation. It is harder to show a financial return for preparedness.

Summary

Mitigation is action taken to eliminate or reduce the impact of a natural-, technological-, or human-caused hazard or undesirable event. Mitigation and a mitigation plan, while important elements of the comprehensive Emergency Management cycle, are required by many emerging standards and guidelines. Before a mitigation strategy can be developed, the hazards must be identified. An examination of historical records and events, hazard identification inspections and process analysis, in addition to an understanding of the potential threats, consequences, and delivery mode, will provide a reasonable basis for mitigation planning. Cost-effective strategies are developed and presented to management for approval and funding, and systems are maintained in a ready state.

Preparedness is the steps taken ahead of time to better enable the organization to respond to and continue its operations when faced with a catastrophic event. Training, communications systems, resource acquisition and management, and drills and exercises, represent examples of preparedness activities. These activities should extend outside of the organization and include the homes and families of the organization's staff.

14 Response Planning

Worrying about possible disasters can cause many sleepless nights for senior executives and board members, and prompt the creation of comprehensive contingency plans to help ease these worries. It is important, however, to maintain a delicate balance between overplanning for events that may never happen and being adequately prepared to respond if disaster does strike.
—Jack E. Cox and Robert L. Barber, "Practical Contingency Planning," *Risk Management Magazine*, March 1996

The response to a fire, evacuation, or the injury of an employee or guest will catch the attention of the organization, its workers, and potentially the outside world. One misstep by an emergency response team member, or the lack of a response, may likely catch the attention of regulatory agencies and stockholder attorneys as well. This visibility and the resulting scrutiny of the organization's emergency planning can have a profound effect on the Security, Safety, or Business Continuity Manager's credibility and career, not to mention the serious impact on the victim's well-being and the survivability of the organization if the incident is of sufficient magnitude.

Today's risk environment has expanded the focus that organizations must apply to identify, prevent, prepare, and react to situations that not too long ago would gain little attention from emergency managers. Bomb-making plans proliferate on the Internet, access to firearms is commonplace, and the killing of innocent bystanders is justified if it suits your political or religious views. Those who have not adapted to these increased challenges are the same ones who are caught off guard when a shooter walks into their building, or they succumb to the intent of terrorists to cause fear and overreaction when the region's hazmat team with television cameras in tow enters your cafeteria to clean up spilled artificial sweetener on a café table, fearing that someone had introduced anthrax into the facility. Those who have adapted and face certain risks are now incorporating physical security measures into the design of their buildings to mitigate the effects of explosive devices or to prevent the introduction of an aerosol into their ventilation system, just to mention a few. Emergency response plans now include instructions to shelter-in-place.

Organizations must be able to respond to situations they can reasonably anticipate during normal conditions and after disasters. Effective action taken to control an emergency will reduce injuries, protect assets and mitigate their loss, and position the organization for a smooth and rapid recovery—all good reasons why response is an important element of the Comprehensive Emergency Management (CEM) model. Emergency services provided by local jurisdictions (police, fire, ambulance, and hospitals) may not always be available. They will likely become overwhelmed by calls for service during and after a regional disaster, pandemic, or terrorist event. When the demand for their resources must be prioritized, businesses are virtually always at the bottom of the list; government believes that the business community has the resources to be self-sufficient and that infrastructure and other organizations, such as public schools, are more in need of its services. Businesses must therefore develop the capability to care for its population for a period of time during and after a disaster.

Many industry standards are becoming matters of law; accordingly, managers can be held liable in civil and criminal court if the programs they require are not effectively implemented. Additionally, violation of these laws is excluded from liability insurance coverage for "errors and omissions" and "directors and officers." With the increased focus on the prevention and response to terrorist events since September 11, courts are increasingly finding that these events are foreseeable and are holding companies liable for not implementing adequate emergency response plans that protect the well-being of their employees and guests. In the United States, an emergency response capability is required by federal and many state regulations for all organizations. All employers must maintain emergency action and fire prevention plans. The plan must be in writing if there are more than 10 employees. The size and type of the plan and of the response capability is dependent on the type of business (for example, hazardous materials producers have much higher requirements under other federal regulations than do general businesses).

Emergency Response Planning and Response Plans

Emergency response is actions taken to manage, control, or mitigate the immediate effects of an incident.[1] As stated previously, an incident can include a fire, a chemical spill, a bomb threat, an explosion, an earthquake, and a list of others.

An emergency response plan is not a disaster recovery or business continuity plan, a mistake often made by business managers. Those who fall into this trap will have a difficult time in a continuity/recovery/legal situation. NFPA 1600, and the ASIS SPC.1-2009 standards, OSHA, as well as documents found within the Department of Homeland Security and the Federal Emergency Management Agency contain instructions and guidelines about how to construct an emergency planning program and the resultant plans. Their requirements and suggestions vary according to the organization's relationship to these documents, and they dictate the plan's level of complexity.

The degree of planning required for most businesses is relatively simple but must match the applicable regulations and foreseeable hazards. The process used to develop

---

[1] The 2010 version of NFPA 1600 defines Response as the "immediate and ongoing activities, tasks, programs, and systems to manage the effects of an incident that threatens life, property, operations, or the environment."

an emergency response program is very similar to that described in the chapters on Business Impact Analysis and Business Continuity Planning. An overview of this process includes:

- Assign a planner, program manager, or planning team. The name or job title of who is to maintain the plan should be included in the response plan document.
- Identify the foreseeable hazards and understand their characteristics (see Chapter 13, Mitigation and Preparedness).[2] Prevent and/or mitigate these hazards to the extent practical.
- Research applicable codes and regulations and incorporate pertinent requirements into the plan. Some regulations (such as OSHA) will require the name or job title of the person who is responsible to notify outside emergency services. List the names or job titles of who is to perform other necessary duties, such as directing responding resources to the incident location, who is responsible for injury management, and who is tasked with the postincident investigation.
- Establish the appropriate response command or organizational structure. This includes clear lines of management and response authority. Decide who plans, coordinates, and authorizes evacuations.
- Develop and communicate plans and procedures (response guidelines). OSHA and some other regulations require the following plan elements:
    - All United States business are required to have a plan
    - The plan must be kept in the workplace and be available for inspection.
    - It must contain procedures for reporting a fire or other emergency.
    - List in the plan:
        - Evacuation types (full evacuation, partial evacuation, for example) and how to account for evacuees (some jurisdictions will also require the use of evacuation maps).
        - The name or job title of the person who can provide information or explain employee duties under the plan.
        - Rescue and medical duties for those who are to perform them.
        - Employee alarm systems that have a distinctive signal for each purpose.
    - Any procedures for employees who remain to operate critical plant operations before they evacuate must be listed.
- Create and train a response organization (see Emergency Response Team below).
- Orient employees to the content of the plan and their responsibilities in an emergency situation. This orientation (training) should occur when the plan is first introduced, upon initial assignment or reassignment (change in job duties) of a new or transferred employee, or if the plan changes. OSHA requires employers to designate and train a

---

[2] As of this writing, OSHA states that terrorist events are not foreseeable, but the organization must be in a position to react to its consequences if affected by a terrorist act.

sufficient number of persons to assist in the safe and orderly emergency evacuation of employees prior to the implementation of the program.
- Practice, practice, practice.

In spite of what the standards and regulations may require, an effective emergency response plan will include the following elements (see Chapter 16, Plan Documentation).[3]

- Clean and easy to follow (if part of a collection of other plans, consider some type of highlight such as a red border). As with most plans, they must be simple to follow and the necessary guidelines easy to find. Unnecessary verbiage should be excluded from instructions used in the field or in the office. An index may be appropriate.
  - Organized in a logical sequence.
  - Includes emergency contact information.
  - Uses action-oriented instructions.
  - Avoids acronyms.
  - Complete but not overly detailed.
  - Training material or excessive explanation is eliminated.
  - Superfluous information is also eliminated.
  - Does not contain information about the planning process. (This is contrary to some of the new standards. While this may be appropriate for a plan developed under the standards, an operational plan, which should be used by the response team, and a summary of the plan included in an employee action guide, should contain only the instructions they need to follow in an emergency.)

Many emergency response instructions begin with the admonition "Do not panic," "Avoid panic," or "Control panic." People rarely panic before, during, or after an emergency, except under very specific circumstances. Assigning responsibility to someone to prevent panic is a task that few, if any, can accomplish if panic were to develop. These instructions are frivolous and should not be included in emergency response plans. Since research has long supported the assertion that panic does not occur, this language in the plan can demonstrate that the planner has not done his or her research. As mentioned above, it is necessary to identify and understand the characteristics of hazards so that the planner is in the best position to prepare for, mitigate, and control their consequences. Without this understanding, we may face unwelcome surprises during the response phase, and needlessly spend valuable time resolving unanticipated issues during the recovery and continuity phase. Hazards can produce different effects depending on their magnitude, their duration, their time of year, building construction, and the level

---

[3]When required to develop plans that don't contain the above elements, consider the generation of the required plan with more compact subplans or operational plans and guidelines written for employees and emergency responders.

of preparedness and mitigation. On a regional basis, hazard impacts can vary based on demographics, location, building habits, and conditions that may increase or reduce the effects of the hazards (soil conditions that may cause liquefaction, creek or river debris, diseased forests, and so on). Be aware that one type of hazard can produce multiple hazards, so-called secondary hazards. Hurricanes can generate tornados and flooding, flooding can cause fires and bridge washouts, and earthquakes can cause hazmat incidents and tsunamis.

Emergency Response Team

An Emergency Response Team (ERT), originally intended to evacuate employees and to fight fires, is an internal organization of typically volunteer employees designed to respond to emergencies before the arrival of public agencies. After a disaster, it may become your primary source of medical, fire control, search and rescue, and hazardous materials cleanup.

Required in nuclear power plants, hospitals, and many high-rise buildings, an emergency response team can help the company to:

- Intervene and stabilize emergencies before they have the chance to escalate
- Increase the survival of injured or sick employees through the rapid intervention and application of life-saving devices and first-aid protocols
- Assist in the complete and orderly evacuation of employees and guests (account for evacuated employees)
- Prevent adverse publicity
- Demonstrate management concern and support for the safety of employees
- Minimize the impact on the environment
- Help comply with regulatory requirements to mitigate hazardous materials incidents (US OSHA, Uniform Fire Code, Environmental Protection Agency). In remote locations away from normal emergency response services or when a "prompt rescue" is required or the responding agencies do not have the technical capability to quickly effect a rescue, the ERT will serve this purpose.
- Reduce property damage and loss
- Become the sole response in a disaster situation when public agencies do not have the ability to respond

Some basic steps helpful in the formation of an ERT include the following.

Management Acceptance and Support

The effective and safe operation of an ERT involves a large commitment of time, money, and resources for training and equipment. The time required for training even a small

ERT in a low-hazard environment should be no less than 8 hours each quarter—much more in a larger, risk-intensive environment. The type of hazards the team may respond to will also affect the minimum training and equipment requirements. Teams involved in "hazwoper" (hazardous waste operations) require a minimum of 40 hours of training. This specialized training and equipment, and the "lost" time of members away from their normal duties, can add up to a significant expense for most organizations. Management must support the program through sufficient funding, policy, and the appropriate delegation of authority to the team members.

Duties and Responsibilities

Defining the scope, or what type of incidents to which the team will respond, will also define the recruitment standards, equipment requirements, and amount of training necessary. This decision should be a natural result of the hazard identification analysis. The scope can (and in some cases, such as fire response, should) speak to the level of response (i.e., the team respond only to small fires in the incipient stage).

Planning

Develop response plans, team structure (Incident Command System, for example), policies, and procedures. If using the Incident Command System, it may be possible (and desirable) to develop Incident Action Plans ahead of time, since in most cases the response environment will be predictable. Many organizations locate the Command Post in the building lobby. Employees from certain departments are generally necessary to ensure an effective response. This includes security (access control, communication), facilities (resources, equipment shutdown), and environmental health and safety (technical support).

Determine Equipment and Resource Needs

Budget for short-term and long-term requirements. Equipment may include reflective vests, hard hats, spill-kits, and a list of other items. Don't forget to include communications equipment. Two-way radios, pagers, and police and fire scanners form a basic response-communications system.

Recruit Team Members

Include sufficient numbers to field full teams for all shifts, to meet the staffing demands of each type of incident, and to allow for a reserve force to substitute for sick or injured team members.

Although membership in the team is usually voluntary, provide incentives to create interest, but be careful that the incentives do not create a de facto situation where the team is no longer considered volunteer. A team compensated for their participation may require a much higher degree of prescreening, medical examination and inoculation, training, and increased liability.

Develop Training Programs

The training must fit the scope of the team's duties and regulatory requirements. It may be necessary to bring in outside consultants or to have the local fire department assist with training. Neighborhood or Citizens Emergency Response Teams (NERT/CERT) are groups of neighborhood residents trained by local fire departments in emergency preparedness, first aid/CPR/AED, basic urban search and rescue, and other skills useful to ERT members. Many of these jurisdictions are happy to include business organizations in these training programs.

Conduct Regular Drills

Practice, practice, practice. Arrange joint drills with public agencies.

Advertise

Let employees and the community know about the existence, authority, and capabilities of the team. This will increase confidence in the organization's ability to respond to emergencies, and it should elicit better cooperation from employees.

Emergency Procedures

The following sections provide guidelines the reader may use to plan a response to emergencies. The guidelines are generic and are not intended to list all foreseeable emergencies or to list all measures the organization should take to prepare for, respond to, and recover from an incident, natural hazard, or disaster. They are primarily directed toward business in general and to construction companies and do not consider situations in which the activation and notification of other agencies or entities may be required. Many of the guidelines and conditions apply to multiple types of disasters, although they are listed here only with the section with which they are most commonly associated. Additionally, we have listed steps that may be taken to Prevent, Prepare, Mitigate, Respond, and Recover, but in many cases, the steps may be better listed in a different category. This categorization in reality is irrelevant as long as the suggestions are properly considered in light of the potential risk (excuse the redundancy). In all cases, always assess the situation before formulating a response, and do so only when trained and the safety of the responders is considered.

Also not included in the guidelines listed in this chapter are methods to activate emergency responders or the ERT and, except in some cases, the need to notify others within the organization. Planners should identify these needs and document them appropriately.

Arrested Fall Emergencies

Falls from height (as opposed to slips and trips, or a fall down a flight of stairs), typically in construction and maintenance operations, account for one of the top categories of fatalities in the workplace. In the United States, employees who are working at a specific height above the ground or above a lower level are required to have some type of fall restraint or personal fall protection.

Personal fall protection is a common option when working at heights and consists of a harness, a connecting device (lanyard or self-retracting lanyard), and a connection point (anchorage). If properly applied and properly worn, this system will prevent serious injury if the worker falls, hopefully decelerating the significant forces generated by the fall and suspending the worker in the air before contact with the ground. Although these systems have saved many lives, they can give a worker a false sense of security. They are intended as protection against the effects of a fall, not fall prevention. Although the worker is prevented from hitting the ground, the dynamics of the fall can still cause serious injury. If the anchorage point is not overhead, the fall can produce a pendulum motion with the worker potentially striking objects in the swing path. Suspension trauma (properly referred to as orthostatic intolerance) is a condition caused by pooling of blood in the lower extremities through the lack of movement. This condition is especially dangerous if the fall victim is unconscious and can be fatal. It is therefore important that rescue, some sources say in 14 minutes or less, is effected immediately. Crush syndrome (see Trenching Emergencies) and blood clots are also possible medical complications encountered in suspension emergencies.

Prevention and Mitigation

1. Remember that most types of ladders are not intended as work platforms. Understand and train employees on the proper selection and use of ladders.
2. Whenever possible, rely on barriers to prevent a fall. Work in an aerial device such as a scissor lift that has rails around the work platform. Follow all rules for working in these devices.
3. Attach foot straps to the harness that the fallen worker may deploy after the fall. The worker can place one or both feet into the straps and take some of the pressure of the harness off the body and aid in blood circulation by "standing" in the straps.
4. OSHA and ANSI standards require the development and training of fall rescue plans. Keep these plans and any related tools and resources up to date.
5. Know the signs and symptoms of suspension trauma (feeling faint, nausea, dizziness, sweating, paleness, and a narrowing of vision).

Preparedness

1. Have fall rescue equipment appropriate to the type of rescue anticipated available for use. Fall protection suppliers are a good source for this equipment and training in its use.
2. Keep training up to date for fall protection safety and rescue.
3. Understand the capabilities of the local fire department to assist or manage an arrested fall response. Supplement your abilities or that of the department as necessary.

Response
1. Initiate self-rescue if able.
2. Dial 911 as necessary. Inform them of the height involved or if required that a "technical or high-angle rescue" is needed.
3. Stabilize the victim and prevent from falling. Stop, if possible, any pendulum (swinging) motion by the victim.
4. Protect the victim from further injury during extrication or disentanglement. If unable to rescue immediately, encourage the person to periodically move their extremities while suspended.
5. Secure or anchor equipment or machinery that may further injure victims or rescuers.
6. Determine the nature of any injuries. Monitor the victim for signs of suspension trauma.
7. Assess existing and potential conditions for additional hazards.
8. Ensure rescuers wear appropriate safety apparatus (harness, lanyard) and other personal protection equipment. Tie off to an approved anchorage. Ensure that equipment and anchorage will support any additional weight of the rescuers.
9. Develop a specific plan for removal/rescue. Identify resources necessary to conduct a safe and effective rescue operation. Use lifts, lanyards, ladders, or other devices to rescue if safe and appropriate.
10. Keep the rescued victim sitting up (unless CPR is required) and transport the person to the hospital with their upper body raised. Because pooled blood in the lower extremities may be unoxygenated, some sources recommend not moving the victim into a horizontal position too quickly (i.e., don't) to allow the heart to adjust to the sudden increase in load. Too much unoxygenated blood at once could cause cardiac arrest (reflow syndrome). It is noted, however, that although common in the emergency medical response community, this practice may not be well supported by research.

Recovery
1. Activate a crisis management plan if the fall becomes a newsworthy event or if a serious injury is involved.
2. Notify the appropriate regulatory agencies such as OSHA as necessary.
3. Conduct an investigation into the cause of and the response to the incident.
4. Provide posttraumatic stress counseling in the case of a fatality.
5. Retrain the person who fell in the safe work practices of working at heights.

Bomb Incident Management

The days since 9/11 have shown that terrorists are capable of using nonconventional means to carry out their objectives. In our rush to protect against these new forms of attack, we must not lose sight of the more conventional tactics—the simple, tried and true tool of both domestic and international terrorism: bombings.

Although almost all the explosives incidents within the United States and Canada are rather small scale (with a few, very notable exceptions), transportation systems and other critical infrastructure, oil refineries, landmarks, businesses with hazardous materials, public assembly, and other sites whose operations could cause collateral or synergistic damage are prime targets. Many believe the types of devices used, most commonly pipe bombs and incendiary devices, will soon be replaced by the increased use of package bombs, higher-yield automobile and truck bombs, and "dirty bombs" carried by suicide or homicide bombers. Large trucks have been used by terrorists as vehicle-borne impro-vised explosive devices (VBIEDs) with great success, and even ambulances have been used to hide explosives. Partially to distinguish between devices manufactured for the military, terrorist bombs are referred to as improvised explosive devices (IEDs).

Throughout history and throughout the world, bombings are a prime tool used by terrorists and are expected to remain the dominant threat. Materials for their construc-tion are easy to obtain, and terrorist groups have a long history of expertise in their use. The terrorist can be a continent away when the bomb is detonated. In 2006, there were 3,445 explosives incidents, up from a total of 1,685 bombings and attempted bombings in the United States in 1996. These 2006 incidents account for 135 injuries and 24 fatalities. Letter bombs (package bombs), depending on the year examined, typically represent less than 1 percent of the total incidents.

Vandalism consistently leads as the top motivation of bombers (mailboxes are a com-mon target as roughly half are committed by persons under the age of 18), followed by revenge (sour love affairs, ex-employees, and "messages" to our beloved Internal Revenue Service, for example). Homicide, suicide, protest, extortion, insurance fraud, and labor disputes are near the bottom of the list.

The Oklahoma City bombing consisted of a 4,800-pound ammonium nitrate and fuel oil mixture that generated a blast pressure of almost 6,000 pounds per square inch and created a 30-foot-wide, 8-foot-deep hole. The winds spawned to replace the vac-uum caused by the explosion exceeded 1,000 times hurricane force. Windows were blown out 1,000 feet away. Even in a small-scale explosion, structural damage is possi-ble. Nonstructural damage can include fire and smoke damage, water damage, soot and dust, glass, and debris. Electronic equipment and circuit boards can be damaged from the blast, and the psychological effects to those nearby can be significant. A bombing or the discovery of a device can put an organization on legal notice that such incidents are foreseeable and thus increase future liability.

Threat Evaluation

In the United States, the chance of finding an explosive device after a threat has been received is very low. The level of physical security and the degree in which security is enforced are primary considerations when it comes time to evaluate the credibility of, and therefore the response to, a bomb threat.

Threats can be emailed, sent through the postal service, or, most commonly, telephoned into the business. It is important to remember that most telephoned threats are hoaxes. Although there is some variation in these numbers, only 2 to 6 percent or less of bombings are preceded by a threat or warning. Some experts put the probability at 0.5% (1 in 200 chance), whereas others, such as the San Jose, California, police bomb squad, place the odds at "infinitesimal."

Generally, the purpose of a valid threat is to prevent injury or to prove intent in an extortion case. In these instances, familiarity of the site and the increased detail provided by the caller about the device and its placement add to the credibility of the threat. The caller may provide the exact location, time of detonation, and a detailed description of the explosive. Although "bomb threat checklists" are found at many reception stations, it is important for someone in your organization to develop the capability to evaluate the credibility of any threats received. This capability can reside with the security director, risk manager, crisis management team, or other responsible person who can be reached quickly and has the ability to make rapid decisions. Most police officers and police dispatchers are not trained to evaluate threats. Many police agencies are restricted by policy to not offer advice to the business owner about the validity of a threat. The person or people who evaluate the threat must have some training in threat evaluation that includes an understanding of the firm's threat risk profile. Basic components of the profile include:

- Level of security at the threatened location
- Public visibility of the company or controversial business activity
- Recent events (product recall, pending labor actions, reduction in force)
- History of bombings and threats
- Activity of individuals and groups (intelligence) who may have a motivation to attack your organization

The wording of the threat is examined for key words ("device" as opposed to "bomb," terms that describe the firing train, and so forth), conditions and circumstances (background noise of a party may indicate a low credibility, or if the threat is received on a day in history that is significant to a terrorist group, it might indicate a higher credibility), detail of the threat, ability to carry out the threat (security level and profile), and motivation. The results of the evaluation will guide the decision to evacuate, shelter-in-place, close the business, or take other action as appropriate.

Evacuation

The decision to evacuate the building or site after the receipt of a bomb threat is still controversial and difficult, although many security managers now follow the practice of avoiding evacuations unless a device is actually found. There is good reason for this approach: because most threats are false, evacuation will cause needless loss of productivity, decrease employee morale, increase the possibility of injury from the evacuation, and, most often, satisfy the intent of the caller. Once a threat has been communicated to employees through an evacuation, expect additional threats generated internally. Most explosive devices in business settings are placed in the parking lot, building perimeter, or public areas, such as the lobby, hallway, and lockers. Evacuating employees will send them through and to these areas. If the threat is not credible, begin a search. If nothing is found, return to normal operations. If a device is found, evacuate a safe distance away from blast effects, or send employees home. If the threat is credible and a detonation time of less than 30 minutes is given, some may want to delay the search and evacuate if the employees can be cleared from the building and placed in a safe location before detonation. If the threat is credible and a detonation time is greater than 30 minutes, initiate a search if it can be completed in time to safely evacuate afterward. These times are somewhat arbitrary and should be adjusted according to your circumstances. This criterion applies only to threats received in the United States. The characteristics of threats and groups overseas may require a different approach.

Searches

Although you should rarely evacuate after a threat is received, you should almost always begin a search for a device, even if the threat is obviously a hoax. Searches should be conducted by company personnel because most police officers are not trained in search methods or are not allowed to conduct a search of your premises. Calling the police to report a threat or to ask them to conduct a search will often result in the incident becoming part of the public record, open to discovery and publication by the media.

Searches can be conducted by the security force, Emergency Response Team, or the facilities department in conjunction with department managers, or by a combination of these methods. Bomb-sniffing dogs are very effective, but their extended response time often makes their utilization impractical. Training in search methods and bomb recognition is desired but not overly important because a bomb can look like anything; any object that is out of place, such as an unidentified, unclaimed briefcase, a package next to the gas main, or a small pipe sitting outside the data center, is suspect. It is most important to identify search methods and searchers ahead of time.

Any search must be systematic, rapid, and thorough. Sectionalize portions of the site and building and prioritize the order in which these sections are searched. Assign specific search areas if they can be searched concurrently, or assign the sections as personnel become available. A good starting place is any area mentioned in the threat (I always check the bottom of my chair first). Search next the most common, accessible areas devices are found (building perimeter, public areas), then areas that could provide a synergistic effect (hydrogen tank, gas main), and then areas critical to the business operation (backup electrical generator or data center). Search last the noncritical areas or areas difficult to reach. Assign these sections to teams most familiar with the areas, such as department managers, who have the advantage of increased search speed, a greater likelihood to recognize objects out of place, and a better knowledge of critical equipment and hiding places.

Rapid communication when conducting a coordinated search is vital. Most explosives incident procedures carry the admonition to turn off all radio transmission devices, including cellular phones, based on the fear that using your radio or phone will cause the premature detonation of a device. Although you may not want to transmit directly adjacent to a suspicious object, the chance of a predetonation caused by your radio is remote. Nevertheless, you should minimize the use of portable communication devices.

SUSPICIOUS OBJECT

If a suspicious object is located, try to find its owner. If the owner can't be found or it is obviously an IED, isolate (secure) the area from entry. Evacuate people away from the area or send them home. Remember that the area is now a crime scene and that the facility (or immediate area) may be closed for 24 to 36 hours or more. Inform the appropriate emergency personnel (police, fire). If safe, open doors and windows around the blast area and shut off hazardous processes that may be affected by a detonation. Consider shutting off utilities. If safe, continue to search for other devices. Do not touch or move the object (even if it is a "dud"). Do not cover the object or cut any wires. Do not put the object in water or pour water on it. Activate your crisis management plan.

PACKAGE BOMBS

Explosives received through the mail can be delivered in envelopes, boxes, or packages and may have one or more of the following characteristics:

- Unusual postmarks or places of origin
- Excessive postage
- Incorrect addresses or titles of recipients
- Excessive handling, wrapping, taping, or inappropriate bulkiness
- Excess weight, stiffness, bulges, or uneven balance and feel
- Smudges and greasy-looking spots or areas
- An odor of almonds or a chemical odor
- Protruding wire or string
- Pinholes from which a safety-pin arming device may have been withdrawn

When a letter is not suspect, the use of nonmetallic letter openers can help to prevent a detonation, but if there is any doubt, isolate the letter and have it examined. Inexpensive sprays exist to quickly identify any explosive residue on the package. If you are a high-risk enterprise, consider the use of small x-ray machines in your mail room.

SUICIDE BOMBS

Robert Mueller, the current Director of the U.S. Federal Bureau of Investigation at the time of the publication of this manuscript, believes that suicide bombers are inevitable in the United States. Powerful enough to kill almost everyone in a 50-foot radius, they can destroy busses, restaurants, and trains. Suicide bombings are a favorite and effective method of terrorists to penetrate a target. According to former Central Intelligence Agency (CIA) Counterterrorism Chief Vince Cannistraro, "There is no 100% defense against suicide bombers."

These devices usually consist of 10 to 30 pounds of explosives and are strapped to the body, hidden in clothing or other objects like backpacks. Suicide bombers may also be the drivers of cars or trucks laden with explosives who detonate their cargo over a bridge or near a crowded intersection. Suicide bombers can be difficult to identify because they no longer fit the profile of the past (perpetrators were almost exclusively young men, but women and children are now found to have joined the ranks). Potential indicators include a person who is alone or with one other person and may appear nervous, apprehensive, or agitated, wearing loose or bulky clothing, possibly inappropriate for the weather (overdressed for summer heat). Exposed wires may be visible, and their midsection may seem rigid. Their hands may be clenched around a trigger (a second person may trigger the device with a call to a cellular phone).

In the case of a detonation, first responders should not approach the suspect or the suspect's remains. There may be undetonated or partially detonated explosives or secondary devices present. A bomb squad should determine when it is safe to approach. If the incident occurred within close proximity to your location, consider taking initial steps to protect against dispersion of radiation, chemical, or biological agents.

DIRTY BOMBS

IEDs specifically designed to cause injury, such as a pipe bomb, can be filled with nails, pesticides, or other poisons to cause the maximum amount of injury. The combination of radioactive materials with an explosive device produces a "dirty bomb," more properly called a radiological dispersal device (RDD). The intent is to use the force of the explosion to distribute radioactive material throughout the area. Waste by-products from nuclear reactors, medical devices, or other radioactive sources are the likely materials found in these devices. This combination of explosives and the radioactive material available to put into the device cannot produce a nuclear explosion. If dispersed through the air, the resulting contamination could affect several city blocks, but based on the probable type of material available to a terrorist, the dispersion would not significantly cause severe illness, a marked increase in cancer cases, or deaths from exposure to radiation unless the device is very large and uses a credible radiation source, and the victims are very close to it. This is assuming that multiple devices are not detonated in unison. Inhaling radioactive dust from an exploding RDD could represent a health risk. The extent of contamination depends on the size of the explosion, the type and amount of radioactive material, and the local weather conditions at the time of the explosion.

Effective RDDs are difficult to acquire and to make. The former Iraqi government and other military forces have tried and failed in the past to produce a dirty bomb capable of lethal results apart from the initial explosion, although in 1996, Islamic rebels from the break-away province of Chechnya planted, but did not detonate, such a device in Moscow's Izmailovo Park to demonstrate Russia's vulnerability. Today, only the United States and the Soviet Union have reportedly manufactured portable nuclear devices small enough to fit into a large backpack. The existence of "suitcase bombs" that contain nuclear devices manufactured in the Soviet Union is believed by many to be a myth. Many large cities and port facilities, however, have installed radiation detectors to identify such devices.

Apart from deaths caused by the explosion, the greatest impact from an RDD is the costly cleanup and the lingering fear of reentering the exposed area or building. Dirty bombs are primarily designed for fear and intimidation. An example of their effectiveness was the rush of people who purchased potassium iodide (KI) after the early reports of the potential use of dirty bombs in the United States. Unfortunately KI will not protect from the effects of an RDD unless the device contains large quantities of iodine isotopes. KI protects the thyroid gland only from radioactive iodine; it offers no protection to other parts of the body or against other types of radioactive isotopes. There is no guarantee that these isotopes would be used in the device.

Prevention and Mitigation

The degree of prevention and mitigation to establish for your organization should be appropriate to your risk. Consider the following:

1. Develop an intelligence program (especially if your risk is from domestic terrorist groups) and maintain continuous contact with law enforcement.
2. Complete an engineering evaluation of the blast resistance of your facilities and implement measures to mitigate the effects of a detonation by creating a blast-resistant exterior and the introduction of structural elements that can resist blast loads. Explore the use of computer-based modeling and decision support systems to assess the extent of blast damage to your building's structural frame. When designing new construction, incorporate blast-resistant architecture.
3. Locate ventilation systems away from ground-level entrances, parking areas, and streets. If possible, locate them on the roof. Install automatic dampers to close air intakes when not in use.
4. Ensure that fire prevention and protection systems are up to the latest codes and best practices (an expensive proposition in many cases). Use blast-resistant separation for fire system equipment, pipes, and controls.
5. Locate critical facilities, offices, and processes away from the perimeter of a campus. Within a building, locate them away from parking lots and loading docks, and away from the mailroom. If justified by the risk, consider locating the mailroom in a separate building. The president's office of a company that was identified as a potential target of the "Unabomber," who specialized in package bombs, was located one floor directly above the mailroom—poor planning, to say the least.
6. Remove hiding places for IEDs from around the building perimeter and especially near lobbies, emergency exits, or paths evacuees may take. This includes trash receptacles, mailboxes, large planter boxes and hedges, newspaper racks, and so forth.
7. Design streets, campuses, and approaches to buildings to prevent vehicular traffic from having a straight approach to a security checkpoint or to building lobbies. This will preclude vehicles from reaching high rates of speed and crashing through the checkpoint or into building lobbies. Place fountains, speed bumps, bollards, sculptures, or other objects in front of lobbies.

8. Control access to parking areas, especially if underground. Physically inspect vans and other large vehicles; check the undercarriage of vehicles, under the hood, and in the trunk. Provide vehicle inspection training to security personnel.

9. Establish no-parking zones around facilities and enforce with tow-away procedures. Approach all illegally parked vehicles in and around facilities, question drivers, and direct them to move immediately; if the owner cannot be identified, have the vehicle towed by law enforcement.

10. Isolate employee parking from visitor parking. If underground parking exists, attempt to accommodate visitor parking on the street.

11. Create as much offset from the street as possible. Overpressure is inversely proportional to the cube of the distance from the blast. Each additional amount of standoff provides progressively more protection.

12. Terrain, other buildings and barriers, and vegetation can provide some additional protection by absorbing and deflecting blast forces and debris. Be careful that vegetation does not provide a hiding place for a criminal or for a device.

Preparedness

1. Audit the physical security, incoming inspection, access, and internal controls of the facility. Analyze the firm's exposure to bombings.

2. Place bomb threat questionnaires and brief instructions at all security, switchboard, and reception stations.

3. Provide specific training to all security officers, switchboard operators, and receptionists who could receive bomb threats.

4. Establish who will evaluate threats. Train this person or an evaluation committee to assess the credibility of threats.

5. Decide what procedures will be followed when a threat is received or if a device is discovered.

6. Decide what conditions (if any) should trigger an immediate evacuation of the building, and list these conditions in a procedure for security or for an ERT commander. Companies may decide to identify certain elements of a threat as justifying evacuation of employees without approval from the threat assessment committee. Depending on the company's bomb-risk profile, these elements may include disclosure of the location of the bomb, the time of detonation, a motive, or an apparent familiarity of the facility.

7. Identify search methods and searchers before the incident. Train teams on search technique, device recognition, and safety. Most police departments will not search for you!

8. Identify and prioritize search areas before the incident.

9. Consider delaying a search to protect the safety of the searchers if the threat is credible and a short time limit is given.

10. Establish a procedure to track the progress of the search. Develop sectionalized maps and checklists.

11. Test all procedures.

12. Review plans, procedures, and contact phone numbers regularly.

13. Understand your community's capability to analyze postdetonation conditions and to monitor for other hazards, such as chemical agents, gases, radioactive materials, or secondary devices.

Response

1. Write down the exact time a telephoned threat is received. Find out, if possible, whether the call originated from within the facility or from the outside.

2. Write down the caller's *exact* words. Permit the caller to say as much as possible, without interruption.

3. Ask the caller the following, and write down the answers to these questions:
  a. When will it explode?
  b. Where is it located?
  c. What does it look like?
  d. Why was it placed?
  e. Who is calling?

4. Attempt to transfer the call to the head of security or to a member of the threat assessment team.

5. Initiate a search for a device. Use a checklist to ensure that all prioritized areas are searched in the most expedient and thorough manner possible. If the caller tells you where the device is located, obviously check that area first.

6. Make other notifications as appropriate (to management or security, for example).

7. If you find an object you suspect may be an explosive device:
  a. Evacuate people to at least 300 yards away and out of the line of sight or blast effect.
  b. Dial 911.
  c. Identify and evaluate the object.
  d. Do not attempt to touch, move, dismantle, or pour water on any suspicious object.
  e. If it is safe, open doors and windows around the area to reduce the blast effect.
  f. Isolate (secure) the area from entry.
  g. Consider shutting down utilities and hazardous processes.
  h. If it is safe, continue the search for additional devices.

8. Contact the public relations spokesperson or activate the crisis management plan.

9. Stage emergency response equipment and strategic resources.

10. Restrict access to the area.

11. If a device explodes, and people are injured, some experts suggest removing the victims immediately and securing the area, not treating them where they are found. Their rationale is based on the practice of some terrorists to place in the same area additional devices set to explode later to kill and injure rescue workers and police.

12. If an RDD is suspected:

    a. Move far away from the immediate area (upwind if possible). If this is not possible, go inside a building to reduce exposure to any radioactive airborne dust.

    b. Cover or filter your mouth and nose to reduce the chance of breathing radioactive dust.

    c. If exposed and once inside, remove clothing and seal them in a plastic bag (if you used material to cover your mouth, place this in the bag also). This practice can remove the great majority of exposure to dust and alpha radiation.

    d. Shower to remove the remainder of the dust.

    e. Building owners should shut and seal all windows, outside doors, and dampers. Fans and heating, ventilation, and air conditioning (HVAC) should be turned off.

    f. Monitor television and radio stations for news and instructions.

Recovery

1. Care for the injured.

2. Begin rescue operations only if properly trained and equipped, and if there are no other devices in the area.

3. Assess damage, including the structural integrity of the buildings.

4. Begin salvage and cleanup.

5. Test electronic and other sensitive equipment for blast damage.

6. Keep employees informed about the status of cleanup and future risk from any radiological residue.

7. Provide posttraumatic stress counseling for employees and rescue crews.

8. Begin relocation and reconstruction if necessary.

9. Investigate and prosecute every incident.

Chemical or Biological Attack

Experts in national and international security affairs have long warned about the use of biological, chemical, and radiological weapons against our populations. The ability to produce and use these methods of mass destruction is well within the grasp of terrorist organizations that often share a fanatical belief that they are defending their religion and homeland. The present-day super terrorists are well educated, well funded, and better organized. Their intent is to cause violence within our cities and to destroy our businesses and way of life.

Authorities believe our greatest risk from these criminals comes from the deliberate destruction of one or more of the 850,000 facilities in the United States that deal with hazardous or extremely hazardous substances, causing the release of these materials into the community. Others point to the desire to use more exotic and more deadly substances such as nerve agents (tabun, soman, VX), vesicants (arsine, lewisite, phosgene), and biotoxins (Cryptosporidium, Ebola virus, smallpox).

Other experts, however, believe that most of these groups lack the technical and tactical ability to use these nonconventional weapons to cause large numbers of fatalities. The Aum Shinrikyo, a Japanese religious cult, employed numerous scientists and spent $30 million in a failed attempt to cause mass casualties by the release of sarin in their subway system. One quart of this agent that contains roughly 1 million lethal doses can be brewed in a kitchen or garage, but more than a ton of nerve agent would be required to kill 10,000 people outdoors. City water supplies are considered safe because of the filtration, frequent testing, and dilution of any agents introduced into a reservoir, but agents and methods exist that may bring this safety factor into question. Most chemical and biological agents that present an inhalation hazard break down fairly readily when exposed to the sun, diluted with water, or dissipated in high winds. However, agents released inside a building will be less affected by these mitigating conditions. The contamination of food and the spread of infectious agents from one person to another represent additional modes of attack.

The best defense a business can take to mitigate and respond to terrorist attacks is through improved security measures and business continuity planning. The purchase of gas masks by individuals is not recommended because they are often outdated and agent specific, and need to be fitted properly. Stockpiling antibiotics is also not recommended because people who self-medicate themselves or their children could do more harm than good (many individuals and companies have, however, stockpiled antiviral medications in anticipation of a pandemic flu).

Nevertheless, we should watch for the warning signs of an attack. A chemical or biological attack won't always be immediately apparent because many agents are odorless and colorless, and some cause no immediately noticeable symptoms. Hospitals and emergency response personnel may be the first to detect a problem. Sudden high levels of absenteeism may also indicate a problem. Indications of a chemical or biological attack include the following:

- Droplets of oily film on surfaces or water
- Unusual number of dead or dying animals in the area
- Unusual or unauthorized spraying (crop dusting) in the area such as over a city or stadium
- Victims displaying symptoms of nausea, difficulty breathing, convulsions, disorientation, lost coordination, or patterns of illness inconsistent with natural disease, such as blisters and rashes. A burning sensation in the nose, throat, and lungs could be further indication of a chemical attack
- Mass casualties
- Low-lying clouds or fog unrelated to weather, clouds of dust, or suspended particles
- People unusually dressed (long-sleeved shirts or coats in the summertime) or wearing breathing protection in areas where large numbers of people congregate such as subways or stadiums
- Unexplained odors (smell of bitter almonds, peach kernels, newly mown hay, or strong garlic)
- Observation of dispersal devices or unexploded or unreleased material in its container

If an attack or exposure occurs, the best defense is to quickly evacuate the area to as far away from the agent as possible. If you cannot outrun a contamination plume, you may need to shelter-in-place. Protection of breathing airways and the skin is the next most important step a person can take in the event of a chemical or biological exposure. Professional response is similar to a major hazmat incident.

Prevention and Mitigation
1. Implement stringent security control of toxic and hazardous chemicals and biological products, especially if they are used or stored in large quantities, such as chlorine or ammonia, or if they are used to manufacture other harmful agents.
2. Store chemical tankers inside buildings or secured yards. Ideally, buildings should provide complete containment with blast-resistant design and be equipped with a scrubber system sufficient to neutralize the entire contents of the tanker.
3. If possible, substitute less toxic substances for toxic chemicals in hazardous processes.
4. Minimize the amount of dangerous chemicals on hand.
5. Secure access to ground-level air intakes. If possible, elevate or relocate them to the roof. HVAC systems can become entry points for chemical and biological (and also radiological) contaminants and distribute them throughout a building.
6. Install HEPA filters on air systems according to the National Institute of Occupational Safety and Health (NIOSH) guidelines.
7. Protect building utilities from tampering, including water mains. Reroute pipes and other utilities underground and away from the exterior of buildings. Secure any access to these utilities.
8. Install HVAC venting and purging systems.
9. Design new construction with security and terrorism protection components in mind.
10. Isolate mailrooms from other parts of the building. If warranted, consider the installation of a separate air-handling system for this area.

Preparedness
1. Understand the types of biological and chemical agents terrorists may acquire, including their properties, consequences, and response options.
2. Identify businesses in the immediate area that have a high risk for attack that subsequently could affect your operations and employees.
3. Develop procedures to notify employees, the surrounding community, and response and regulatory agencies if the hazard can be created within your organization.
4. If at high risk, develop evacuation routes and procedures to move employees out of the area before exposure.
5. Assess the physical security of buildings.

6. Restrict access to mechanical rooms and building operations systems.
7. Identify safe rooms to use for sheltering-in-place. Develop procedures and training programs for employees.
8. Develop a means to quickly monitor conditions in your surrounding area. This could be assigning responsibility to monitor police and fire broadcasts, news channels, or other forms of rapid communications and information gathering.
9. Work with human resources or other departments within your organization to quickly identify periods of sudden, out-of-the-ordinary absenteeism among employees.

Response
1. Gather as much information as necessary from radio, TV, or other official sources. This information may include the type of agent or type of attack, area affected, and the recommended response. If the agent is known, they may broadcast signs and symptoms of exposure, centers where any medications or vaccinations are distributed, and where to seek medical attention for symptoms.
2. Move upwind from the source of the attack.
3. If evacuation from the area is impossible, move upwind and indoors to an interior room on a higher floor if possible (many agents are heavier than air and will tend to stay close to the ground).
4. Close building dampers, windows, and doors. Turn off HVAC systems.
5. Shelter-in-place. This means that once indoors, close all windows and exterior doors and shut off air conditioning and heating systems. Seal all cracks around windows and doorframes with duct tape. Seal all openings in windows and doors (including keyholes) with cotton wool or wet rags and duct tape. A water-soaked cloth can be used to seal gaps under doors if wide tape is not available. Cover bare arms and legs and make sure any cuts or abrasions are covered or bandaged. Choose a room with access to a bathroom and preferably with a telephone. If possible, store a 3- to 5-day or more supply of food, water, flashlights, radio, prescription medications, and other emergency supplies. In a commercial building, close dampers and shut off HVAC systems (some sources say to leave the system running if it is able to filter out the contaminants; however, HEPA filters cannot filter chemical agents).
6. If splashed with an agent, immediately wash it off using copious amounts of warm soapy water or a diluted 10:1 water to bleach solution. Taking off your clothing can remove roughly 80 percent of the contamination hazard. Place your clothing in a sealed plastic bag. If outdoors and you cannot go inside, attempt to locate a fountain, pool, or other source of water so that you can quickly and thoroughly rinse any skin that may have been exposed. Flush eyes with water. If water is not available, talcum powder or flour are also excellent means of decontamination of liquid agents. Sprinkle the powder liberally over the affected skin area, wait 30 seconds, and brush off thoroughly. If available, rubber gloves should be used when carrying out this procedure.

7. If in a car, shut off outside air intake vents and roll up windows if no gas has entered the vehicle.
8. In any case of suspected exposure to chemical or biological agents, medical assistance should be sought as soon as possible, even if no symptoms are immediately evident.
9. If contact with anthrax through the mail or a shipment is suspected:
   a. Do not handle, shake, or empty a letter or package suspected of containing anthrax. Place it in plastic such as a zipper-lock bag or plastic sheet protector.
   b. Wash your hands with soap and warm water for 30 to 60 seconds, and then wash your face. Blow and wipe your nose.
   c. Close the doors and windows of the room where the package or letter is located and turn off air conditioning, heating, and fans.
   d. Collect the names of all people who have had contact with the letter or package.
   e. If you develop flu-like symptoms, see your health care provider immediately.
10. Call the police and fire department.
11. Monitor news broadcasts for additional information such as locations for vaccinations.
12. Maintain resource information for decontamination and the cleanup of biohazards.

Recovery
1. A chemical, biological, or radiological attack can have a lasting effect. The effects of a biological agent or pandemic could persist for many months and spread around the world. Plan for an extended relocation.
2. Track employees who may be quarantined, killed, or evacuated. The temporary or permanent loss of personnel may become the biggest issue from a continuity standpoint. This will include the employees of vendors, suppliers, and customers.
3. Develop a mass casualty plan for the organization that considers the biohazard risk.
4. Establish work-at-home capability or alternate worksites (this is applicable to most all recovery planning situations).
5. Cross-train employees where possible.
6. In the case of widespread sickness, minimize direct contact with others. Set up video conferences, web casts, and so forth for normal meetings.
7. Disinfect phones and other work surfaces on a regular basis.
8. Maintain sufficient out-of-region resources to meet continuity objectives.
9. Keep copies of building design and engineering data off-site and secured to aid in the decontamination of the building.
10. Once an affected building has been decontaminated, educate the users about the process and its safety to reoccupy.

Civil Disturbance

An organization's risk from a civil disturbance can include a range of exposures from peaceful protest to the direct action against its workers and facilities, or from the result of being located in "the wrong place at the wrong time" by suffering the consequences of a violent and destructive street protest, too often sparked by the infection of an anarchist element. Organized labor activity can also pose its own set of problems, and it demands its own set of protective measures. While the need to protect people and facilities from these destructive elements is easily recognized, the workers who need to respond must keep in mind the potential to create legal liability and media exposure if an encounter with a protester is not handled properly.

Prevention and Mitigation
1. Audit and improve physical security and access control.
2. Remove objects such as decorative rocks that can be used by demonstrators, activists, or rioters to break windows, injure employees, or damage other property.
3. Isolate lobbies from the remainder of the building, offices, or facility.
4. Develop intelligence programs if prone to civil unrest and demonstrations.

Preparedness
1. Train receptionists or security officers to understand arrest policies, and not to:
   a. Be provoked by name calling or derogatory remarks
   b. Discuss or argue the merits or issues of the dispute with protesters or picketers
   c. Antagonize demonstrators or picketers
   d. Throw back objects thrown at them
   e. Attempt to take anything from a protester unless in self-defense
   f. Make physical contact unless blocking a doorway.

Response
1. Dial 911 to report any indication of a civil disturbance such as a riot, demonstration, or picketing. Report any disturbance originating in an office suite to security or to 911 and to building management.
2. Remain in the building and close windows and drapes.
3. Lock exterior or suite doors and monitor who is attempting to gain entry.
4. Avoid confrontation with demonstrators.
5. Take elevators out of service to limit unauthorized access if necessary.
6. Warn other employees.
7. Decide with police, security, and local or building management if unauthorized persons need to be removed.

Recovery
1. Immediately clean up biohazards and physical damage.
2. Inventory and document any losses and submit to insurance.

3. Provide stress counseling for employees.
4. Review security and response procedures.

Confined Space Emergency

Construction and maintenance workers are often required to enter a confined space to perform work, but if they become injured or trapped, a rescue attempt by the untrained and unequipped too often causes additional victims. A *confined space* is generally defined as a space that has limited opening for entry and exit, poor natural ventilation, or that contains or could (as the result of welding operations, for example) contain dangerous air contaminants and is not intended for continuous worker occupancy. Check with the local safety enforcement agency for an exact definition used in your location. These safety agencies have very specific rules for the entry into a confined space and for the training and equipping of rescue operations. Examples of a confined space include pipelines, ventilation and exhaust ducts, storage or process tanks, pits, silos, vats, sewers, and tunnels. Annual fatalities in the United States average just short of two each week.

Emergency response to a confined space emergency can be difficult due to the environment in which they occur or because of the difficulty of access to the injured worker. The three types of rescue are *self-rescue* (the person recognizes or is told of the hazard or condition and escapes on their own), *nonentry rescue* (extrication of an incapacitated worker by the use of a lifeline or other mechanical means without entering the space), and *entry rescue* (removal of the worker or workers by entry into the space by properly trained and equipped rescue personnel using appropriate precautions).

Prevention and Mitigation
1. Ensure that workers understand how to identify a confined space.
2. Post warning signs on permanent confined spaces and inform employees of their locations and restrictions.
3. Where possible, avoid the need to enter a confined space. Use long-handled tools for shoveling or cleaning, install automatic cleaning systems, or other devices that may prevent the need to enter—that is, use an engineering means to avoid the hazard.
4. Follow all atmosphere testing procedures and entry protocols prior to going into the space.
5. Devote a sufficient number of workers to the operation to ensure the safety of those in the confined space.
6. Follow applicable safety regulations and good practices related to confined space entry as well as other hazard mitigation such as lockout/tagout of hazardous energy.

Preparedness
1. When a confined space entry is anticipated, implement a confined space safety and permitting program that at a minimum meets the requirements of local safety regulations.

2. Have all testing and rescue equipment appropriate to the risk at hand. Some equipment may need periodic calibration.
3. Train workers in their duties and responsibilities, as well as safe operating and rescue procedures appropriate to the risk. Ensure that this training is current and includes any other requirements such as medical examinations for respiratory protection.
4. Understand the capabilities of the local fire department to assist or manage a confined space response. Supplement your abilities or that of the department as necessary.

Response

1. Dial 911 and report injury/victim trapped within a confined space.
2. Do not rush into a rescue attempt; enter only if trained, equipped, and not alone. Consider the following:
   a. Location, magnitude, and nature of the incident
   b. Risk versus benefit (danger to rescuers)
   c. Available resources, personal protection equipment necessary to conduct safe and effective operations
   d. Potential changes in conditions, environment;
   e. Additional hazards
3. Assist the entrant who needs to self-rescue without breaking the plane of the entry opening.
4. Without entry, determine the number of victims and their locations, and, if possible, attempt rescue without entry by mechanical means using pulleys, winches, or lifelines.
5. If applicable, provide ventilation to victims in confined spaces (from outside) with a blower.
6. Assess the perimeter surrounding the confined space to determine the presence of, or the potential for, hazards or risks to rescuers and the victims. Control hazards as practicable including shutting off nearby operations or utilities that may pose a hazard.
7. Monitor changes in conditions and environment (without entry).
8. Determine if rescuer and equipment capabilities are appropriate for rescue operations.
9. Only those trained and properly equipped in confined space rescue should enter.
10. Ensure rescuers wear appropriate safety equipment (harness, lifeline, etc.).
11. Determine if special needs or conditions exist—that is, the need to decontaminate the victim if hazardous materials are present.
12. Stage equipment as necessary (first aid, air/oxygen). If hazardous materials are involved, be prepared to present a copy of the MSDS to the fire department or ambulance crew.
13. If time permits, develop a specific rescue plan. Consider the type, size, access, and internal configuration of the confined space, and provide this information to the fire department.

Recovery
1. Activate the crisis management plan if a serious injury is involved.
2. Notify the appropriate regulatory agencies such as OSHA as necessary.
3. Conduct investigation into the cause of and the response to the incident.
4. Provide posttraumatic stress counseling in the case of a fatality.

Earthquake

The release of energy caused by resistance to the continual shifting of large segments of the earth's crust (tectonic plates) is responsible for most of the world's earthquakes. The boundaries of these plates are called fault zones. Although most major earthquakes occur near the fault zones, few places in the world are totally immune.

Major earthquakes can occur many thousands of miles away from these boundaries; they are thought to be related to interplate crustal weakness. Such was the case for the largest earthquakes in the continental United States, along the New Madrid (Missouri) fault in 1811 and 1812. No portion of the United States or southern Canada is immune from the effects of earthquakes. The entire West Coast (including Canada, Alaska, Nevada, and Utah), the Midwest near the Mississippi River, and the East Coast north of Florida up to southeastern Canada and New England are susceptible to significant ground movement.

Damage from earthquakes is related to the amount of energy released, length of the fault rupture, depth and type of fault, velocity and acceleration, distance from the fault, soil types and conditions, and type of building construction. Various methods are used to measure earthquakes. The Richter scale indirectly measures the energy released from an earthquake by recording needle deflections on a seismograph. An earthquake with a Richter magnitude of 3 is barely felt unless you are very close to it. A magnitude 6 earthquake can cause major damage. The Richter scale uses a logarithmic progression. That is, the energy released from a magnitude 6 earthquake is not twice as big as that from a magnitude 3 but about 100 times greater. The Great Sumatra earthquake of 2004 that generated the tsunami responsible for almost 300,000 deaths was estimated to be a magnitude 9.15, one of the largest since 1900. Some scientists placed the magnitude as high as 9.3. This was caused by a sea-bed rupture of nearly 800 miles, with estimates of vertical movement in some locations of 50 feet, vibrating the planet by as much as 1/2 inch (1 centimeter). The shaking lasted for about 10 minutes. The earthquake released energy equivalent to 100 gigatons of TNT and caused a slight change in the earth's rotation.

The Modified Mercalli scale is used to map areas of intensity, based on personal reports by victims and by inspection of the damage. It uses a scale from I through XII, based on the type of damage observed. Although not as popular as the Richter scale, it provides a better indication of actual damage.

Certain soil conditions amplify or attenuate seismic forces. Soil and bedrock in the eastern United States tend to transmit this energy over a wider area than that in the West, allowing an earthquake in Bolivia to be felt in Minneapolis. Sand and silt near unstable bay and river areas can cause liquefaction and soil failure. Liquefaction occurs when the ground motion causes sandy materials saturated with water to behave like a liquid.

Most modern structures are designed to withstand earthquakes. They may sustain heavy damage, but they should not collapse. The greatest danger is from objects such as parapets, signs, bricks, and glass falling off buildings. Older buildings, nonretrofitted buildings of concrete tilt-up, and nonreinforced masonry construction can partially or completely collapse. The Uniform Building Code in the United States and the National Building Code in Canada contain maps that assign a seismic risk (based on expected ground movement, not probability) to various parts of the countries and define building codes to balance the known level of risk in these areas.

Earthquakes can cause a complete collapse of transportation systems: highways, overpasses, bridges, shipping ports, and airport runways can be damaged or destroyed. Infrastructure can fail; expect to be without water, sewer, and utilities for 3 days or more; police, fire, and hospital services can be destroyed and will be likely overloaded.[4] Thousands can be made homeless or refuse to return to their homes, even if damage is minor. Dam and levee failures, train derailments, landslides, hazardous material releases and spills, uncontrolled fires, avalanches, people trapped in buildings, and people injured in falls could all result from an earthquake. The destruction generated by tsunamis produced by underwater faults is well known. The 2011 earthquake in Japan generated a tsunami that took tens of thousands of lives locally and all the way across the Pacific Ocean caused millions of dollars in damage (and claimed one life) on the west coast of the United States. The results of this earthquake further illustrates the cascade of disasters that one event can trigger (tsunami, the release of nuclear materials, fires, homelessness) in a country with modern infrastructure and response capability. The possibility of multiple events must always be considered in emergency preparedness and planning.

Presently, forecasting is based on known fault locations and the historical recurrence rate of earthquakes on these faults. These predictions are difficult, in part, because many faults are not discovered until they cause the earth to move. A fault under Los Angeles, only identified in 1999, is expected to cause losses of up to $250 billion and result in between 3,000 and 18,000 fatalities.

Prevention and Mitigation
1. Understand the seismic risk for your area and plan accordingly.
2. If your business processes are at risk of unacceptable damage from a large earthquake, consider relocating or diversifying facilities to areas less susceptible to seismic events.

---

[4] Subsequent to the effects produced by Hurricane Katrina, and the response to it, disaster researchers and disaster organizations now recommend that people and organizations prepare to be "self-sufficient" for a period of 5–7 days and some suggest a period of up to 14 days.

3. Identify the structural and nonstructural hazards to your facilities, occupants, equipment, and processes.
4. Ensure that automatic fire sprinkler systems are supported with earthquake sway bracing (see National Fire Protection Association [NFPA] Standard 13, Standard for the Installation of Sprinkler Systems). Pipes can break during an earthquake, causing water damage or the inability to extinguish a fire.
5. Bolt bookcases, file cabinets, display racks, workstations, flammable liquid storage cabinets, wire or equipment racks, and other tall, heavy objects to the floor or wall. These objects can topple over during an earthquake, causing injuries to employees, blocking escape routes, damaging equipment, and delaying the cleanup and recovery process.
6. Strap sensitive, critical, or expensive equipment (including computers and servers) to desks, workbenches, or equipment racks. Place equipment that cannot be strapped or that must be moved often on seismic damping mats. Base-isolate (prevent the base from moving, or attach devices that allow the base to move independently) large equipment such as boilers, pumps, chillers, and backup electrical generators.
7. Identify and mitigate other nonstructural hazards, including light fixtures, storage racks, hanging objects, mirrors, and fireplaces.
8. Strap water heaters to the wall and install flexible connections to the gas line. Water heaters can topple over during an earthquake, causing fires and the loss of a potential source of drinking water.
9. Consider the installation of seismic switches (valves) that shut off hazardous process gasses or equipment when a preset acceleration is achieved.
10. Arrange for backup power generation.
11. Ensure you have a plan for redundant communications.
12. Prepare a mass casualty plan.

Preparedness
1. Stockpile supplies of food, water, lighting, and disaster first-aid supplies. Employees may need to remain on-site temporarily while transportation systems are being restored. Encourage employees to keep emergency supplies in their desk or car. At a minimum, this should include the following:
   a. First-aid supplies and prescription medications
   b. High-energy, low-salt packaged, dried, or canned food
   c. Nonelectric can opener
   d. Portable radio with batteries
   e. Flashlight or light sticks
   f. Extra cash
   g. Tennis shoes
   h. Blanket

      i. Pocket knife

      j. Safety whistle (this should remain in the employee's pocket or purse)

      k. Picture of kids, family

      l. Out-of-town contact phone numbers

      m. Change of underwear and socks

      n. Toiletries and personal hygiene items

      o. Proof of residence, such as a water bill

      p. Drinking water—at least 1 gallon per person per day. Water supplies may not be available for 3 to 5 or longer days after an earthquake

2. Encourage individual and family preparedness. Employees who know that their families and loved ones are safe at home are more willing to remain at work or to return to work soon. Conduct earthquake safety workshops and obtain information from the American Red Cross and the Federal Emergency Management Agency (FEMA) to distribute to employees.

3. Consult with a structural engineer to ensure that your buildings meet seismic safety standards. Arrange to have these engineers inspect the buildings after an earthquake. See #7 below.

4. Train personnel in the location of emergency shutoff valves for gas, electrical, water, and hazardous gas or chemical lines. Install flow valves on hazardous material lines to prevent uncontrolled discharge if the pipes are broken.

5. Train employees in first aid and cardiopulmonary resuscitation (CPR).

6. Maintain an adequate supply of cash.

7. Set up an agreement with the local authorities to accept a contracted structural engineer's judgment about the safety of your facilities after a major earthquake. Local inspectors may not check your building in a timely manner and, because of time constraints and workload, may be less likely to allow occupancy.

8. Consider purchasing earthquake insurance.

Response

If you are inside—DUCK, COVER, and HOLD:

- Take cover under a desk or table, or sit or stand against an inside wall (not inside a doorway).
- Hold tightly to the desk or table until the shaking stops.
- Move away from windows or objects that may fall on you.
- Do not run outside during the shaking.
- After the shaking has stopped, evacuate the building if structural damage is apparent.
- Avoid use of the telephone. Replace telephone handsets shaken off the hook.
- Do not use open flames.

If you are outside:

- Do not enter a building during the shaking.

- Move clear of buildings, falling glass, utility poles, wires, and large trees.
- Get on the ground—DUCK, COVER, and HOLD.
- After the shaking stops, watch for falling glass, electrical wires, poles, and other debris.

If you are driving:

- Drive away from overpasses and underpasses.
- Stop in a safe place.
- Set the parking brake.
- Stay in the vehicle. If wires fall onto the vehicle, stay inside until rescued.

If located in a tsunami inundation zone, move to higher ground after a large earthquake, especially if the epicenter is off shore. Monitor media for tsunami warnings and reports.

Recovery

1. Check for injuries.
2. Check the structural integrity of the building.
3. Rescue victims, if safe, trained and equipped to do so.
4. Check for fires and other damage. Turn off gas only if a leak is detected.
5. Clean up hazardous material spills.
6. Turn off noncritical electrical equipment if power is out. When power is restored, turn equipment back on gradually.
7. Avoid the use of open flames. Do not use camp stoves or charcoal in an enclosed area, such as a tent.
8. Use caution when reentering damaged buildings; aftershocks can cause further damage.
9. Check that vents and exhaust pipes have not separated from water heaters or other machinery.
10. Listen to radio and television stations for official information.
11. Begin relocation and reconstruction if necessary. If employees are to occupy a partially damaged building, concentrate first on cosmetic repairs. No one wants to look at cracks in the wall during an aftershock.

Evacuation Planning

Since the collapse of the Twin Towers in New York on September 11, evacuation planning is no longer as simple as listening for the fire alarm and finding the nearest exit. Past practices, such as relocating two floors up and three below in a high-rise building equipped with a sprinkler system and pressurized stairwells, are coming into question. The admonishment to not use elevators for evacuation is under review, with much discussion directed toward changing this policy. The installation of elevator systems composed of hardened smokeproof enclosures, backup power, and other protections may make them serviceable in a fire situation. Greater emphasis is placed on planning for the full and complete evacuation of high-rise buildings, despite the time this may require (estimated at about 10 seconds per floor to move down a stairwell, assuming everybody keeps moving). According to the draft Federal Building and Fire Safety Investigation of the World Trade Center Disaster Occupant Behavior, Egress and Emergency Communications report, "Success in evacuating a building in an emergency can be characterized by two quantities: the time people needed to evacuate and the time available for them to do so. To the extent the first time exceeded the second, it follows that there will be casualties. When the second time exceeds the first, perhaps by some suitable margin, nearly all should be able to evacuate the building."[5]

Fire and smoke, as well as other hazards, can spread very rapidly, necessitating quick action to protect building occupants and move them away from the danger. However, the decision to evacuate a building should not be taken lightly; injuries can occur, productivity is lost, morale can suffer, and people may be less likely to treat seriously the need to evacuate during an emergency if preceded by a number of "false alarms." Planners must analyze their hazards and decide how they need to structure evacuations for their environment and for each foreseeable situation.

Once evacuation plans are in place, employees must be made aware of their responsibilities—that is, how to recognize the need to evacuate, what is expected of them, where to go, and what to do when they get there, and markers indicating where evacuees are to report once outside the building. This is often accomplished by the publication of an Emergency Action Guide issued to new employees and reviewed during new hire training. Visitors must also be provided with this knowledge, usually accomplished by maps that contain brief instructions and evacuation routes posted inside entrances and elevator lobbies, instructions on the back of visitor badges, or brochures distributed during the visitor registration process. In high-rise buildings, color-coded symbols can be painted on the walls of stairwell entrances with instructions to proceed down the stairs until they find a like symbol (if not calling for a full evacuation). That floor becomes their evacuation relocation floor (examples include a green star, yellow triangle, red circle, blue square). The same symbol is located every three floors below, or the number of floors mandated by local regulations.

A primary responsibility after an evacuation is to verify that no one is left in the building or in the area vacated. In days past, this was done by taking a head count of those who were present at the assembly point (roll call). Although this may prove effective when dealing with small groups of employees, it is nearly impossible with larger and more mobile populations. In general, the better approach is to assign primary and alternate "floor wardens" to search their segmented areas of responsibility and to report to the incident commander (or equivalent) that no one is left behind in their section. It is their

---

[5] *National Institute of Standards and Technology National Construction Safety Team Act Report 1-7* (Draft) (September 2005) U.S. Government Printing Office.

responsibility to ensure the safe and complete evacuation of all occupants from their area. The size of the search area must be manageable so that the warden is not in excessive danger due to a delay in evacuating (obviously, the warden should be the last out). In high-rise buildings, additional wardens may be assigned to monitor stairwells to keep people moving, stationed at elevator lobbies, or assigned to help the disabled. Wardens should be clearly identified by wearing distinctive hard hats, reflective vests, armbands, or other means of identification. Many large cities will mandate the number and type of wardens for buildings above a certain height and may require them to perform additional duties. Floor wardens must be thoroughly trained in their responsibilities: in the emergency plan, first aid, CPR, and Automated External Defibrillator (AED) use; and in their own safety. They should understand the requirements of the ICS. Employees must be trained to acknowledge their existence and authority.

All evacuation planning should include visitors, vendors, contractors, and guests. This is especially true for the disabled; an increasing number of court cases are awarding damages to the disabled who were not adequately cared for during an evacuation.

Experience has shown that people tend to leave by the door from which they entered the building, even if they are just a few feet away from a more appropriate exit. They may also delay evacuating, even when they know there is danger. Although people tend not to panic in emergency or disaster situations, evacuation planning and drills will reduce injury and confusion.

Prevention and Mitigation

1. Ensure that the evacuation warning system and fire department communication panels (if applicable) are connected to backup power and that the systems will not be damaged in a fire or earthquake. Absent local regulations, panels should be located away from evacuation routes, noisy areas, or areas subject to blast damage.
2. Ensure that your evacuation planning conforms with codes and standards and with local regulations. In the United States, OSHA has very specific guidance and regulations for evacuations.
3. In high-rise buildings, assign a dedicated fire safety director whose sole function is life safety for the building occupants. This is required in certain jurisdictions.
4. To the greatest extent practical, upgrade life safety elements of the building. Strengthen evacuation stairwells, make them wider to accommodate a greater volume of people, improve lighting, and make them smokeproof.
5. In high-rise buildings (or single-story buildings, for that matter), increase the evacuation discharge area to ensure it is large enough to accommodate a full evacuation.
6. Remain current on the latest standards and thinking about evacuation procedures.
7. Have backup (redundant) communication systems in place (public address system, elevator telephone system, and telephone system).
8. Ensure that stairwell doors unlock during an evacuation or fire emergency.

9. Maintain close ties with the fire department, and ensure that response plans are coordinated.

10. Ensure that part of the fire department's response plan includes a search of stairwells to ensure that all people have left the building.

Preparedness

1. Determine who will plan, coordinate, and authorize evacuations. Assign an alternate for this person. Employees should be trained to activate fire alarms if this method can be used to initiate an evacuation.

2. Develop a script to use when announcing an evacuation. For example: "Attention. All occupants of floors 10, 11, 12, 13, and 14 are to immediately enter the nearest stairwell and proceed down to the floor with the same evacuation symbol as the one for your current floor." Repeat the message several times.

3. Develop unobstructed emergency escape routes that comply with local regulations and that do not lead to or through hazardous areas. Ensure that your plan is consistent with legal requirements, such as proper signage (illuminated exit signs), posted evacuation route maps, and number of exits. Never nail doors shut; clearly mark any doors and hallways that are not exits. If the emergency is minor, it may be necessary only to relocate employees to a safe location within the building. In high-rise buildings, it may be appropriate to evacuate only the three floors above the incident and the two floors below (some jurisdictions require five-floor relocation: two floors above a fire floor and two below are moved down five floors).

4. Devise special procedures for employees to stay behind and shut down equipment or perform critical operations. However, delaying the evacuation of employees should be minimized, if not eliminated.

5. Select and train evacuation monitors or floor wardens and clearly define their duties. Floor wardens typically confirm that everyone in their assigned area evacuates, ensure that the escape route is safe, and assist where needed.

6. List methods used to notify employees of an evacuation. These can include activation of the fire alarm or other distinctive audiovisual warnings (be certain that warnings can be heard in all locations of the facility), public address system announcements (develop scripted warning messages that can be adapted to the immediate situation—see earlier), or word of mouth (use floor wardens or supervisors). Consider repeating the notification in a different language.

7. Identify employees with handicaps who may need assistance with evacuations. Assign "buddies" or floor wardens to assist these people. Purchase special equipment, such as evacuation chairs, to assist with their removal from the building. Train potential users of this equipment on its location and operation. Wheelchairs should not be taken into stairwells (part of the discussion mentioned above includes the use of elevators for the disabled).

8. Establish assembly points. Show the assembly points in the Emergency Action Guide (if required) or on wall maps.[6] If feasible, place a sign, placard, or brightly painted markings on the ground to indicate where employees are to assemble. Assembly points should be a reasonable distance from the building so that employees and guests are not injured by the emergency or by other dangers. They should not be located where they could impede emergency response operations or equipment.

9. Devise a method to account for employees and guests. This can be accomplished through a sign-in log, by having area supervisors taking roll calls of their employees and guests, or through the use of floor wardens.

10. Establish "all clear" procedures and guidelines to follow when authorizing the return to the building. These should include a method to check whether the building is safe to reoccupy. Designate who has the authority to notify evacuees when it is safe to reenter.

11. Document your plan. Summarize the procedures in an Emergency Action Guide and distribute the guide to each employee.

12. Conduct drills often, no less than annually. Local regulations may require drills more frequently. Drills should always be announced ahead of time, and no one should be excused from the drill. Drills should actually simulate an evacuation; that is, if evacuees are to relocate to a different floor, actually go to that floor.

Response

1. If it can be done safely, shut down hazardous processes before evacuation.

2. Train all employees to recognize the evacuation signal and to follow evacuation procedures. Include the following instructions:

   a. All employees are to leave immediately by the nearest exit.

   b. Do not return to the work area to obtain personal effects (if you are at your work area when the evacuation signal is heard, take your car keys with you).

   c. Do not use elevators (see above).

   d. Assist the disabled (if there are not floor wardens assigned to this task).

   e. Go directly to the assembly point unless instructed otherwise (do not stop to talk with friends or to get something from a vehicle).

   f. Do not reenter the building until the "all clear" is given. Your plan and instructions should designate who has the authority to issue the reentry order.

3. Ensure that all employees and guests evacuate. Floor wardens will sweep an assigned area to check that all have evacuated and to help the disabled evacuate.

4. Notify emergency officials if anyone is not accounted for or if injured are left behind.

5. Keep employees informed often until a decision is made to release employees to go home or to reoccupy the building.

6. Close stairwell doors after you enter; do not prop them open.

7. Remove high-heel or flip-flop shoes if descending stairs.

---

[6] Some states require employers to provide employees with written instructions to follow during an emergency.

Recovery

1. Establish several methods to disseminate return-to-work instructions with employees if they are released to go home.
2. Review the plan and procedures at least annually or after any major change in the configuration of the building. Provide for the training of newly hired or transferred employees.
3. Consider instructing business continuity team members to take laptop computers with them when they evacuate as long as they don't need to return to the workstation to get the computer. Some organizations issue laptop computers to business continuity team members so that they can continue to operate at an alternate location. This is a controversial task because evacuees should not do anything that delays evacuation or carry anything with them that could cause injury or delay others.

Excavation or Trench Collapse

The entrapment of a worker in an excavation or trench collapse accounts for just over one death each week in the United States. Just one cubic yard of dirt or sand can weigh 1 ton (dirt) to 1.3 tons (sand) or more. This weight will obviously make breathing difficult if buried up to chest level, but most of these fatalities occur when victims are only partially buried. A common story found in these cases (and also in cases of building collapse from earthquakes or other types of entrapment) is that of the victim in relatively good spirits, speaking with the rescuers, and even helping to dig himself out, but dies in the ambulance or hospital after extrication. This is due to crush syndrome (traumatic rhabdomyolysis) that results in the sudden reperfusion of toxins produced by the breakdown of muscle tissue that has been crushed or compressed and as a result without blood or oxygen for a period of time (usually one to four hours), primarily causing kidney failure and potentially heart arrhythmias and other complications. Treatment once this occurs includes the administration of intravenous fluids that is beyond the capability of most emergency response teams, so prevention and rapid rescue are important.

Prevention and Mitigation

1. Ensure all hazardous utilities are identified and exposed prior to digging.
2. Ensure the trenching contractor has a "competent person" present during all phases of the trench work to evaluate hazards (soil conditions, poor air quality, fall protection, stability of adjoining structures, access and egress) and to implement the appropriate protective systems (proper benching or sloping of excavation walls, insertion of shoring or trench boxes).
3. Empower workers with the knowledge and expectation they must stop work and remove themselves from a trench or excavation if they believe conditions pose a hazard.
4. Monitor the work of contractors or maintenance workers on your property to help ensure that safe work practices are followed.

Preparedness
1. When trenching or excavation is anticipated, train workers in safe work practices for these operations.
2. Train emergency response workers in response protocols.
3. Maintain tools appropriate for a rescue on-site or ensure the local fire department is trained and equipped for these types of emergencies.

Response
1. Dial 911 and inform the dispatcher that a victim is trapped in a cave-in.
2. Remove other exposed individuals from the excavation.
3. Cease any operations or processes that may cause ground vibration (with the exception of dewatering equipment). Restrict vehicle traffic.
4. Determine the number of victims and their location. Obtain probing poles for use, if applicable.
5. Quickly determine why the trench collapsed, the potential for secondary collapse, and how it can be made safe for rescuers.
6. Stabilize trench/excavation walls, ensure ground pads are placed at the lip of the trench, ventilate the trench and monitor atmosphere, dewater soil, support unbroken utilities, as applicable.
7. In appropriate soil and environmental conditions (sand, grain, pea gravel, or any type of running product), protect the victim from the surrounding soil with the use of concrete or steel pipe, or pre-engineered structures that physically isolate the victim.
8. Stabilize adjoining buildings, walls, other structures, or vehicles endangered by excavation operations. Find out if the failure damaged a utility. If so, take appropriate action.
9. Remove or mitigate other potential hazards to victim or rescuers.
10. Perform non-entry-type rescue, such as placing a ladder over the trench to allow the victim to self-rescue if possible and safe.
11. When safe, use hand digging for extraction.
12. Avoid the use of heavy equipment or mechanical winches to physically lift, pull, or dig out (extricate) victims. Do not attempt to pull a partially buried victim out by a rope or sling.
13. List special circumstances of the trench construction and pass this information on to rescuers/fire department personnel. This should include trench geometry, soil types, type of collapse, shoring, the presence of utilities and hazardous substances.
14. Stage emergency equipment (oxygen, stretchers, hand tools, shoring, pumps, ladders, etc.).
15. Once extricated, check for injuries such as broken bones, and if possible, some sources suggest maintaining pressure on or above crushed or compressed body parts so that reperfusion can occur gradually and not overload the renal system. Inform ambulance or paramedic crews of the possibility of crush syndrome injury.

Recovery
1. Activate the crisis management plan if a serious injury is involved.
2. Notify the appropriate regulatory agencies such as OSHA as necessary.
3. Conduct investigation into the cause of and the response to the incident.
4. Provide posttraumatic stress counseling in the case of a fatality.

Fires

Fires account for the greatest number of losses to business, exceeding that from natural disasters. Sixty-four to 70 percent of businesses that experience a major fire never recover. They go out of business for good, primarily because of the loss of vital business records, particularly their accounts receivable files. Fire is also a major cause of accidental death in the United States. America's fire death rate is one of the highest per capita in the industrialized world, with over 1.3 million fires causing the death of just over 3,000 and injuries to more than 17,000 people in 2009. These figures represent a downward trend since 1977 but typically account for a leading cause of death in the home. Direct property loss due to fires in the United States is estimated at $12.5 billion. Injuries and burns from fires are often so hideous that their victims become reclusive; they do not want to be seen in public. The healing process from serious burns is said to be the most painful of any trauma.

Many fires in business are caused by intentional or careless acts and by equipment failure, but bear in mind that arson is the preferred tool of many domestic terrorists, such as animal rights groups. Labor disputes can result in, and disgruntled employees have resorted to, the burning of facilities. Terrorist plots have revealed plans to drive gasoline tankers into the lobbies of buildings. Fires in public assembly areas and nightclubs have in the past and continue into the present to cost too many lives. Fires can accompany natural and manmade disasters, during times when the ability to control them is severely compromised.

Fires can develop when a combustible fuel comes in contact with an ignition source. This contact does not need to be direct because radiant energy can be sufficient to bring a fuel up to its ignition temperature. Because we are surrounded by combustible materials, it is generally easier to prevent fires by controlling their ignition sources.

Prevention and Mitigation
1. Understand your fire risk, the fire codes, and fire prevention engineering to prevent the ignition and spread of fires. Retrofit as much as possible to bring buildings up to the latest fire codes.
2. Install fire extinguishers on every floor and near hazardous areas, a code requirement most everywhere in the world.
3. Conduct regular fire prevention inspections and immediately correct any hazards identified (see later).

4. Maintain all fire control devices and equipment in good working order and up to local codes and standards.
5. Establish a vegetation "clear zone" extending at least 100 feet from brush, fields, or forests. Use fire-resistant landscaping and building materials.
6. Ensure that the address of the business is clearly marked and that the fire department is aware of the locations of shutoff valves and hazardous materials. Work with the fire department ahead of time to develop "pre-fire" plans.
7. Establish no-smoking policies and permit smoking only in a fire-safe area.
8. Implement and enforce a fire permit system for "hot work" such as welding.

Preparedness
1. Establish a written fire prevention plan. The plan should at a minimum include the following elements:[7]
   a. A list of potential fire hazards and their proper handling and storage procedures, potential ignition sources (such as welding, smoking, and others) and control procedures, and the type of fire protection equipment or systems that can control a fire involving the hazards
   b. Names or regular job titles of those responsible for maintenance of equipment and systems installed to prevent or control ignitions or fires
   c. Names or regular job titles of those responsible for the control of accumulation of flammable or combustible waste materials
   d. Housekeeping. The company must control accumulations of flammable and combustible waste materials and residues so that they do not contribute to a fire emergency. The housekeeping procedures shall be included in the written fire prevention plan.
   e. The company shall apprise employees of the fire hazards of the materials and processes to which they are exposed.
   f. The company shall review with each employee upon initial assignment those parts of the fire prevention plan which the employee must know to protect the employee in the event of an emergency. The written plan shall be kept in the workplace and made available for employee review.
   g. Maintenance. The company shall regularly and properly maintain, according to established procedures, equipment and systems installed in the workplace to prevent accidental ignition of combustible materials.
2. Install automatic fire detection, suppression, and warning systems (UL-listed monitored fire-alarm systems with smoke detectors and sprinklers).
3. Establish guidelines and other written material for employees. Train employees in fire safety and evacuation.

---

[7] Adapted from the California Code of Regulations, Title 8.

4. Shut off utilities, processes, and electronic systems if evacuation is anticipated. Be aware that some utilities cannot simply be turned back on without inspection. Never delay evacuation if this cannot be done safely.
5. In the case of a wild land-fire, place escape vehicles in position before evacuation is ordered.
6. Ensure that electrical systems are properly maintained. Is lightning protection required?
7. Inspect gas lines, furnaces, and boilers for good condition and the absence of leaks; protect processes that use flammable materials from damage, both accidental and intentional.
8. Assign someone to ensure that soldering irons, hot plates, and other ignition sources are turned off at the end of the day. Security officers should be trained and instructed to check for fire hazards while on their rounds.
9. Control traffic lanes and parking so that emergency vehicles have unencumbered access to the facility.
10. Preregister with restoration service companies that specialize in the cleaning, dehumidification, and odor removal of electronic systems and the recovery of paper documents. The components of smoke can severely damage electronic equipment even if not in contact with heat and flames.

Response
1. Obtain basic information about the location, type, and size of the fire if it is reported to you. Record the time, name, and extension number of the person reporting the fire. Ask if personnel have evacuated the area and if anyone is trapped or injured by the fire or smoke.
2. Evacuate the immediate area. If it can be done safely and quickly, open windows and close doors. Don't use elevators (see above). If floor wardens are used to verify that everybody has evacuated, leave your office door open so they can check that everybody has left. Stay low, near to the floor, if smoke and heat are strong.
3. Dial 911 (or your nation's emergency number) or the number that directly accesses the fire department dispatch center. Remain on the line until the dispatcher hangs up. Some organizations will first send a member of management, security, or the ERT to evaluate the need to notify the fire department. This can be a dangerous and costly approach because fires can get out of control quickly.
4. Fight the fire, if you are qualified (trained) and it is safe to do so. In many jurisdictions, the equipment and training required to fight a fire beyond the incipient stage (nonstructural, preflashover) is extensive.
5. Shut off heating, ventilation, and air conditioning systems if a potential for bringing smoke from the outside into the building exists.
6. Dispatch one or more people to meet the fire department and direct them to the location of the fire. The less obvious the route, the more personnel should be assigned to this task.

7. Remove any vehicles or other impediments to responding equipment.
8. If the building has a sprinkler system, assign a person to monitor the post indicator valve (PIV). This person should verify that the valve is functional and in the open position. Remain at this position to ensure that the valve is not prematurely shut.
9. Assign a person to check or start the fire pump, if one is utilized.
10. Determine whether equipment or hazardous processes should be shut down. Do so only if it is safe.
11. Move flammable or hazardous materials away from the area. Do so only if it is safe.

Recovery
1. Contact a restoration service that can repair smoke-damaged equipment, clean up water, and control mold and odor.
2. Prevent further damage to facilities and equipment from rain, theft, and other fire-related problems.
3. Restrict access to the damaged area.
4. Contact the insurance carrier or broker.
5. Determine whether the fire will affect production schedules or the functions dependent on any equipment, materials, or product destroyed by the fire.
6. Ask the fire department if it or disaster recovery teams can remove critical equipment or records not damaged in the fire. Some companies place large red or yellow dots on this equipment so it can be quickly located and removed if access to the building is restricted.
7. Recreate any lost vital records.
8. Provide status reports to major customers and employees.
9. Quickly replace or replenish used fire control equipment or devices.
10. Begin relocation and reconstruction.
11. Determine the cause of the fire and what actions are necessary to prevent a reoccurrence of the fire.

Floods and Heavy Rain

Second only to fires, floods are the most common and widespread of all natural disasters in the United States. On a worldwide basis, floods have accounted for between 30 and 50 percent more loss of life than earthquakes. According to FEMA, flooding has caused the deaths of more than 10,000 people in the United States since 1900, with more than $1 billion in property damage each year (this figure varies greatly from year to year; flood damage from Hurricane Katrina is estimated at $125 billion). Flooding is a major component of almost 9 of every 10 presidential disaster declarations that result from natural phenomena.

The onset of a flood can be gradual or sudden, and it can occur in areas where it is not normally expected. Heavy rains are usually the root cause. A summer storm in Colorado in 1976 was responsible for the deaths of 100 campers and vacationers; water came rushing down a river without warning. This type of tragedy has been repeated a number of times since this date. These floods can reach heights of 30 feet or more, and they can result from the collapse of a dam formed by a landslide, ice, or a debris jam or from the runoff of water from the walls of a canyon. Flooding can also result from the structural failure of permanent dams during an earthquake. High seas produced by major storms (storm surge) can also cause flooding, especially in low-lying areas, as experienced in many hurricanes. Tropical storms can carry huge amounts of water.

Floods are usually easy to predict. Flash floods generally recur in the same location and can happen within minutes of a storm, even if heavy rain is not apparent in the immediate area of the flash flood. Where floods are prevalent, the National Oceanic and Atmospheric Administration (NOAA) maintains River Forecast Centers that report on river levels, rainfall, and predicted weather. Up-to-the-minute information is also obtained from firsthand observations, monitoring police and emergency services radio communications, and river-level websites. Some jurisdictions prone to flash flooding have installed horns and sirens to warn people of impending danger.

Area-wide flooding causes many more problems than simply getting things wet. The force of the water in a river at flood stage can move vehicles and destroy buildings and bridges by battering them with an incredible force from the flowing water and with the debris it contains. It can isolate areas from fire and police response and sever communications lines. Floods can contaminate drinking water; sewage treatment facilities and pumping stations can be knocked out of service due to power failures.

Despite all the water, fires can be a big problem during floods. Underground fuel and storage tanks containing hazardous or flammable liquids can be forced to the surface. Equipment can be damaged or destroyed by water and mud, which often are deposited everywhere. Finished product in your warehouse can be contaminated and vital records damaged or destroyed. The foundations of buildings can be undermined by the erosive effect of rapidly moving water. The growth of certain molds after the flood can be deadly, and the odor is difficult to remove.

Determine the flood risk for your site, for access to your site, and for any infrastructure that feeds your site. List the elevations of facilities, critical processes, power sources, or other items that may be affected by flooding and understand what actions must occur at differing water levels. FEMA, as part of the Flood Insurance Rate Map (FIRM) program, issues maps that show the flood hazard area of your community. These maps are used to establish your flood risk and are different from inundation maps that show the areas of potential flooding due to dam or reservoir breaks. Use caution, however, because the maps do not always consider the maximum flooding or indicate all areas where flooding has, or can, occur. The maps outline the areas subject to flooding based on 100-year and 500-year flood potentials. Other terms for a 100-year flood potential include a *base flood, baseline flood*, and a *1 percent chance flood*. The term *base flood* is increasingly used because it does not carry the misnomer that a 100-year flood is a rare event or that water levels will not be exceeded. A 100-year flood does not mean that this level of flooding occurs only every 100 years! The 100-year flood has a 26 percent overall chance of occurring every 30 years and can (and has) repeated itself in the same location within the same year. A 1 percent flood means that there is a 1 percent or greater probability of the mapped area being equaled or exceeded during any given year. A 500-year flood indicates a 0.2 percent chance of being equaled or exceeded in any given year. FEMA's HAZUS software mentioned in Chapter 12 also contains a flood mapping component.

Prevention and Mitigation

1. Maintain a flood mitigation and response plan and list the necessary resources (sandbags, pumps, dikes, etc. to implement the plan, appropriate to the risk). This may be required before you can apply for federal aid. Coordinate your planning with local agencies and include plans to assist employees who may need transportation to evacuate.
2. Keep streams, culverts, and other waterways adjacent to your property or to escape routes clean of debris or silt deposits that could restrict drainage or cause it to dam. Work with local governments to assist in the removal of vegetation or other impediments. Monitor obstructions such as ice jams.
3. Install flood barriers, flood panels and shields, diversion dikes, or other architectural barricades and maintain them in good condition. Floodproof at-risk structures (water-seal walls, install flood doors, reinforce exposed walls). If located in an area prone to flash flooding, consider erecting diversion walls.
4. Don't develop or expand into flood-prone areas.
5. Ensure that power shutdown instructions are distributed to the appropriate personnel and that these people are trained in the shutdown authority and procedures.
6. Determine whether underground tanks need to be drained and refilled with water to prevent damage from flotation.
7. Develop a plan to evacuate domestic or farm animals or equipment if appropriate.
8. If critical materials and supplies need to be delivered to keep the site in operation, develop a contingency delivery plan for this purpose.
9. Consider the purchase of flood insurance.
10. Build diversion channels; straighten river channels.
11. Increase sewer and storm-water drainage capacity.
12. Anchor any items or equipment that may float away. (Large tanks of flammable or hazardous materials can and do float away.) Consider off loading storage tanks of hazardous materials and separating chemicals that can react if mixed together. Also consider filling emptied tanks with water to add weight, if this will not contaminate, or cause an adverse reaction with, its contents when refilled.
13. Install check-valves in sewer traps and other piping systems to prevent flood water from backing up into the building.

Preparedness
1. Review flood exposures that could affect your facility during a flood—that is, a neighboring facility's yard storage, service interruptions, and access and escape routes. Check ground-level openings or windows that have been bricked over to make sure they are waterproof.
2. Install monitoring devices or arrange to directly monitor local and upstream water levels. Monitor river levels at locations that may affect your facility or site, at access points to your site, and at locations that may affect infrastructure. Monitoring should include physical and weather service or water resources predictions. Have plans in place with up-to-date phone numbers, radio frequencies, or other resources needed to scrutinize water levels.
3. Know the flood-warning signs and your community's alert signals.
4. Review the contents of your community's flood plan, evacuation routes, and high-ground points.
5. Preregister with restoration service companies that specialize in the cleaning, dehumidification, and odor removal from facilities, electronic systems, and the recovery of paper documents. Often, these services will provide a list of actions the company should take to prepare, collect, and preserve documents and equipment to prevent further damage. Write these instructions into your plan.
6. Be prepared for erosion or landslides.
7. Keep roof and parking lot drainage systems free of dirt, leaves, and other debris or obstructions.
8. Inspect roofs for water stains, ponding, plant growth, or other signs of potential weakness.
9. Remove any yard storage from drainage or other low-lying areas. Coat exposed metal with grease if it cannot be moved to higher ground. Evaluate the need for additional flood-proofing of vital equipment.
10. Maintain an adequate supply of sandbags, if located in an area susceptible to flooding.
11. Maintain a supply of waterproofing materials to protect equipment in the case of water leaks from the roof.
12. Park vehicles on the escape side of bridges in case they are weakened by rushing water. Keep vehicles fueled.
13. Monitor conditions by listening to AM/FM radio, television, or NOAA weather radio broadcasts.[8] Flood watches (flooding is possible) and flood warnings (flooding is imminent or occurring) are issued by the National Weather Service. Some amateur

---

[8] The NOAA Weather Radio All Hazards (NWR) in the United States is a nationwide network of radio stations that broadcast continuous weather information directly from the nearest National Weather Service office. This service will also broadcast information on non weather related hazards such as chemical spills, earthquakes, and Amber alerts. These broadcasts are found on one of the following frequencies:

162.400 MHz	162.425 MHz	162.450 MHz	162.475 MHz
162.500 MHz	162.525 MHz	162.550 MHz	

television clubs have rented helicopters to transmit live video of flood conditions. Amateur radio clubs are good sources of communications and information during emergencies.

14. Stockpile other materials in a secure, accessible location away from flood level such as an upper floor or higher ground. These should include:
    a. Cleanup and salvage equipment
    b. Emergency food and water
    c. Flashlights, batteries, and light sticks
    d. Portable water pumps and hoses
    e. Portable generator and fuel
    f. First aid supplies and snakebite kits
    g. Shovels and tools (including chainsaw and crowbars)
    h. Rubber boots and gloves
15. Inspect fire protection equipment. Test fire pump and ensure that it has sufficient fuel.
16. Sandbag protection equipment and backup generators if they are susceptible to flooding. This includes fire sprinkler risers, valves, and hydrants to protect them from the force of floating debris.
17. Move equipment and documents to upper levels or to higher ground. Preidentify the most important documents to move if not duplicated and stored off-site.
18. Back up data systems, and transfer operations to an alternate location.
19. Test and confirm that sump pumps are working and well maintained.

Response
1. Immediately de-energize equipment if the flooding is internal to the facility because of sprinkler system activation, broken pipes, and the like. Cover equipment and product with waterproof sheeting.
2. Monitor conditions and escape routes.
3. Shut off utilities (gas, water, and electric) to protect equipment and to help prevent fires (power to a fire pump should be maintained).
4. Immediately evacuate to higher ground—floodwaters often rise rapidly.
5. Watch for and avoid low-lying areas. Don't drive through flooded areas. If your car stalls, abandon it immediately. Six inches of rushing water can knock people off of their feet and a foot of water can float away a vehicle. Almost half of all flash flood fatalities involve the occupants of vehicles.
6. Don't attempt to cross flowing streams or to swim to safety.
7. Beware of snakes and other animals.

Recovery
1. Use caution when returning to the area of a flood because roadways and bridges may be weakened as a result of the flood.

2. Before entering a building, inspect foundations for cracks or other structural damage. Avoid entry until water surrounding the building has receded.
3. Assess nonstructural damage to the building. Beware of electrical hazards outside the building.
4. Contact the restoration service provider. Ensure that facilities and equipment are cleaned, dehumidified, sanitized, and deodorized before allowing the reentry of employees and guests.
5. Do not turn on utilities until the structure, appliances, and utilities are dry and the building is checked for natural gas or propane leaks.
6. Be sure water supplies are safe to drink. Be prepared to use bottled water until the community's water system is declared safe to drink. Avoid physical contact with floodwater because it may contain hazardous chemicals, sewage, or other contaminants. Dispose of any food or consumables that may have been in contact with floodwaters or mold.
7. Obtain permit and inspection procedures for repairs and reconstruction.
8. Retrofit structures during repair and reconstruction.
9. Search for properties that should be acquired in order to remove at-risk structures from the floodplain.
10. Begin mitigation planning to avoid a repeat of the same problems in future flooding.
11. If relocating temporarily, inform customers of the new location.
12. Inform employees when they can expect to return.

Hazardous Materials Incidents

The response to a hazardous materials incident, typically a chemical spill or airborne release, can range from a relatively simple cleanup procedure to one that is complicated, dangerous, and requires extensive training and equipment depending on the type and amount of the substance. A hazardous material is simply defined as a substance capable of creating harm to people, property, and the environment. The United States Environmental Protection Agency (EPA) and the Resource Conservation and Recovery Act (RCRA) have more definitive definitions.

Companies that use hazardous materials or that generate hazardous waste are in some form regulated by a number of federal and local laws. Additionally, civil and criminal sanctions are available to authorities if the regulations are willfully violated. The mitigation and response procedures for companies that use large volumes of hazardous materials can go far beyond what we have listed here.

The following are basic procedures the firm can use to prepare and respond to chemical spills, toxic releases, and other hazardous materials emergencies. The reader should check with local regulations and match or exceed the level of response and training they require. Before attempting to clean up chemical spills, be sure to use the proper protective equipment. If there is doubt, call the fire department.

Prevention and Mitigation

1. Construct secondary containment around hazardous materials storage areas and connect the area to a scrubbing system that will neutralize any leaks if applicable.
2. Separate incompatible materials that may react if mixed and store inside approved cabinets or containers.
3. Train staff to the appropriate level in the proper storage, use, hazards, and cleanup of hazardous materials.
4. Minimize the amount of hazardous materials stored on-site.
5. Find alternate chemicals that can be substituted for hazardous materials.
6. Ensure that hazardous materials and hazardous waste are properly marked, stored, and removed from the property in an expedient manner.
7. Equip chemical storage tanks and hazardous processes with seismic shutoff switches if located in earthquake-prone areas.
8. Build dykes or diversion around drains or implement some method to prevent hazardous materials from entering storm drains.
9. Establish the appropriate level of security to prevent theft or the intentional release of hazardous materials into the community or environment.
10. Identify potential shelter-in-place rooms or locations.

Preparedness

1. Identify the hazardous materials used on site and maintain material data safety sheets in strategic locations (EOC, spill carts, and the like). Include a list of their quantities and their locations. Be aware of hazardous materials located at neighboring facilities or transportation systems.
2. Determine what federal, state, and local regulations the company must follow. Train staff to the highest level appropriate to the level of response and cleanup that may be required.
3. Work with local officials, such as the fire department, to coordinate pre-emergency plans.
4. Place detection equipment, spill kits, safety showers, and equipment at strategic locations.
5. Maintain the proper equipment for first aid and for the cleanup and decontamination of materials used in your operations.
6. Install panic alarms and closed-circuit television (CCTV) in hazardous areas.
7. Install wind socks and weather stations if appropriate. Calculate release and dispersion models if appropriate and understand how those from surrounding sources (including transportation systems) can affect your location.
8. Conduct joint training drills with city or county services.
9. Investigate the need to install a community warning system. Identify the appropriate authorities to notify in case of a release or spill.
10. Inspect containers and piping for damage or leaks.

Response

1. Evacuate the immediate area (upwind if possible). If required, shelter-in-place (see later).
2. Isolate the area and deny access to any unauthorized personnel.
3. Decide whether outside assistance (fire department or hazmat cleanup) is required.
4. Wear the highest level of protective equipment appropriate to the situation.
5. Remove injured victims or personnel overcome by fumes if it can be accomplished safety. If contaminated, provide immediate eye and body drenching for at least 15 minutes. Seek medical attention.
6. Apply other first aid appropriate to the injured.
7. Identify the materials, their properties and hazards, cleanup procedures, and toxicology. Assume the substance is hazardous until you know otherwise. Look for identification on containers (don't walk into the secured area), placards, or material data safety sheets. Smell is not always a reliable way to identify chemicals. Never taste an unidentified substance (during the early days of the semiconductor industry, a Ph.D. engineer could not speak for several weeks because of an acid burn on his tongue caused by an unsuccessful attempt to identify a spilled substance).
8. Eliminate ignition sources if a flammable liquid or explosive gas is involved. Stage fire control and emergency equipment at appropriate locations (upwind away from the area of contamination).
9. If it is safe to do so, mitigate or eliminate the source of the spill—that is, close valves, cap bottles, patch leaks.
10. Determine the size of the area affected and whether additional evacuations are warranted.
11. Contain the spill if properly trained and equipped. Do not let material go down a drain or into a waterway, basement, or confined space.[9]
12. Use only properly trained, equipped, and certified personnel to clean up the hazard. Never work alone.
13. Make notifications to regulatory agencies as required.
14. Shelter-in-place procedures (see the previous section on Chemical or Biological Attack for additional measures):
    - Turn off HVAC systems.
    - Close all doors, windows, shades, blinds, and transoms.
    - Place wet towels in spaces under doors.
    - Block air vents with plastic, and place tape around doors and windows.
    - Cover your body with clothing as much as possible.
    - Avoid eating anything uncovered that may be contaminated.
    - Listen to informational broadcasts.

---

[9]Tap water that is disinfected with chloramines may be toxic to fish.

Recovery

1. Activate the crisis management plan if not done earlier.
2. Decontaminate responders, facilities, and equipment.
3. Repair damage.
4. File any required reports with regulatory agencies.
5. Monitor health and environmental problems caused by the incident.
6. Investigate the cause and take actions to prevent a recurrence.

Hurricanes

The combined energy of the atmosphere and the ocean can turn a tropical storm into a massive center of power and destruction. These cyclonic storms, intense low-pressure centers with swirling arms of clouds and winds over 75 miles per hour, cause billions of dollars in damage to the coastal areas of the Gulf of Mexico, the Atlantic Ocean, and many other regions in the middle latitudes. Hurricanes are considered the most destructive type of weather condition because they are capable of spawning tornados in addition to high winds and flooding.

Although the center, or "eye," of these storms is clear and calm, the winds across the rest of their 50- to 500-mile diameters can range from 75 to 150 miles per hour or more. Hurricanes are classified according to their wind speed; however, the storm surge, moisture content, and other damaging factors (damage generally raises by a factor of four for each increase in category) are not precisely related to these classifications. The Saffir-Simpson hurricane scale (Table 14-1) lists these classifications.

Despite the power of a hurricane, the storm itself usually progresses at a speed of only up to 30 miles per hour, at times stopping entirely. Although the strong winds account for many of the problems associated with hurricanes, high seas and flooding cause the most damage and loss of life. During storms of this magnitude, the level of the sea may rise many feet higher than normal (a phenomenon known as storm surge). Hurricanes can drop tremendous amounts of water. Hurricane Floyd in 1999 dropped more than 20 inches of rain in North Carolina, causing the worst flooding in their recorded history, after it was downgraded to a tropical storm. This flooding contributed to almost 50 deaths and more than $1.3 billion in damage.

An average of seven named storms per season form during the months of June through November and last for a week or more. This life span allows forecasters to spot and track a storm and predict its landfall. The loss of life in hurricanes in the United States is decreasing.[10] More accurate forecasting and a likelihood of people to take the warnings seriously

---

[10] Hurricane Mitch, which devastated Central America on October 22, 1998, was a category 5 storm with 180 mph sustained winds. This hurricane reportedly killed more than 10,000 people in Central America and will serve for generations to show what hurricanes can do. "These deaths are not just numbers; they were children, moms, dads, and friends." Jerry Jarrell, Former Director, National Hurricane Center. Hurricane Katrina, responsible for 1,836 deaths and $81 billion in property damage, reminds us that we cannot let our guard down.

Table 14-1 The Saffir-Simpson Hurricane Scale

Category	Sustained Winds	Description	Effects
1	74–95 mph	Minimal	Very dangerous winds will produce some damage. People, livestock, and pets struck by flying or falling debris could be injured or killed. Older (mainly pre-1994 construction) mobile homes could be destroyed, especially if they are not anchored properly as they tend to shift or roll off their foundations. Newer mobile homes that are anchored properly can sustain damage involving the removal of shingle or metal roof coverings, and loss of vinyl siding, as well as damage to carports, sunrooms, or lanais. Some poorly constructed frame homes can experience major damage, involving loss of the roof covering and damage to gable ends as well as the removal of porch coverings and awnings. Unprotected windows may break if struck by flying debris. Masonry chimneys can be toppled. Well-constructed frame homes could have damage to roof shingles, vinyl siding, soffit panels, and gutters. Failure of aluminum, screened-in, swimming pool enclosures can occur. Some apartment building and shopping center roof coverings could be partially removed. Industrial buildings can lose roofing and siding especially from windward corners, rakes, and eaves. Failures to overhead doors and unprotected windows will be common. Windows in high-rise buildings can be broken by flying debris. Falling and broken glass will pose a significant danger even after the storm. There will be occasional damage to commercial signage, fences, and canopies. Large branches of trees will snap, and shallow-rooted trees can be toppled. Extensive damage to power lines and poles will likely result in power outages that could last a few to several days.
2	96–110 mph	Moderate	Extremely dangerous winds will cause extensive damage. There is a substantial risk of injury or death to people, livestock, and pets due to flying and falling debris. Older (mainly pre-1994 construction) mobile homes have a very high chance of being destroyed, and the flying debris generated can shred nearby mobile homes. Newer mobile homes can also be destroyed. Poorly constructed frame homes have a high chance of having their roof structures removed, especially if they are not anchored properly. Unprotected windows will have a high probability of being broken by flying debris. Well-constructed frame homes could sustain major roof and siding damage. Failure of aluminum, screened-in, swimming pool enclosures will be common. There will be a substantial percentage of roof and siding damage to apartment buildings and industrial buildings. Unreinforced masonry walls can collapse. Windows in high-rise buildings can be broken by flying debris. Falling and broken glass will pose a significant danger even after the storm. Commercial signage, fences, and canopies will be damaged and often destroyed. Many shallowly rooted trees will be snapped or uprooted and block numerous roads. Near-total power loss is expected, with outages that could last from several days to weeks. Potable water could become scarce as filtration systems begin to fail.

Table 14-1 (Continued)

3	111–130 mph	Extensive	Devastating damage will occur. There is a high risk of injury or death to people, livestock, and pets due to flying and falling debris. Nearly all older (pre-1994) mobile homes will be destroyed. Most newer mobile homes will sustain severe damage, with the potential for complete roof failure and wall collapse. Poorly constructed frame homes can be destroyed by the removal of the roof and exterior walls. Unprotected windows will be broken by flying debris. Well-built frame homes can experience major damage involving the removal of roof decking and gable ends. There will be a high percentage of roof covering and siding damage to apartment buildings and industrial buildings. Isolated structural damage to wood or steel framing can occur. Complete failure of older metal buildings is possible, and older unreinforced masonry buildings can collapse. Numerous windows will be blown out of high-rise buildings, resulting in falling glass, which will pose a threat for days to weeks after the storm. Most commercial signage, fences, and canopies will be destroyed. Many trees will be snapped or uprooted, blocking numerous roads. Electricity and water will be unavailable for several days to a few weeks after the storm passes.
4	131–155 mph	Extreme	There is a very high risk of injury or death to people, livestock, and pets due to flying and falling debris. Nearly all older (pre-1994) mobile homes will be destroyed. A high percentage of newer mobile homes also will be destroyed. Poorly constructed homes can sustain complete collapse of all walls as well as the loss of the roof structure. Well-built homes also can sustain severe damage, with loss of most of the roof structure and/or some exterior walls. Extensive damage to roof coverings, windows, and doors will occur. Large amounts of windborne debris will be lofted into the air. Windborne debris damage will break most unprotected windows and penetrate some protected windows. There will be a high percentage of structural damage to the top floors of apartment buildings. Steel frames in older industrial buildings can collapse. There will be a high percentage of collapse to older unreinforced masonry buildings. Most windows will be blown out of high-rise buildings, resulting in falling glass, which will pose a threat for days to weeks after the storm. Nearly all commercial signage, fences, and canopies will be destroyed. Most trees will be snapped or uprooted and power poles downed. Fallen trees and power poles will isolate residential areas. Power outages will last for weeks to possibly months. Long-term water shortages will increase human suffering. Most of the area will be uninhabitable for weeks or months.
5	>155 mph	Catastrophic	People, livestock, and pets are at very high risk of injury or death from flying or falling debris, even if indoors in mobile homes or framed homes. Almost complete destruction of all mobile homes will occur, regardless of age or construction. A high percentage of frame homes will be destroyed, with total roof failure and wall collapse. Extensive damage to roof covers, windows, and doors will occur. Large amounts of windborne debris will be lofted into the air. Windborne debris damage will occur to nearly all unprotected windows and many protected windows. Significant damage to wood roof commercial buildings will occur due to loss of roof sheathing. Complete collapse of many older metal buildings can occur. Most unreinforced masonry walls will fail, which can lead to the collapse of the buildings. A high percentage of industrial buildings and low-rise apartment buildings will be destroyed. Nearly all windows will be blown out of high-rise buildings, resulting in falling glass, which will pose a threat for days to weeks after the storm. Nearly all commercial signage, fences, and canopies will be destroyed. Nearly all trees will be snapped or uprooted and power poles downed. Fallen trees and power poles will isolate residential areas. Power outages will last for weeks to possibly months. Long-term water shortages will increase human suffering. Most of the area will be uninhabitable for weeks or months.

account for this improvement. A hurricane watch means that hurricane conditions may occur; a hurricane warning means that a storm is expected within 24 hours.

Prevention and Mitigation

1. Develop facilities, equipment, and manufacturing shutdown procedures and ensure workers are trained on these procedures.
2. Keep trees in good health and trim branches that may fall into the building or onto equipment.
3. Reduce the potential for windborne debris. Have straps or other means on hand to brace/anchor yard storage, signs, cranes and roof-mounted equipment. Remove loose debris. Anchor all portable buildings to the ground. Brace or remove outdoor signs.
4. Inspect and repair roof coverings and edges a few months before the hurricane season. Clean drains and catch-basins. Inspect and test fire systems.
5. Upgrade existing structures and ensure that new construction incorporates high-wind building standards. Provide information from FEMA and other sources that employees can use to strengthen the wind resistance of their homes.
6. Avoid construction in areas subject to the greatest damage.
7. Strengthen life lines (utilities, roadways, and vital equipment). Elevate equipment such as fuel tanks, fire pumps, and electrical generators. Enclose this equipment in some type of protection. Strap or anchor to the roof deck support assembly (e.g., the joists) all roof-mounted equipment such as HVAC units and exhaust vents.
8. Set up alternate work sites outside the affected zone and provide access to data systems.
9. Provide prefitted hurricane shutters and/or plywood for windows and doorways where practical. Practice the installation of the protections annually.

Preparedness

1. See the section on Floods and Heavy Rain for additional guidelines. Prepare for hurricane-related flooding with sandbags and an ample supply of brooms, squeegees, and absorbents.
2. Check flood insurance policies for adequacy of coverage and exclusions.
3. Before the storm season, stockpile such extra supplies as food, water, battery-powered weather radio, cash, plywood and nails, ropes, sandbags, emergency lighting, and power. These items may be in short supply when a storm is forecast.
4. Determine where to relocate business operations and employees if ordered to evacuate.
5. Inspect the integrity of roof edging strips, drains, pipe racks, and sign and stack supports.
6. Cover sensitive equipment and finished goods with waterproof covers.
7. Train employees to prepack clothes, medications, pictures, cash, extra glasses, and baby supplies, in case they are required to evacuate from their homes.

8. Maintain ongoing agreements with contractors for supplies and repairs that may be needed after a windstorm. If possible, use contractors who are from outside potential hurricane areas. Prequalify restoration and building contractors for the business and employees.
9. Ensure that any workers who stay on-site have proper supplies and equipment (drinkable water, nonperishable food, medical supplies, flashlights, and communications).

Response
1. Monitor the progress of the storm and estimate the amount of time required to perform all essential tasks.
2. If ordered to evacuate, do so immediately.
3. Move valuables out of the area or to upper floors if located in a flood or storm surge zone.
4. Secure or store furniture, planters, or other objects located outside the building.
5. Board up windows with 5/8-inch marine plywood (taping does not work).
6. Turn off utilities and HVAC. Shut down operations that depend on outside power sources in an orderly manner, following established guidelines.
7. Arrange for extra security if necessary.
8. If required to remain, stay inside and away from windows, doors, and outside walls. (Employees should never remain alone. Do as much as possible to ensure their safety.)

Recovery
1. Wait until the storm has passed before beginning repairs and restoration. Continue to monitor news sources.
2. Remain alert for extended rainfall and subsequent flooding even after the winds have subsided.
3. Evaluate the structural integrity of the building and utilities.
4. Open windows and doors to ventilate and dry the facility. Prevent the development of and protect workers from mold.
5. Ensure that the site is safe for cleanup, salvage, and reoccupation. Ensure workers wear the proper protective equipment. Avoid loose or dangling power lines, broken or loose gas connections, or other hazards.
6. Account for employees as they return from evacuation.
7. Begin damage assessment.
8. Begin relocation and reconstruction if necessary. Facilities, electrical, and other equipment should be checked and dried before returned to service.
9. If vital records or other resources were moved to alternate locations, return these documents as appropriate.

Lightning

Lightning develops as a result of interactions in the atmosphere between charged particles, interactions that produce an intense electrical field (up to 100,000,000 volts) within a thunderstorm. These bolts produce so much heat (50,000°F) that the air explodes into rumbling shock waves. Lightning kills more people across the United States each year than tornados; it is second only to flooding in weather-related deaths. In the United States, most deaths occur in Florida, Missouri, Texas, New York, and Tennessee. Lightning is one of the leading causes of interruptions and other power-quality disturbances. The amount of lightning activity an area receives is measured in terms of flash density, the number of cloud-to-ground lightning strikes per square mile during a year.

Of all weather-related phenomena, lightning is associated with the greatest number of myths. Lightning can and often does strike in the same place twice, and it seeks the best conductor to the ground, not necessarily the tallest object. The tires of an automobile do not protect it from lightning; the metal body will conduct the charge away from its occupants—convertibles excepted, but why would someone drive with the top down in the rain? You can be struck by lightning from a storm many miles away, even if the sky overhead is clear. A software engineer in Silicon Valley, an area of California where lightning is uncommon, was injured when his computer exploded. The engineer was on the ground floor of a high-rise building that was struck by lightning.

In addition to injuries, lightning can cause fires, structural damage, and electrical power outages. Lightning caused a fire and explosion at the Jim Beam distillery that caused more than $1 million in damage and the loss of 800,000 gallons of bourbon. Included in the loss figure is the payment of a fine of almost $30,000 for the contamination of a nearby river when the bourbon was allowed to spill into it. Lightning can cause brick or concrete walls to explode.

Prevention and Mitigation
1. Understand the frequency of severe thunderstorms in your geographical area and plan accordingly. If located in a high-risk area, have your facilities evaluated by engineers, and implement the appropriate mitigation.
2. Build structures that incorporate lightning protection if located in a high-risk area.
3. Install line conditioners that reduce the effects of power surges on sensitive and critical equipment, and connect critical equipment to uninterruptible power supplies.
4. Consider the purchase and installation of a backup generator.
5. Upgrade fire protection systems.
6. Establish 24-hour guaranteed service or equipment replacement agreements with vendors.

Preparedness
1. Monitor the storm's conditions. The National Weather Service broadcasts continuous weather and warning information over the NOAA weather radio stations. Internet sites provide weather conditions and information, and local TV stations often broadcast severe weather information.

2. Instruct employees to stay indoors. Outdoor activities should be discontinued immediately. Use the 30/30 rule: If there is less than a 30-second period between the lightning flash and thunder, remain indoors for 30 minutes.
3. Avoid contact with pipes, railings, wire fences, telephones, and other electrical equipment and appliances, including faucets and showers.
4. Train employees about lightning safety, to avoid open areas if outside, and to seek low ground in a wooded area (not under a single tree). Employees should crouch close to the ground on the balls of their feet, but they should not lie down. Stay clear of metal objects; if in a group, move away from one another. Include first aid as a part of the training.
5. Maintain scheduled backups of computer, telecommunications, and other data systems.
6. Shut down electronic systems if feasible.

Response
1. Bring employees inside the building; warn employees who may be about to leave that they should remain inside.
2. Contact emergency services if someone is struck by lightning.
3. Treat injured victims. Remove them from danger. Cardiopulmonary distress, broken bones, and burns are the most common lightning-related injuries. Look for entry and exit wounds.
4. Stand clear of windows, doors, and electrical appliances.
5. Unplug appliances well before a storm nears, never during.
6. Avoid contact with piping, including sinks, baths, and faucets.
7. Avoid the use of the telephone except for emergencies (lightning can strike telephone lines and subsequently injure people speaking on the phone, but this accounts for a small percentage of lightning injuries).
8. Check for and extinguish fires.
9. If power is lost, shut down equipment.

Recovery
1. Log all events, actions, decisions, and expenses.
2. Activate emergency response teams as required.
3. Assess and document damage.
4. Reestablish utilities.
5. Test and replace damaged equipment and connectivity. Restore data if any are lost.

Serious Injury or Illness

Injury and illness are probably the most common emergencies found in the workplace. Businesses lose billions of dollars annually in lost productivity, recruiting, fines, and worker compensation costs because of injuries. The firm's response to an injury is very visible and if handled incorrectly may become an issue, producing poor employee morale, a labor dispute, or negative publicity, not to mention the well-being of the injured.

As we have often mentioned, during and after a disaster situation, businesses may not be able to rely on the availability of ambulances, paramedic response, or even fully staffed hospitals. These services can become quickly overloaded: first responders and doctors can be injured or killed, and medical facilities can be damaged. Even if services are available, their response may be delayed as a result of traffic congestion, damaged bridges and overpasses, lack of fuel, and other obstacles. All employers have a legal and moral responsibility to protect the safety and health of their employees. They must be prepared to deal with first-aid issues, especially if located in areas subject to catastrophic disasters that cause injuries beyond the everyday cut finger. It is therefore important to train employees to a commensurate level in first aid and CPR. Because of their effectiveness, businesses should also consider the purchase of Automated External Defibrillators (AED) for use in cardiac emergencies.

Prevention and Mitigation
1. Implement effective injury and illness prevention plans (safety programs).
2. Understand the risk and type of injuries possible in your environment. Develop prevention programs, response procedures, and stock any special first-aid supplies required to provide emergency treatment.
3. Purchase biohazard cleanup kits and train employees in their use
4. Ensure that the location address is prominently and clearly visible to the fire department and ambulance companies. If appropriate, know and provide your Global Positioning System (GPS) coordinates to responding agencies if they are equipped to use this technology in the event that street markers may be destroyed.
5. Sponsor flu inoculation programs and health fairs.

Preparedness
1. If employees are exposed to hazardous chemicals, install eye-wash stations and emergency showers.
2. Inoculate employees at risk of bloodborne diseases, if necessary.
3. Know any special conditions or needs of employees and make this information available to first responders or the ERT.
4. Train employees in first aid and CPR. Match or exceed the level of training required by regulations to compensate for any extended response times of emergency medical personnel.
5. Purchase and train employees in the use of AEDs.
6. Maintain first-aid supplies at strategic locations. Most first-aid kits are not designed for the type of injuries or delays in care that may occur after a disaster. At least one kit should contain disaster first-aid supplies.
7. Control traffic and parking so that emergency vehicles have access to the facility.
8. Keep a supply of blankets and a stretcher or other transportation device.

Response

1. Contact, or request someone to contact, emergency services (fire department, ambulance) and report:
   a. Your name
   b. A description of the illness or injury
   c. Your location
2. Stay on the line until the dispatcher hangs up.
3. Determine the extent of the injuries.
4. Provide first aid if qualified:
   a. Control bleeding.
   b. Check breathing.
   c. Check circulation.
   d. Treat for shock.
5. Remain with the victim:
   a. Ensure that the victim is not moved (unless to protect from further hazard).
   b. Obtain as much information about the victim's condition, history, and needs as required.
6. Keep those not involved in the emergency away from the area.
7. Send people to meet the fire department, paramedics, or ambulance at the driveway and front door of the building to escort them to the victim.
8. Assign someone to hold elevators at the ground floor.

Recovery

1. Follow biohazard procedures and regulations when cleaning up body fluids.
2. Notify regulatory agencies (such as the Occupational Safety and Health Administration [OSHA]), legal advisors, and insurance carriers if required.
3. Completely investigate the cause of all injuries and take steps to prevent their recurrence.
4. Retrain injured employees in safety procedures.
5. Activate the crisis management plan if injuries are major, if illness is widespread or controversial, or if fatalities are involved.
6. Provide posttraumatic stress counseling if necessary.

Structural Collapse

Construction failures, earthquakes, fires, weather, and terrorist events are just a few reasons that structures unexpectedly (but sometimes not unpredictably) collapse, entrapping their unfortunate victims. Unless specially trained and equipped, most response will be geared toward steps to take while professional emergency services are on their way. CERT/NERT classes will train students in basic Urban Search and Rescue measures that go beyond what we have listed here that may be useful subsequent to a regional disaster. Since many collapse situations will occur on construction sites, the following guidelines are geared toward this environment.

Prevention and Mitigation

1. Ensure proper bracing for all types of construction operations and wind loads.
2. Follow all regulations, engineering plans, and best practices when engaged in demolition operations.
3. Develop preventative maintenance programs and frequent inspections of structures.
4. Do not rush into a collapsed structure; always ensure your safety first and evaluate the situation.
5. Respect weight and occupancy loads.
6. Follow the latest or most stringent seismic building codes.

Preparedness

1. Identify and understand the risk and prepare accordingly.
2. Cribbing or pneumatic air shores, jacks, rigging, ventilation, and other similar types of equipment should be available when this hazard is a potential.
3. Train response teams in Urban Search and Rescue techniques.

Response

1. Dial 911, or if using a cell phone, dial the direct fire/rescue dispatch number and report injury/victim trapped within a structural collapse.
2. Remove nonessential workers from the area, and control access to the area. The restricted area should be considered a danger zone for further collapse or falling debris.
3. Fight any fires within the scope of your training, turn off utilities, or protect from adverse weather conditions.
4. Evaluate the stability of the collapsed structure and surrounding supporting structures including ground, equipment, remaining structures or formwork, and so forth.
5. Shut down all operations that cause vibration or that could cause secondary collapse.
6. Account for all workers. Without entry, determine the number of victims and their locations. Mark the location of trapped workers with a brightly colored "V."
7. Assist survivors, and triage rescued victims if necessary.
8. Stabilize structure to prevent secondary collapse or rescue operations from causing additional injury or collapse.
9. Begin rescue of injured or trapped workers. If victim cannot be readily rescued from an entrapment, if safe, notify incident command of location and place shoring around victim.
10. Identify and obtain necessary resources.
11. Establish staging for rescue equipment and debris removal area.
12. Inform responding rescue authorities of any hazardous materials or other potential hazards within the collapse are.
13. Monitor changes in conditions and watch for signs of secondary collapse.
14. Ensure appropriate PPE is worn and maintain normal safety practices throughout search and rescue operations.

Recovery
1. Activate a crisis management plan if a serious injury is involved.
2. Notify the appropriate regulatory agencies such as OSHA as necessary.
3. Conduct investigation into the cause of and the response to the incident.
4. Provide posttraumatic stress counseling in the case of a fatality.

Tornados

A tornado is a violently rotating column of air that makes contact with the ground. (A tornado that does not touch the ground is referred to as a funnel cloud; a tornado over water is a waterspout.) Tornados usually develop from severe thunderstorms. They generally originate in the right-rear quadrant of the storm cell or at the leading edge of a line of thunderstorms. Their paths are unpredictable, cutting straight lines of destruction, zigzagging back and forth, or hopping and skipping around, even reversing their general direction. They usually travel from the southwest to the northeast at an average speed of 30 miles per hour, but they have been known to remain stationary. The National Weather Service warns against attempts to outrun a tornado in a vehicle because some tornados can travel at speeds of nearly 70 miles per hour.

Although most tornados occur between 3:00 and 7:00 PM during the months of April, May, and June, they can occur any time of the day and anywhere in the world. The Texas Panhandle, Oklahoma, Kansas, Nebraska, Iowa, Missouri, parts of Arkansas, Illinois, and Indiana have the most tornados, although they form in every state and at greater frequencies than previously thought. Southern Ontario accounts for one-third of Canada's total tornados, with seven of the nine strongest occurring in this region.

Most fall into the "weak" or "Minor Damage" category, with rotational wind speeds of 100 miles per hour or less. Only a very small percent of tornados are classified as "violent" or "Massive Damage," with wind speeds reaching 300 miles per hour or more. These violent tornados account for almost 70 percent of the fatalities. It is the wind speed that accounts for most of the destruction, not a sudden drop in air pressure. Wind speed is greatest at the upper portions of the tornado. Flying debris is the cause of most injuries.

The Fujita scale (Table 14-2) is still used in Canada and other parts of the world to rate the intensity of a tornado by examining the damage it caused. It is interesting to note that the size of a tornado is not necessarily related to its intensity. Adopted for use in 2007, the United States uses the Enhanced Fujita Scale (Table 14-3) to better reflect damage patterns and wind speeds.

A tornado watch is issued when severe thunderstorms or tornados are most likely to occur and when conditions are favorable for their formation. A warning is issued when a tornado is detected on radar or reported by observers. Warnings and firsthand reports are received by the following means:

- NOAA Weather Radio broadcasts
- Community warning sirens

Table 14-2   The Fujita Tornado Scale

F-Scale	Name	Wind Speed	Damage
F0	Gale tornado	40–72 mph	Some damage to chimneys; breaks branches off trees; pushes over shallow-rooted trees; damages sign boards
F1	Moderate	73–112 mph	The lower limit is the beginning of hurricane wind speed; peels surface off roofs; mobile homes pushed off foundations or overturned; moving autos pushed off the roads; attached garages possibly destroyed
F2	Significant	113–157 mph	Considerable damage. Roofs torn off frame houses; mobile homes demolished; boxcars pushed over; large trees snapped or uprooted; light object missiles generated
F3	Severe	158–206 mph	Roof and some walls torn off well-constructed houses; trains overturned; most trees in forest uprooted
F4	Devastating	207–260 mph	Well-constructed houses leveled; structures with weak foundations blown off some distance; cars thrown and large missiles generated
F5	Incredible	261–318 mph	Strong frame houses lifted off foundations and carried considerable distances to disintegrate; automobile- sized missiles fly through the air in excess of 100 meters; trees debarked; steel reinforced concrete structures badly damaged
F6	Inconceivable	319–379 mph	These winds are very unlikely. The small area of damage they might produce would probably not be recognizable along with the mess produced by F4 and F5 wind that would surround the F6 winds. Missiles, such as cars and refrigerators, would do serious secondary damage that could not be directly identified as F6 damage. If this level is ever achieved, evidence for it might only be found in some manner of ground swirl pattern, for it may never be identifiable through engineering studies.

Table 14-3   The Enhanced Fujita Tornado Scale

Scale	Wind Speed (mph)	Relative Frequency	Potential Damage
EF0	65–85	53.5%	Minor damage.
			Peels surface off some roofs; some damage to gutters or siding; branches broken off trees; shallow-rooted trees pushed over.
			Confirmed tornadoes with no reported damage (i.e., those that remain in open fields) are always rated EF0.
EF1	86–110	31.6%	Moderate damage.
			Roofs severely stripped; mobile homes overturned or badly damaged; loss of exterior doors; windows and other glass broken.
EF2	111–135	10.7%	Considerable damage.
			Roofs torn off well-constructed houses; foundations of frame homes shifted; mobile homes completely destroyed; large trees snapped or uprooted; light-object missiles generated; cars lifted off ground.

(Continued)

Table 14-3 (Continued)

EF3	136–165	3.4%	Severe damage.
			Entire stories of well-constructed houses destroyed; severe damage to large buildings such as shopping malls; trains overturned; trees debarked; heavy cars lifted off the ground and thrown; structures with weak foundations blown away some distance.
EF4	166–200	0.7%	Extreme damage.
			Well-constructed houses and whole frame houses completely leveled; cars thrown and small missiles generated.
EF5	>200	<0.1%	Massive damage.
			Strong frame houses leveled off foundations and swept away; steel-reinforced concrete structures critically damaged; high-rise buildings have severe structural deformation. Incredible phenomena will occur.

- The Internet
- Real-time Doppler radar
- CCTV
- Citizens band, amateur radio, and police communications frequencies
- Local television and commercial radio channels

The key to surviving a tornado is early warning and getting to an appropriate shelter.

Prevention and Mitigation

1. Keep trees and shrubbery trimmed. Make trees more wind resistant by removing diseased or damaged limbs, then strategically remove branches so that wind can blow through.
2. Strengthen the wind resistance of buildings and consider the installation of shutters on windows.
3. Work with the local jurisdiction to establish tornado warning systems if none exist in the community.
4. Build safe rooms or tornado shelters. FEMA has publications that describe how to construct these shelters.

Preparedness

1. Monitor weather broadcast stations and conditions. Obtain and use NOAA weather radios.
2. Learn what tornado conditions look and sound like (often described as the sound of a freight train or airplane). Tornados frequently emerge from near the hail-producing portion of the storm; wind may die down, and the air may become very still. An approaching cloud

of debris can indicate the location of a tornado even if a funnel is not visible. A rotating protrusion of the cloud base may indicate a developing tornado, especially if the upper portion of the cloud is rotating. The cloud may turn a greenish color (a phenomenon caused by hail). Tornados generally occur near the trailing edge of a thunderstorm. It is not uncommon to see clear, sunlit skies behind a tornado.

3. Know the meaning of watches and warnings.
4. Identify a shelter room. A storm cellar or basement is best, but an interior room without windows on the lowest floor is the next-best alternative (excluding mobile homes).
5. Instruct employees in the meanings of local warning systems (sirens, a red bar on a television screen, and the like).
6. Construct storm shelters if none are close by.
7. Maintain supplies of food, water, and emergency lighting in case employees cannot leave immediately after. Power may be lost, so plan accordingly.
8. Keep plastic sheeting and other materials on hand to cover equipment or a damaged roof.

Response

1. Seek shelter in a storm cellar, basement, or interior room without windows. Get under something sturdy, such as a desk or table. If in a modular office, seek shelter inside a sturdy building if there is no storm shelter nearby.
2. Occupants of a high-rise building should go to an interior room, hallway, or interior stairwell on a lower floor (if you have time). Center hallways are often structurally the most reinforced part of a building. Sit or kneel down in the corridor to protect yourself from a possible head injury.
3. Stay clear of windows (do not open them), doors, and outside walls. Close exterior office doors. Take shelter under a desk if there is no stairwell, basement, or other protected location.
4. Continue to monitor radio or television for information.
5. If outside, seek a safe place in a sturdy nearby building. As a last resort, take cover in a ditch or low-lying area; lie flat with your hands covering your head.
6. If in a vehicle, seek shelter in a ditch as described previously.
7. Leave mobile homes for a storm shelter or a safe place in a sturdy nearby building.

Recovery

1. Assess damage to facilities.
2. Begin search and rescue operations only if properly trained and equipped.
3. Treat injuries.
4. Avoid the use of open flames.
5. Monitor news broadcasts and the location of disaster assistance centers.
6. Form teams of volunteers to assist employees affected by the tornado.

7. Document damage for insurance reimbursement.
8. Clear debris.
9. Begin relocation and reconstruction.

Workplace Violence

Violence is generally the leading cause of death for women and the second-leading cause of death for men in the workplace. In 1994, robbery-homicide represented 9 percent of total crime but accounted for 75 percent of workplace homicide. With more than 2 million workers physically attacked, more than 6 million threatened, and more than 16 million workers harassed, it is not surprising that two-thirds of workers did not feel safe at work.[11] In this same year, there were 1,080 workplace homicides in the United States, but this number declined to 540 in 2006 and continues to trend downward. The cost of workplace violence reported in 2004 was $121 billion. Direct legal and medical costs total hundreds of thousands of dollars per incident, and jury awards exceed $5 million. Firms that suffer an incident of workplace violence experience an 80 percent loss of productivity in the subsequent week, and some have failed as a result.

Workplace violence is defined as violent acts, including physical assaults and threats of assault, directed toward people at work or on duty. It is classified by the relationship of the perpetrator to the workplace:

- Type I—Criminal Intent. The perpetrator has no legitimate relationship to workplace, usually an act of robbery or a political or terrorist act. These are the most common, accounting for about 85% of workplace homicides.
- Type II—Customer/Client Related. Accounting for about 3% of the total, there is a relationship that becomes violent while interacting with the business or entity. While most of this type of workplace violence is believed to occur within health services (hospitals, nursing homes, and psychiatric facilities), customers or clients, students, and passengers are examples of others that fall within this category.
- Type III—Worker versus Worker. Violence committed by an employee or ex-employee against other employees account for about 7% of workplace homicide but typically accounts for a large percent of total multiple homicides from a single act.
- Type IV—Domestic violence "spillover" into the workplace. This category pertains to victims of domestic violence assaulted or threatened while at work, and accounts for about 5% of all workplace homicides.

A person who is engaged in a mass homicide attempt is referred to as an "Active Shooter" because most of these incidents involve the use of one or more firearms and can involve more than one perpetrator.

---

[11] National Institute of Safety and Health, Bulletin 57 (Washington, D.C., June 1996); 1994 figures.

Prevention and Mitigation
1. Observe good hiring practices, conducting criminal background verification and investigation where appropriate.
2. Establish policy that strictly forbids any form of workplace violence, intimidation, racism, sexism, or possession of weapons and that requires employees to report threats and potential problems to management immediately. Communicate this policy to all employees. Consistently enforce sanctions for noncompliance.
3. Foster a good working environment, with consistent discipline, a problem-solving rather than a blaming attitude, open communication, respect, and a nonauthoritarian management style.
4. Never terminate an employee in a manner that is degrading or does not allow him or her the full use of social services available.
5. Audit and improve physical security and access control. Ensure there is sufficient lighting, good visibility, CCTV, and posted signs, as well as other protections such as minimizing cash, avoiding high-crime areas, and using multiple workers for late-night operations. Access control and panic (or robbery) alarms are helpful.
6. Consider the construction or designation of a hardened "safe room."
7. Isolate lobbies from the remainder of the building, offices, or facility.

Preparedness
1. Develop a comprehensive workplace violence prevention and management program.
2. Train supervisory and management personnel in the following areas:
   a. Recognition of potentially aggressive behavior
   b. Recognition of signs of domestic violence
   c. Diffusion of aggressive behavior
   d. Recognition of the importance of an immediate response to complaints
   e. Elements that may be required by OSHA, such as crime awareness, location of and operation of alarm systems, communications procedures, and late-night procedures for Type I violence that includes how to react to a robbery
3. Protect potential victims. Obtain corporate and personal restraining orders, provide special parking and escort privileges, change their shift or work location, and consider providing the employees with personal protection devices.
4. Form a committee to evaluate and manage threats or acts of violence. The committee should include the employee's manager; human resources; and legal, security, and medical (employee assistance program representative, staff, or contract psychologist) personnel.
5. Train security officers to handle potentially violent situations.
6. Train drivers or installers to:
   a. Know their destination and have a planned route of travel. Don't use alleys or isolated side streets as shortcuts.
   b. Keep car doors locked, especially the passenger door, when driving.

c. If you see someone on the road who indicates they need help, don't stop. Use your radio or cellular phone, if available, to call the police.

d. If your vehicle is bumped from behind, do not stop. Call for assistance or drive to the nearest police station.

e. If you are told that something is wrong with your vehicle, do not stop immediately. Drive to the nearest service station or another well-lighted public area.

f. Do not pick up hitchhikers.

g. If you become lost, find a public place, like a service station, to read your map or ask for directions.

h. Valuables should be kept out of sight whenever traveling or leaving a vehicle parked.

i. Always park in well-lighted areas. Don't park in isolated areas for lunch.

j. Be alert for suspicious persons loitering in the area of your destination. Always be alert and aware of what is around you. Stop and observe the sidewalk, street, and doorways for potential problems before exiting.

k. Glance under your truck—someone may be hiding there. Look into the cab before entering.

l. Know your vehicle license and company truck number so that you can furnish this information to police in the event the criminal hijacks or escapes in your truck.

m. If you observe anyone tampering with your truck, do not endanger your personal safety. Not all crimes are committed against the driver. They may try to break into or steal the truck or van, or to remove product. Call the police immediately.

Response

TYPE I (ROBBERY)

1. Follow the robber's directions.
2. Do not argue or fight with the robber, and offer no resistance whatsoever.
3. Speak to the robber in a cooperative tone. Let the robber know you intend to follow his or her instructions.
4. Never produce a weapon during the event.
5. Move slowly and explain each move to the robber before you make it. Avoid surprising the robber.
6. Do not follow or chase the robber.
7. Do not touch anything the robber has handled.
8. Close and lock all doors.
9. Call the police immediately.
10. Write down everything you remember about the suspect and the robbery.
11. Protect evidence.

TYPE III – IV (ACTIVE SHOOTER)

1. Assess the situation (if safe to do so), but do not delay calling 911.
2. If possible:
   a. Leave the building or area if safe to do so. Keep hands visible and notify responding officers of the location of any injured victims. Avoid the rescue of the wounded—the shooter may return.
   b. If you cannot evacuate, find a place to hide that is out of the shooter's view and that can provide protection against gunshots.
   c. If in an office or other room, lock and barricade the door with a desk or other heavy object.
   d. Place cell phones in silent mode, turn off radios or other indications the area is occupied.
   e. Close blinds; turn off lights if shooter is outside.
   f. Take cover.
   g. Do not respond to any voice commands until you can verify they are issued by responding officers and not the shooter.
3. Summon help:
   a. Dial 911. If you can't speak because of the proximity of a shooter, leave the line open so the dispatcher can listen to what is taking place.
   b. Activate panic alarm if available.
   c. Pull fire alarm (except in the case of protesters or if you don't want to place employees outside of the building). Some argue against this practice because it could cause stairwell doors to unlock.
4. Warn other employees if safe.

Recovery

1. Account for employees and guests when able.
2. Activate the crisis management plan.
3. Immediately clean up biohazards and physical damage.
4. Provide stress counseling for employees, victims, and families.
5. Review security and response procedures.

Summary

Emergency response is actions taken to manage, control, or mitigate the immediate effects of an incident such as a fire, illness or injury, chemical spill, or terrorist event. Because of the increased emphasis on standards, regulations, and litigation, managers must identify their foreseeable hazards and design plans and programs to deal with these very visible events. Effective action taken to control an emergency will reduce injuries, protect assets and mitigate their loss, and position the organization for a smooth and rapid recovery. The training of Emergency Response Teams and the stockpiling of survival resources will help protect the organization and its employees and guests when a disaster overwhelms local police, fire, and medical services and forces the organization to be self-sustaining for a period of time. Employees must understand their responsibilities under the emergency response plan, and those who develop these plans must become knowledgeable about the conditions that an incident may produce. Response plans that are difficult to use, that are used as a training guide, or that contain useless information should either be discarded or supplemented with very brief, direct, action-oriented instructions. Emergency response actions should almost never be taken if the responders are not trained or properly equipped, or if the action places them in an unsafe position.

15

Business Impact Analysis

The great enemy of the truth is very often not the lie—deliberate, contrived, and dishonest—but the myth—persistent, persuasive, and unrealistic. Belief in myths allows the comfort of opinion without the discomfort of thought.
—**John F. Kennedy**

A business continuity plan that is not predicated on or guided by the results of a business impact analysis (BIA) is at best guesswork, is incomplete, and may not function as it should during an actual recovery.

The BIA will help the company establish the value of each functional unit and business process as it relates to the organization and not to itself, illustrating which functions need to be recovered and in what order they may need to be recovered. It identifies the financial and subjective consequences to the organization of the loss of its functions over time, highlights interdependencies, and establishes the function's "outage tolerance" or recovery time objectives (RTOs). Its results are used to determine which functions are the most critical, and at what times they are critical. It is fundamental in the understanding of the amount of risk to retain, transfer, or mitigate and will assist management to make timely decisions about future business issues, as well as guiding management's decision as to the scope and breath of the continuity effort. This is accomplished by examining impacts over significant blocks of time (hours or days) on service objectives, cash flow and financial position, regulatory requirements, contractual issues, and competitive advantage. It presents management with a financial basis for selecting the most cost-effective continuity strategies.

A BIA will also help to accomplish the following:

- Identify which processes and computer applications are important to the survival of the organization and/or to the continuation of the organization's key business objectives
- Develop an understanding of how interdependencies between functions and other organizational units, processes, and outside (third-party relationships) will affect continuity management
- Identify critical resource requirements of the organization
- Gain support for the continuity process from senior management
- Increase management's awareness of the issues and resources required for a workable program, as well as introduce a basic planning structure to the management group
- Satisfy regulatory requirements and standards
- Potentially reveal inefficiencies in normal operations
- Help to justify or better allocate continuity planning budgets (cost/benefit)

A major objective of the BIA is to establish the RTOs and recovery point objectives (RPOs) of business functions and data processes. Outage tolerances or RTOs are established based on the results of the analysis and the continuity priorities that are assigned. Management will then understand how best to allocate recovery resources to these functions. A shorter RTO will require the most expensive strategies; a longer RTO will allow for the selection of lower-cost availability options. Outage tolerance is the amount of time the organization can be without the use of a function before it has a detrimental effect on the company. The Recovery Time Objective is the amount of time by which the organization would like to have the process or function back in service. These terms are often incorrectly synonymous. The Outage Tolerance is the maximum downtime before the function or process is critically affected or the deadline in which the function or process must be restored to prevent severe impact to the business.[1]

The RTO is a decision based on the results of the BIA and on the judgment and agreement of the process or business unit owner with the concurrence of senior management. The definitions refer to critical or severe impacts, but the organization can make its own decision about what level of loss or impact is tolerable (which may be that no loss is acceptable, hence the terms "continuity," "always on," and "0.9999% up-time"). Because major recovery strategies are dependent on and should be driven by the RTO, senior management acceptance is vital to the funding of the strategies.

RTOs can also assist in the escalation of a problem to the declaration of a disaster[2] or to a higher level of response. If the time required to recover a process is 4 hours and the RTO is 6 hours, you have about 2 hours to fix the problem or to decide what to do about it.

The RPO is the point in time to which systems and data must be recovered after an outage (*Disaster Recovery Journal* definition) or the acceptable amount of data that can be lost before a function or process is critically affected. In other words, how much data loss are you willing to accept in your recovery, or how current does your data need to be to start your recovery? It may be acceptable to lose 4 hours of data and then recover to that point and, perhaps, manually recreate the data that were lost. The RPO is often used in the selection of a backup strategy. Tape backup, for example, could mean a day or more of lost data.

Once systems and resources are recovered or repaired, the time available to recover any lost data and/or any work backlog is referred to by some as the Work Recovery Time (WRT). When developing the business continuity plan, keep in mind that a data process that supports the RTO of a business function may need to be recovered sooner. If the function cannot begin recovery until the data are ready, that time differential must be considered in the dependent RTO and RPO.

---

[1] The term "Maximum Tolerable Downtime" (MTD) is often used and can be defined as the maximum downtime the organization can tolerate for a business process or the length of time a process can be unavailable before the organization experiences significant losses.

[2] A term used by hot-site vendors to indicate a customer needs to use their facility.

Another objective of the BIA is to show the business cycle criticality of various functions and refine strategies or reallocate resources accordingly during a continuity situation. Companies are composed of individual business functions that work together to deliver services or products. Although all functions are important during normal operations, some are more critical or time dependent than others after a disaster. In many cases, not all business processes need to be recovered at the same time, although functions initially less critical can become more important over time, and some functions can have more importance depending on the timing of the business cycle or retail season. Companies that survive after a disaster do so by focusing on the recovery of their most critical or time-dependent functions, by implementing the most cost-effective recovery strategies, and by making the best use of resources that become strained or scarce after a disaster. Be careful, however, that the prioritization of functions is not based solely on their outage tolerances. Their relationship to the organization's mission is the most important factor.

It is obviously important to decide which continuity strategies to use before scripting instructions or steps in the plan. Often, the selection of a strategy necessitates a change in policy or even the strategic plan of the company. Management may have difficulty making these decisions if the impact, solutions, and costs are not presented in a complete, positive (avoid a "savior versus doom-and-gloom") package.

A purist will advise the planner to treat the BIA as a project separate from the continuity planning phase. He or she will argue that the planner's only objective at this point in the process is to demonstrate the cost (impact) of a loss over time. As a practical matter, however, this is difficult, and it comes with some disadvantages. The BIA will generate interest, support, and momentum in the continuity project as a whole. These advantages will diminish over time, so the planner must complete the initial phases of the project as quickly as possible. Separating the BIA from other elements of the continuity planning process will often add unnecessary time and expense, which can result in failure to complete the project. Management time is often at a premium; thus, it is difficult to arrange more than one meeting to conduct the impact analysis, identify critical functions, and discuss recovery strategies and resource requirements. The planner may have only one opportunity to meet with these individuals. While a top-down approach is generally the best technique, much of the resource information will usually come from a lower level of management, but the provision of this information should be via the senior manager. As a result, planners, especially those who use questionnaires extensively, include questions related to mitigation, preparedness, hazard identification, and resource requirements as part of the BIA meeting. Critical functions, the identification of single points of failure, and the initial selection of recovery strategies are often discussed at this juncture.

Risk Analysis versus Business Impact Analysis

A BIA is often thought of as another name for a risk analysis. Some contingency planners believe a risk analysis is a process that focuses solely on physical assets and that a BIA focuses solely on business processes. A close examination will reveal that this belief is not correct.

A BIA is a means of assessing the impact of a disruption in any functional area or on the operations of the enterprise as a whole. It can be considered a subset of a risk analysis, in that it places an "asset value" on business functions and focuses on the criticality of a disruption over various time periods. The source or cause of the disruption, or a detailed understanding of the probability of its occurrence, is relatively unimportant when conducting an impact analysis and is not considered.

Therefore, the BIA focuses much less on hazard identification (some say there should be no focus on hazard identification). Listing all the hazards that might befall an operation may be useful in understanding the conditions, environment, and special needs required when selecting mitigation or continuity strategies (such as the inability to move large pieces of equipment across town after an earthquake or the amount of time required to remove bodies, complete an on-site police investigation, and remove the carnage after a workplace homicide), but it adds little to the understanding of financial or subjective loss to the operation over time. Outage tolerance is exclusive of its cause.

The approach taken to initiate and manage a BIA is very similar to that used in a risk analysis. It must begin with senior management's commitment. Because a top-down approach provides the fastest and often most accurate results, management must emphasize that this is a task not to be delegated downward in the organization. The planners' discussions with these managers in the process of collecting impact data will help to build relationships that will be useful later on if the business continuity planning process becomes stalled or if resistance is encountered.

"The BIA is all about estimating the impact that *could* occur if a business process is disrupted for *any reason*. Risk assessment identifies *specific risks*, estimates the likelihood of those risks impacting the business, and puts controls into place to mitigate the risks, identifying any residual risk that still remains" (Cardoza, K&M Publishers, Inc.).[3] The risk analysis is a more or less solitary experience, whereas the BIA is a top-down partnership with senior management.

Business Impact Analysis Methodology

Plans based on intuitive analysis are often too generalized or miss details important to an effective recovery. A comprehensive analysis often reveals interdependencies and outage tolerances that are not obvious even to those with intimate knowledge of the company's operations. The BIA, if conducted in a structured manner, can help guarantee the success of the entire business continuity process. Different methods exist to accomplish this task. Whatever method is used, certain steps must be completed if the planner is to avoid

---

[3] Residual Risk is any risk that remains after one or more controls have been implemented.

obstacles commonly encountered in the analysis. See Chapters 13 and 14 for additional information on the development, practical application, and project management of the BIA. The major elements of a BIA include the following:

- Project planning
- Data collection
- Data analysis
- Presentation of data
- Reanalysis

Project Planning

It is important to understand at the beginning of the analysis what goals and milestones to establish, what regulatory requirements or standards (if any) must be satisfied, and what methods return the best results within your organization. The following initial steps should be taken in the development of a BIA:

- Obtain management commitment
- Define the scope of the analysis
- Identify the participants
- Decide how to collect the data
- Arrange interviews or distribute questionnaires

As stated previously, the biggest single predictor of the success or failure of the business continuity planning process is the level of senior management commitment. Most often, this commitment is gained only after management becomes fully aware of the potential harm to the company from the loss of its ability to deliver service or products and to maintain its revenue streams. These losses can be fully demonstrated through BIA. The BIA owes a great deal of its success to senior management support!

If the planner or security manager is at a lower-level managerial position, bound by a tight reporting structure (as is often the case), partnership with senior management is an exceptional opportunity to familiarize oneself with their concerns. It allows senior management to recognize the value of your department and you as its leader.

But which comes first, and how? Planners who have the ear of the Chief Executive Officer (CEO) or an influential board member have little difficulty getting the needed support. Unfortunately, most planners are not blessed with this degree of leverage. To get the process started, the planner may find or convince someone in senior management to sponsor the project. Often, this is the Chief Financial Officer (CFO). Comprehensive financial audits increasingly mention the lack of a continuity plan. The CFO is more readily convinced of the importance of the project, owing to his or her intuitive understanding of potential financial impacts on the organization. This is the person who also may be legally responsible for the protection of certain corporate documents. The CFO is usually high enough within the organization to drive the project and to help the planner both collect and interpret the data. If the planner has difficulty getting financial data from other managers, a "down and dirty" analysis can be derived from the CFO's knowledge alone.

Because the BIA examines a great deal of financial data, the planner would normally maintain a close relationship with the CFO. The results of the analysis must be acceptable to management; everyone must believe that the analysis represents the true impact to the organization. The CFO should provide good insight into how the data in the final report should be represented. The data collected in this way are more often accepted by management. Data presented to management by an outside consultant or developed by a bottom-up approach may be met with skepticism and subject to increased scrutiny. Unless the consultant or you are intimately familiar with the financial position of the company, either may have difficulty responding to senior management's questions in a meaningful way. When the analysis is developed using a top-down approach, the management group will have had a hand in developing the data and will tend to answer each other's difficult questions themselves.

The CFO can convince the CEO and other senior managers of the need to commit themselves to the project and of the value of a top-down approach. Beginning the analysis at this level will help to ensure the accuracy of data by avoiding the tendency of department managers to inflate the importance of their business units or functions. Lower managers usually do not have the "big picture" of the impact that loss of their functions could have on the objectives of the organization.

An important reminder is this: the partnership with senior management should generate the support needed to drive the project through the remainder of the process. Without this support, continuity planning tends to bog down, die, or take so long that a great deal of time must be wasted revising information that has become outdated before the plan can gain acceptance.

The BIA should be completed in the shortest time possible in order to produce the most accurate results. The financial position (impact), even the organization of the company, can change rapidly, and data may become outdated generally if more than 3 months old. Consider the use of outside consultants to assist with the interviews and analysis if the scope of the project is sufficiently large to justify their use. The interest and momentum gained for the continuity planning process will be diminished if this phase is extended for any length of time. Enlist senior management's help in arranging to have all participants available for the study. A senior executive or the project sponsor should issue a memo to all affected personnel describing the importance of the project and introducing the project team or project leader.

Senior management should also agree on the scope of the analysis, timelines for the project, type of outcome expected (for instance, the format of the presentation), and who is to participate (such as an outside auditor). Product introductions, or a pending Food and Drug Administration inspection in a biotechnology firm, are not times when full cooperation can be assumed. You will have better success when all participants are otherwise available.

Normally, the scope of the analysis will mirror that of the business continuity planning project, but management or the planner can tighten the scope to an individual division, site, or location. Many planners make the mistake of including in the analysis only those departments or business functions they believe are critical without the understanding that a major goal of the BIA is to determine what is and what is not critical. At this juncture, assume that all functions are critical at some point in time, or under certain circumstances, and include them in the analysis.

Meet with the CFO or your sponsor to determine who should participate in the analysis. Interview the highest-level manager in each functional business unit. Try to maintain consistency among the positions of the participants; that is, interview only at the director or vice president level. Organization charts are useful tools to guide the planner to the proper levels within the organization for interviews. As with continuity planning, many planners insist that a planning group, composed of the planner and members of management, must develop questionnaires and determine what impact the organization will suffer from disasters. They believe this is the best and only method to ensure management support and an effective planning process. In fact, planning groups can be useful if the time allotted to do the analysis is small or the number of business units participating is very large.

Prepare an informal list of potential financial and subjective impacts that may affect the operations of the company (see Box 15.1, Sample Questions). In the meeting with the project sponsor, review the types of questions and assumptions you intend to use. Discuss any additional impacts or risks the sponsor may wish to add to the list. If you are a consultant, the sponsor can ensure that terminology used in the questionnaires is consistent with that used in the organization because this will help to avoid confusion or inconsistencies in the data.

Determine how best to manage the collection of the data. Data collection is accomplished through the use of interviews, questionnaires, or both. In large organizations, a workshop is sometimes conducted to outline instructions and explain expectations. Questionnaires distributed directly to managers tend to be delegated, so avoid their exclusive use unless they are to be completed during the interview. Questionnaires are best used in a bottom-up approach to collect resource information (team and vendor contact numbers and addresses, for example) that you will incorporate in the plan, but not for the BIA.

Schedule interviews. Give a brief presentation to the participants outlining the purpose of the analysis to give them an opportunity to collect the information you need. It can be difficult to get managers to think in terms of the financial and subjective impact of their operations and to discuss recovery strategies, especially if they believe they will be responsible for selecting the best strategy then and there. This belief may overwhelm some managers and cause them to mentally give up on the project before it starts. Avoid this problem by clearly explaining to the manager the goals and expectations for the project before the interview.

In addition to the CFO, the risk manager is an important resource for the analysis. Meet with the risk manager to review insurance requirements. Determine what perils

BOX 15.1 SAMPLE QUESTIONS

The following are sample questions the planner might ask to obtain the type of information useful to a complete and meaningful impact analysis, through interviews, questionnaires, or a combination of both.

1. If your department or function generates revenue for the organization, what are the sources of this income? Sources of income can include:
   a. Product sales (list by product lines)
   b. Services rendered to outside clients
   c. Discounts or commissions
   d. Interest from investments or floats
   e. Incentives for on-time or ahead-of-schedule completion dates or milestones
   f. Tax base, if a government agency
   g. License and use fees
   h. Maintenance fees
   i. Other _____

2. In addition to the loss of revenue, what other types of financial impact are realized if your department or function is lost? These may include:
   a. Canceled orders due to late delivery
   b. Penalties for late payments
   c. Regulatory requirements, late filings, and fines
   d. Contractual obligations delayed or not met
   e. Interest on borrowed funds
   f. Wages paid to idle staff
   g. Other _____

3. What is your estimate of the total exposure for each item or product if the inability to function or deliver the service lasted for a 30-day period, assuming the loss occurred during the period of its greatest negative impact? For example, if the greatest percentage of sales for your Christmas products occurs during December, use the sales figures from this month to estimate the exposure. If your greatest percentage of sales for swimwear occurs during June, use the sales figures for June to estimate loss of sales potential for swimwear.
   List the cumulative minimum and maximum loss for the days 1, 2, 3, 4, and 5 of the outage for each exposure. If your organization includes retail sales, list the daily sales volume for a 7-day period. Continue by calculating and listing the exposure for weeks 2, 3, and 4 of the outage. Again, use a time period that is significant to your business. Recovery times are often thought of in weeks and months, but impacts in certain institutions, such as a bank, can increase dramatically in a few hours and days. In banking, the organization may cease to exist long before the end of a 30-day period. These organizations must therefore calculate their losses by hours, not days.
   It is useful to list these figures in table format (Table 15-1), for ease of input onto a spreadsheet or into a database program that can combine and report the totals.

4. What types of extraordinary expenses are necessary to implement expected or projected recovery strategies for your business functions? These expenses may include:
   a. Transportation costs
   b. Rent for alternative space

    c. Contract services
    d. Emergency requisitions
    e. Temporary relocation of employees
    f. Temporary employees to catch up backlogged work
    g. Equipment rental
    h. Additional supplies
    i. Other _____
    Determine the minimum and maximum extraordinary expenses over time and list them as described in question 3.

5. At what times of the week, month, year, or business cycle is processing or production especially critical? In other words, when can an outage hurt you the most? This information is useful to the management team in making strategic "corrections" or modifications to the recovery plan made necessary by the timing of the disaster. It can illustrate changing priorities and allow decision makers to redirect resources to where they are needed most and when.

    Critical times are typically the end of each quarter or year, or some significant product-related event. A toy manufacturer may list the months of September, October, and November as its most critical time, preceding the Christmas season. When using a questionnaire, allow respondents to provide free-form answers. Some questionnaires and BIA software programs only allow for the selection of individual months. Although this monthly period may be sufficient for the analysis, it may not be significantly focused for recovery purposes. Later, this information is matrixed in the continuity plan.

6. What other business impacts might result from the loss of your business functions? The financial loss due to reduced customer service or technical support, tarnished public image, or the loss of future business is often difficult, if not impossible, to predict or measure.

    Some impacts may be purely subjective or difficult to state in direct financial terms. These impacts can have the greatest effect on the survivability of the company, and they should always be included in the analysis. This question asks respondents to list what these impacts may be. Examples include:
    a. Loss of competitive advantage or market position
    b. Loss of shareholder confidence
    c. Increased liability
    d. Decreased employee morale
    e. Cash-flow difficulties
    f. Reduced public image or confidence
    g. Reduced customer service
    h. Contractual consequences
    i. Regulatory violations and consequences
    j. Loss of key personnel

7. How would you rate the severity of these impacts (Question 6 above) if your business functions were lost? On a five-tiered scale, rate the estimated severity of the impacts if the outage were to occur during the worst possible moment. A value of 1 represents a minor impact, whereas a value of 5 represents a severe or fatal impact on the function or the organization. The use of a wider range of choices, such as a scale of 1 to 10, may add ambiguity to the evaluation.

Some analysts attempt to assign specific definitions to each value. Although the definitions vary, the following is typical:

1 = Minor or no impact (the problem is easily handled, functions are not affected, or it is not an issue)

2 = Somewhat critical (although the problem is still easily handled, some degradation in the image, service provided by the function, or ability to meet requirements will occur)

3 = Moderate (nearly all of the functionality or ability to meet requirements is degraded)

4 = Serious (continuation of the function or ability to meet requirements is extremely difficult)

5 = Severe (the function or ability to meet requirements ceases)

The criticality categories listed elsewhere in this publication can also be used to express the severity of the impact. Use these impacts to help to determine which functions are critical to the organization and to demonstrate to management the consequences of the loss of certain functions. The severity ratings are combined by type of impact for each function, and then by type of impact for the entire company. These are often reported in a table or matrix format, but a graphical representation is easier to understand.

8. How much time is required to reconstruct records or backlogged work once your function is back in operation? This question seems redundant to the calculation of certain extraordinary recovery expenses, but it will help determine RTOs. Based on a reasonable or expected duration of an outage, estimate the amount of time required to catch up, considering that normal (recovery) operations may be concurrent.

   Allow the respondent to give an open-ended response to this question because the range of possibilities can be great. For some functions, the time required can be a few hours; for others, it can be a number of years (and yes, you should discuss proper backup procedures in this instance).

9. What short-term and long-term resources will your function require to operate while returning to normal operations? This question is intended to give the analyst an overview of critical systems and equipment to help determine outage tolerances and interdependencies. The respondents should detail their resource needs on resource forms or questionnaires after the interview. Questions such as this, and others that ask about the criticality of systems, equipment requirements, application tolerances, and dependencies, are also useful to the issues of critical function recovery and recovery prioritization. Keep in mind that software applications and systems (servers) are not critical functions for the purpose of a BIA, but they support critical functions. If a server or application is lost, its impact is the loss of the critical functions it supports.

10. What are the lead times for the replacement and installation of the equipment and systems listed previously? This is an important question, one that can drastically affect the selection of recovery strategies. High-technology equipment can have lead times of 6 months or more. Even if the equipment can be quickly replaced, the re creation of its environment and support systems can take a long time.

and property are excluded in existing policies. Ascertain the period of indemnity for business interruption and contingent business interruption, and estimate the time delay in reimbursement. Calculate the value of the physical assets and identify lead times for the replacement of assets. These values are often not considered because they are subject to insurance reimbursement; however, it is important to know whether the policies call

Table 15-1   Example of Business Impact Analysis Table

Exposure	Day 1		Day 2		Day 3		Day 4	
	Min	**Max**	**Min**	**Max**	**Min**	**Max**	**Min**	**Max**
Lost Sales	0	0	5,000	5,000	7,500	10,000	10,000	20,000
Canceled orders	0	0	1,000	1,000	2,500	3,000	75,000	100,000

Exposure	Day 5		Week 2		Week 3		Week 4	
	Min	**Max**	**Min**	**Max**	**Min**	**Max**	**Min**	**Max**
Lost Sales	20,000	30,000	75,000	125,000	350,000	500,000	750,000	950,000
Canceled orders	100,000	250,000	450,000	750,000	750,000	750,000	750,000	750,000

for reimbursement at replacement or depreciated values. Although we will not reduce the financial impact in the analysis by the amount of reimbursement, we can consider the difference in the replacement versus depreciated cost. The loss of interest on cash reserves or on loans to secure temporary or replacement property, and extraordinary expenses during the delay between purchase and reimbursement, should also be added to the calculations later in the analysis.

Although it is more effective to concentrate on the restoration or continuation of critical business functions rather than on individual computer systems and applications, as was common in the past, critical functions can include "mission critical" processes, equipment, and applications if they support a critical function; however, this approach is still debated. Technology Departments (IT) should not be directly considered in the BIA, since the impact of the loss of their systems are indirectly accounted for in the business functions. However, it is extremely important at this point to also meet with the information systems director to determine the impact of losing these functions and equipment as well as other vital communication links within the company. The cost of replacing equipment, software, and the cost and time anticipated to recreate the IT environment should be considered. This may also include the cost of controls and recovery/continuity strategies such as the build-out of a redundant data center. How these costs are represented in the BIA must be given careful consideration and should be listed separately from loss/impact data, but more appropriately reported as part of the risk analysis.

Data Collection

To obtain data that are valid, examine all current business functions and operations. Some planners suggest that only critical business functions be analyzed, but again, the purpose of this analysis is to determine which functions are critical. Any designation of a function, operation, or process as critical or noncritical before the analysis is complete is merely intuitive. Intuitive estimates of potential acceptable outage tolerances or downtime can misdirect many thousands of dollars to recovery strategies that cause overplanning or unnecessary loss during a recovery situation. Data for the analysis are best collected through interviews with unit or department leaders. This approach helps to gain their buy-in for the project more effectively than if they were simply to respond in a questionnaire. During a personal interview, managers can ask questions and better understand what is expected from them.

Start the main part of the interview with the difficult financial and subjective (operational) impact questions (see Box 15.1, questions #6 and 7) so that neither the planner nor manager feels the need to rush through the process as the meeting time begins to expire. If planners cannot schedule sufficient time to cover these topics adequately, they should consider postponing meetings for another time. Allow 1 to 1 1/2 hours for each meeting, longer if nonimpact issues are discussed or if the manager is not experienced with putting impacts in financial terms. Although this is a top-down data collection procedure, it may be necessary to meet later with the next-lower level of management to validate impacts and strategies developed.

During the interview, determine the following:

- How the business unit or function fits in with the overall mission statement of the organization
- The primary service objectives of the business unit or function
- The business unit or function processes and dependencies. Do the function(s) cross over to other departments?

The analyst will also guide the manager to accomplish the following:

- Estimate the maximum loss if the function is out of service, out of operation, or access to it is denied for 30 days or more. This 30-day period can be adjusted to fit expected outages, but it should remain consistent for all functions being evaluated. Assume that the function, or access to it, is lost at the worst possible time of the month, year, business cycle, or production schedule. Be careful that evaluation of functions is not duplicated. If one process impacts multiple functions or if a function impacts other functions, Barry Cardoza (Cardoza, K&M Publishers, Inc.) says to record the impact at the first point it occurs. This should help to avoid duplicative loss.
- Determine how this maximum loss is allocated over the following times: day 1, day 2, day 3, day 4, day 5, week 2, week 3, and week 4. Again, these times can be adjusted to fit the specific environment. For example, it may be more realistic for a financial institution to track its losses by the hour instead of by the day. Consult with the CFO to decide what time periods are best for the calculations, but again, keep them consistent throughout.
- Estimate any extraordinary expenses the unit may incur if the function is lost. As before, indicate the maximum amount, as well as the details over a specified time. Extraordinary expenses can include costs associated with idle staff, wages paid to extra staff to handle backlogs, equipment rental, outside services, and transportation.
- List the impact on any contractual or regulatory obligations the outage will cause. What fees, fines, penalties, or missed milestone payments will the company face?

- Indicate how cash flow will be affected by the outage.
- Rate the loss of goodwill or damage to the corporate image.

When seeking answers to these and other questions, be careful not to put participants in a defensive situation. Form questions in such a way as not to cause managers to believe they need to justify their positions within the company. Don't ask, "How valuable is your function to the organization?" Instead, ask:

- What would be the impact to the organization if this function were lost?
- How would this impact change over time?
- How would the loss of this function affect other functions within the organizations (that is, what other functions are dependent on the input to or output from this function)?

Some planners and software developers believe the best method for completing the analysis is through the simple distribution of questionnaires to the appropriate participants. Although many planners use questionnaires exclusively, questionnaires without personal interviews will not maximize the planner's understanding of how each function fits into the overall organization and how its interdependencies relate to the whole. Managers are reluctant to put sensitive financial and other impact information "in writing" and may need additional time to answer the questions adequately. Another difficulty with the use of questionnaires is some analysts tend to list a predetermined scale to record the financial impact. While in many cases simply checking the appropriate range of values is easier for the responding manager, it can allow overinflated estimates, and the ranges selected may be too broad (or contain too many smaller categories) to make the most meaningful comparisons. Allowing managers to place a checkmark next to a range of figures does, however, return a greater number of responses in a shorter period of time, but this method forces the analyst to predetermine the ranges—a practice counter to the goals of the BIA (it becomes intuitive). Face-to-face interviews can stimulate managers' understanding of the process and, more importantly, their ability to think through strategies and resource needs. A simple compilation of financial information from questionnaires will miss subjective information that could be important to an accurate analysis of the data.

During an interview, the analyst can ask the questions listed on the questionnaire, with the exception of items that are simply resource oriented. Have a printed or electronic (diskette, shared drive file, or web) version of the questionnaire available to give to the manager after the interview. Many of the business continuity planning and BIA software programs on the market have this capability. The manager fills out the questionnaire electronically, and the analyst simply imports the data into their software, thus avoiding the need to input data twice.

If a questionnaire is used without an interview, meet briefly with the recipients and explain both the purpose and importance of the project. Recipients rarely complete all sections of questionnaires, necessitating follow-up questions.

Other Questions for the Impact Analysis

The following questions are commonly listed on questionnaires or asked during the business impact interview. They can help the analyst maximize the use of time while gaining information for an effective analysis and continuity plan. Some of the answers to these questions are verified by comparison to the financial data. Remember that from a practical point of view, you are also collecting information to be used later in the Business Continuity Process.

- What is the name of your department, unit, or function? Please describe its function. Include an overview of what your department or team must do to recover from a disaster (you will be asked to list detailed instructions later).
- How long can your department, unit, or function be out of service without adversely affecting the overall operation of the company?
- What are the most critical functions of your department or unit? Which of these functions would need to be recovered first, either at an alternative site or reconstructed at the original site?
- What tasks would be necessary to recover these functions?
- If these tasks for the recovery of the various functions are dissimilar, how many separate recovery teams would be necessary to implement these tasks?
- Who would you select as team leaders and alternate leaders for these teams?
- What customers (internal or external) or other business activities would be affected by the inability to deliver or perform your function? Describe the extent of the impact and estimate the amount of time before the impact would affect the customer.
- What other areas or functions of the company are you dependent on to operate effectively, and how would their inability to deliver services or products affect your operations? How would an area-wide disaster affect the physical delivery of these services or products?
- Where can you obtain other required services, products, or raw materials if access to your present supplier ceased?
- Is your function directly involved with billing, collection, or the processing of revenue? If yes, please describe.
- Have any arrangements or agreements been made with your vendors for emergency delivery of critical resources?
- Where would or could the organization go to perform this function if employees were unable to gain access to the current location?
- Can you depend on the resources of similar functions at other company locations? Please state the names of the units, sites, and office locations.
- Indicate the estimated square footage required to house disaster recovery operations for both a short-term and long-term (or permanent) relocation of your business unit. Attach a copy of any special floor plans needed.
- What computer support systems (that is, mainframe, stand-alone PC, or LAN) are necessary for the continued operation of your unit? What is the maximum amount

of time these support systems could be unavailable before their loss would have a negative impact on the company? Rank these systems in order of priority and indicate what configurations are necessary. Are these systems maintained by the Information Systems Department or by your department?

- Do your critical vendors and suppliers have tested recovery plans in place, and is their recovery time consistent with your needs?

Resource Questionnaires and Forms

Ideally, the resources required to implement the recovery strategies are best determined subsequent to the selection of the strategies themselves. Most often, however, question-naires are distributed after the interview to give the respondent time to list the short-term and long-term resources they would need to recover or to continue their operations based on only the preliminary discussion of strategies. Resources are included for the following:

- Employees and consultants
- Internal and external contacts
- Customers
- Software and applications
- Equipment
- Forms and supplies
- Vital records
- Other _____

Employees and Consultants

List the names, titles, addresses, and contact phone numbers for the employees assigned to your team. Set up call trees if desired. A call tree is when one team member is respon-sible for contacting a fixed number of team members, who in turn have a list of other team members they are responsible for calling. Call trees can expedite the notification of large groups of people. Outside vendors exist to provide this service and can be useful if there are a large number of calls to complete. List as many numbers as practical (home, office, cellular, pager, fax, and e-mail addresses) and use a consistent format, such as (415) 555-1212.

Not all employees on your staff or team may be needed during the initial recovery phase. After a disaster, space and resources may be limited. Identify at what point during the recovery process the individual is required, according to the following:

1 = Required within the first 24 hours
2 = Required after 24 to 72 hours
3 = Required after 3 to 5 days
4 = Remain on standby until advised

If plans are maintained electronically, as most now are, the address section should be separated into street, city, state, and zip code columns so that a database sort can be achieved (Table 15-2). Such action may be useful to determine who lives closest to another or for identifying who lives inside an affected area. By the same token, separate first and last names.

Internal and External Contacts

List the names, addresses, and phone numbers of vendors, suppliers, consultants, or other people and groups, either internal or external to the organization, that you may need to contact to assist with the recovery or with replacement of supplies, equipment, or raw material, or to provide technical support (Table 15-3). This list should be complete but as short as possible. Bear in mind that your original list of contacts (phone books, vendor lists, etc.) may be buried under tons of debris after a disaster.

Briefly describe (one or two words, if possible) the type of service each contact provides and list any account numbers, passwords, or other type of authorizations required. If there are contact people within a vendor company, you may list those names also. If the vendor has a corporate office or other location outside the potential disaster impact area, consider including those numbers also because the local office may suffer damage from the same disaster. Examples of these contacts include the following:

- Plumbers
- Electricians
- Seismic or structural engineers
- Janitorial service
- Security service
- Food service vendor
- Payroll processing

Table 15–2 Sample Employee Resource Questionnaire Form

Function:							
			Contact Information				
Employee	Title*	Address	Home	Cellular	Pager	Priority	Called by:

*Title, position, or recovery function.

Table 15–3 Sample Internal/External Contact Resource Questionnaire Form

Function:						
Contact or Vendor Name	Representative	Address	Primary Phone	Alternate Phone	Type of Service	Account Number or Password

- Hot-site providers
- Portable toilet/shower providers
- Insurance broker
- Bank manager
- Federal or regulatory agencies
- Media contacts
- Equipment/systems suppliers

Customers

Although customers may be listed under "Internal/External Contacts," you may want to list separately certain customers or contacts you may wish to inform of a disaster, in order to let them know of major delays, when you expect to return to normal operations, or that you have relocated to an alternate site. If your organization is unaffected by a regional disaster that has received some publicity, you may need to affirm to these customers your ability to continue to serve them. The questionnaire is similar to the Internal/External Contact list, but without the need for account numbers or passwords (unless this information is necessary).

Software and Applications

List all software applications utilized by the unit or team that would be required in a recovery situation. List the name and version number of the application or operating system, the number of licenses required, and at what point during the recovery it is required (in the first 24 hours, by day 2, by the end of week 2, for example). Record the location of backup copies (this may be a task for the IT department). Be sure to list the platform it requires (PC, mainframe) (Table 15-4).

Equipment

List the critical equipment your unit will need to function after the disaster. This should include, but is not limited to, desktop or personal computers (if possible, state configuration or hardware requirements—that is, amount of random access memory (RAM), hard disk, processor speed, CD-ROM drives, special attachments), terminals, printers, fax machines, calculators, phones, modems, tables, chairs, soldering irons, microscopes,

Table 15–4   Sample Software Resource Questionnaire Form

Function:						
Software Description	Version Number	Serial Number	No. of Licenses	Backup Location	Time Required	Platform

shovels, and like equipment. Include model numbers and enter the quantity required for a short-term outage (usually defined as less than 5 days) and also for a long-term outage. Estimate lead time for the replacement of equipment (Table 15-5).

Forms and Supplies

List all forms and supplies required to conduct business for both a short-term and long-term outage (for instance, letterheads, vouchers, diskettes, and special forms). Regular office supplies, such as paper, pens, or staplers, do not have to be included because they should be readily available at an alternate location unless the amount is extraordinary. Include such supplies as food, glassware, shipping cartons, and raw materials. Indicate where these supplies are stored or from where they can be obtained (Table 15-6).

Vital Records

List the critical records, reports, or documents on which your unit depends—records that must be available to send to regulators or that are necessary to conduct normal operations. Examples include accounts receivable, floor plans, corporate minutes, technical manuals, and the disaster recovery plan (Table 15-7).

Indicate where these records are stored or located within the company and on what type of media (paper, microfiche, or computer tape, etc.). If the record is in off-site storage, or at a law firm, identify the company and its location on this form and in the plan.

Data Analysis

The information gained from the interviews and questionnaires is analyzed according to the scope and goals of the project. Data are combined from all business functions to allow the planner and management to decide which are critical to the continued

Table 15–5 Sample Equipment Resource Questionnaire Form

Function:

Equipment Description	Make or Model	Configuration or Other Spec.	Short-Term Requirement	Long-Term Requirement	Lead Time

Table 15–6 Sample Forms and Supplies Resource Questionnaire Form

Function:

Description	Make, Model, or Lot Number	Location or Source	Short-Term Requirement	Long-Term Requirement	Lead Time	Quantity on Hand

Table 15-7   Sample Vital Records Resource Questionnaire Form

Function:				
Description	Media Type	Internal Location	External Location	Form or Other Number

operation of the organization and which are dependent on others. Their outage toler-ances/RTO are determined, and recovery priorities for both the individual business func-tions and support systems, such as computer applications, are assigned. When drawing these conclusions, the planner must keep the following guidelines in mind.

Outage tolerances (RTOs) and critical functions are not determined solely on the numerical data. For one thing, their designations may change. Also, whereas the analy-sis may indicate that product A generates the most income, it may also carry the highest recovery cost; management may decide to recover product B first because it has the low-est recovery cost or is a product they wish to emphasize at the time. In most cases, it is readily apparent from the data which functions are the most important, but the planner must obtain early verification of this from the sponsor or senior management.

Again, account for impacts only once each; be careful of duplicates. In a large or com-plex organization, it is easy to add the loss of a dependent function more than once when calculating the overall impact.

Do not deduct insurance coverage or expected claims reimbursement from the loss figures. Although it is possible to predict (assume) the maximum amount of reimburse-ment for claims, it is more difficult to predict when the reimbursements will actually be paid or how the total amount will be distributed over the life of the claim. During a recov-ery situation, claims are usually not filed all at once—the extent of the loss, the amount of documentation, the varying procedures between insurance companies, and the added workload on the claims departments during an area-wide disaster will delay or extend the process. Small and medium-sized companies have little tolerance of an interruption in cash flow caused by a disruption; for them, anything less than a rapid reimbursement may have fatal consequences for the organization. Figures adjusted for expected reimburse-ment may not adequately warn of this danger. Also, disputed claims may be ultimately lost.

After the data from the interviews and the questionnaires are analyzed, verify the results with business unit management and with the CFO. This verification is important for keeping the data credible throughout the process.

Establish RTOs both within the critical processing time windows and during "normal process times"—that is, not end of month or end of quarter. List both in the recovery plan.

Presentation of the Data

Distribute a written report that includes a description of the BIA program and process, supporting data, the impacts for the individual functions or units, and the combined impacts for the organization. Depending on the practices of the organization, or on regulatory requirements, the findings are illustrated in a manner that satisfies the goals and scope of the BIA. By collecting the data in an iterative manner as suggested (as opposed to simply asking, "What is our total loss after 30 days?"), information can be presented in a meaningful way. Functions can be ranked by the greatest loss and/or RTO and RPO. The categories of loss (such as revenue streams, penalties and fines, for example) are listed with their corresponding totals and subtotals.

Operational loss is also listed, ranked and discussed by category and function. Some standards and regulations mandate the presentation of the results in a matrix format. Some managers argue this type of representation of the data is meaningless. It may be best, in this case, to include both the matrix and financial/operational loss tables that list actual dollar figures and other rankings. Some analysts suggest converting financial loss figures to impact ranks by assigning severity levels of 0–4 (no impact, minor impact, immediate impact, major impact). Others may argue that ranking financial impact in this manner is too simplistic.

Critical processes (software applications) can be mapped and tied to their IT System platforms. These applications are often assigned "values" such as Critical, Vital, Sensitive, or Noncritical, but keep in mind that the assignment of these categories should be supported by the data and by management decision. Other ranking methods include a "Tier" rating. For example, Tier 1 could include all processes that are the highest level of criticality whose continuity strategy includes real or near-real-time mirroring to a second data center; Tier 2–4 could be data and programs of lesser criticality whose recovery is affected by less expensive, increased time-delayed means. These rankings are, of course, established after recovery/continuity strategies are agreed upon.

The report is often combined with other information derived from the questionnaires and interviews (resource information, however, is better left for the plan itself, but occasionally appears in the report). Existing plan maturity, the results of a Gap Analysis, a discussion of the initial continuity/recovery strategies, and mitigation recommendations often find residence in the report.

Because the BIA is often used to gain management support for the continuity planning program or to justify program budgets, their results are presented to senior management. This presentation could be the most important step in the impact analysis. If it is successful, the planner will gain the political and financial support needed to complete the project. If not, the planner will have wasted many hours of valuable time and will face needless delay and frustration and may produce a plan that is less than effective.

The analysis presented to management must be credible. As stated earlier, a top-down approach will result in the greatest degree of success. If this approach is used, the planner will have few problems with the acceptance of the analysis. When presenting the results, ensure that figures are accurate and not misleading and that the data and the presentation are short and simple. Matrices, tables, and statistics tend to bore and confuse people, even at the senior management level. Slides that contain spectacular pictures of disasters and natural phenomena for background tend either to distract or cause the graphics to be remembered only for their scenery. Relationships and financial data are better represented graphically, either by pie, bar, or other types of charts. Whenever possible don't present expected occurrences in terms of probability—state them as a fact. For example, if the local Office of Emergency Services or the U.S. Geological Survey predicts that the probability of a major earthquake is 82 percent in the next 10 years, report, "I believe there will be a major earthquake during the life of our strategic plan" (assuming it is a 10-year plan).

Unless it is important to the understanding of interdependencies or other relationships, discuss only the impacts of the major functions and the impact to the organization as a whole. During the presentation, outline why you are there and what you expect the group to decide or to do as a result of the analysis.

Reanalysis

Finally, BIA should not be a "one-shot deal." The information in continuity plans must be updated regularly and the basic strategic framework of the plan reviewed annually. It is logical that the impact to the organization will change as the structure and strategic direction of the business changes.

Whenever a new project is initiated such as an IT upgrade in equipment or software, or an existing function is restructured, BIA should be built into the beginning stages of the project. Not only does it keep portions of the analysis fresh, it helps to perpetuate a culture of Business Continuity Planning.

Summary

The Business Impact Analysis identifies the financial and operational loss of the organization's business functions over time periods significant to the individual organization regardless of what caused the loss by examining their impact on service objectives, financial position and cash flow, regulatory and contractual issues, and market share/competitive risks. It allows management to decide which functions and processes are critical and what recovery time objectives (and potentially recovery sequence) to assign to these processes or business functions. The BIA identifies interdependencies and establishes a financial justification for continuity strategies. It also helps to provide an understanding of the amount of risk to assume, transfer, or mitigate.

A BIA differs from a risk analysis in important ways. The Federal Finance Institutions Examination Council (FFIEC) has said that the BIA is the first step in developing a Business Continuity Plan and that the Risk Analysis is the second step. This is contrary to other planning models that place the Risk Analysis before the BIA.

The manner in which the BIA is conducted can help guarantee the success of the entire business continuity process or doom it to failure. Effective project planning, a data

16

Business Continuity Planning

A plan is nothing, planning is everything.
—**Dwight Eisenhower**

Business continuity planning is defined in many different ways, reflecting its author's particular slant, background, or experience with the process. Many of these definitions attempt to combine the meaning of continuity planning and of a continuity plan. There is an important distinction between the two.

Business continuity planning is a process that identifies the critical functions of an organization and that develops strategies to continue these functions without interruption or to minimize the effects of an outage or loss of service provided by these functions. The most common strategies involve some type of third-party data center or alternate, off-site processing location and alternate workspace to restore operations to a minimally acceptable level. In today's business environment, it is less desirable to return to or to achieve a minimum level of service after a disaster. These companies wish to, or need to, maintain operations at their current level or to take advantage of the disaster by utilizing their business continuity plan to gain market share over the competition. NFPA 1600 defines business continuity as "an ongoing process to ensure that the necessary steps are taken to identify the impact of potential losses and to maintain viable recovery strategies, recovery plans, and continuity of services." According to the ASIS SPC.1-2009 Organizational Resilience Standard's Glossary, it is the "strategic and tactical capability, preapproved by management, of an organization to plan for and respond to conditions, situations, and events in order to continue operations at an acceptable predefined level." It goes on to say that business continuity is to ensure an organization's ability to continue operating outside of normal operating conditions.

Disaster recovery planning is another term used either synonymously with business continuity planning or to represent the idea that planning is important only to telecommunications and data centers. Business continuity planning implies planning for all the critical functions or business units of an organization. In other words, disaster recovery refers to the reestablishment or continuity of information technology and data systems; business continuity refers to the recovery or continuity of business unit operations (systems versus people).

A *business continuity* plan is a comprehensive statement of consistent action taken before, during, and after a disaster or outage. The plan is designed for a worst-case scenario[1] but should be flexible enough to address the more common, localized emergencies, such as power outages, server crashes, and fires. Although the actions listed in the plan contain sufficient detail to implement strategies designed to recover critical functions, they are more guides than inflexible dictates. Because it is not practical to plan for every type of contingency, and because each disaster has its own set of conditions, the ability to modify the plan must be incorporated.

Although a continuity plan is important, it is the planning process that returns the greatest value. This distinction is often missed by both planners and end users of continuity plans. The identification of critical functions, the thought and analysis behind the development of the strategies designed to recover or to continue the functions, and the knowledge of why one particular strategy was selected over another are not always apparent from simply reading the plan. This is valuable knowledge when last-minute decisions are required to adapt the plan to a particular situation. The planning process is also a training exercise. The participants must think through contingencies so that the actions required to recover from them will be already familiar. Reading the plan for the first or second time just after the disaster will provide for a less than effective recovery. This is assuming, of course, that the plan is not buried under a hundred tons of rubble.

Why Plan?

Responsibility for continuity planning often resides with the risk manager, the chief financial officer (CFO), or the data center manager. Security managers are, however, increasingly taking the role of plan developers. Their experience with the protection of assets, involvement in the identification and the mitigation of risk, and their emergency response duties makes them logical choices for this role. The ability to work effectively with all levels of management is a required trait for security managers, a trait that all successful continuity planners must possess.

Some types of businesses, such as healthcare, financial institutions, and industries regulated by toxics laws, are required to maintain continuity plans. Businesses are increasingly regulated by other laws, regulations, and standards, many differing widely in their approach and requirements. Some are intended to be industry specific and others

---

[1] While this is mostly a true statement, one should use caution when describing this concept to stakeholders, since some may adopt the attitude (as briefly described in Chapter 12) that planning for the worst-case scenario equates to an asteroid hitting the earth and therefore mentally shut down; they believe you are giving them a project that is impossible or pointless to complete. Indeed, based on engineering studies, a worst-case scenario for a multifacility site familiar to the author located only thousands of feet from the Hayward Fault, which is capable of (and "overdue" for) a magnitude 7.9 earthquake, predicts that a number of its buildings will suffer catastrophic damage, with one multistory building sliding (intact) down a hill. These buildings contain highly toxic materials, asbestos, and radiological hazards that will make damage assessment and reentry problematical. The challenges of planning for this degree of destruction can be overwhelming. (Some would give up and simply say "scrape it," i.e., just tear it down and start over, which may be a viable strategy, but what do you do if there is a lesser (and more likely) magnitude earthquake that causes major, but survivable damage?)

broadbased. Some use differing terminology or try to package the same methodologies in different-looking boxes.

In any case, even in the absence of regulatory requirements, it makes good business sense to maintain a continuity plan. The cost of downtime, the cost of reconstructing lost data, and the loss of cash flow can severely damage many organizations, even beyond their ability to recover. If they are unable to operate, retail and transportation operations can lose an average of more than $100,000 per hour, high-technology manufacturing $200,000 per hour, and financial brokerages more than $6 million per hour. A Business Impact Analysis can help to pinpoint your organization's exact loss potential.

Without continuity planning, the organization may lose its competitive advantage, valuable employees, and future research. Organizations cannot insure against lost customers or a diminished public (customer) image. History consistently shows that between 35 and 50 percent of businesses never recover after major disasters.[2]

Other rationales for continuity planning include the following:

- Fulfill requirement by financial auditors or by potential customers
- Prevent the loss of market share
- Capitalize on the lack of planning by the competition
- Uphold fiscal responsibility
- Avoid stockholder liability
- Fulfill regulatory requirement
- Retain key employees
- Prevent the loss of research
- Help ensure the safety of employees
- Preserve customer confidence
- Assist in the overall economic recovery of the community
- Assist in a quick and orderly recovery after a disaster
- Minimize the economic loss (devaluation) to the firm

The Planning Process

The basic steps involved in business continuity planning are simple, although their implementation can be complex and time-consuming. The critical functions of the organization are identified and ranked according to their value to the organization or to their interdependencies with other critical components. Cost-effective strategies for continuing or recovering the critical functions to an acceptable level are evaluated. Once the recovery strategies are chosen, a plan is developed to implement the strategies. Users are trained and the plan is "tested" (the proper term is *exercised* or *simulated*), and provision for maintenance of the plan is established. A process for continuous improvement and program effectiveness is put into operation.

---

[2]These figures are often reported in the industry and vary widely. There is some discussion about their validity. In any event, almost every major disaster has more than one example of a business that did not recover.

Before these steps commence, it is important to identify physical or procedural hazards that could cause an outage or delay the recovery process. When dealing with multiple sites, the planner should visit each location (if within the scope of the project) and conduct an inspection for these hazards. This inspection should identify single points of failure in critical systems, and it should produce a set of recommendations to prevent or mitigate the hazards identified. This is often included as part of the risk and business impact analysis.

Next, the organization must prepare to respond to the disaster or to the emergency when it happens. The goals of emergency response are to protect the health and safety of employees, guests, and the community and to minimize damage to the organization and the environment by stabilizing the situation as quickly as possible. Response planning is not continuity planning, but the two plans can be integrated.

Although the focus is normally on the continuity or continuance of operations, once the disaster or emergency is stabilized, recovery and restoration will begin. The terms *recovery, resumption,* and *restoration* refer to separate phases of the organization's return to predisaster service levels (although some planners use them interchangeably). Resumption embraces the initial, short-term strategies and steps necessary to get back into production as quickly as possible. Moving to a hot site (a separate building, office area, or third-party data center with duplicate or equivalent equipment already installed, waiting for emergency use) and transferring production to a satellite facility are examples. Recovery and restoration refer to the long-term strategies and steps the company will follow to reestablish its normal goals, service, or production levels. The replacement of a production line, installation and testing of replacement equipment, and the construction of new facilities are examples. In this text, we will use these terms interchangeably.

Project Management

Business continuity planning projects, if not properly managed, will lose momentum, languish, and die, or assume such a negative tone that the participants become hesitant to continue with the project. Information and the strategic mission of an organization can rapidly change, so that once the project is started, any significant delay will cause the end product—a business continuity plan—to be outdated before it is finished.

Project management is a major skill, and it is required of anyone who undertakes responsibility for business continuity planning. It is a partnership between members of management, outside services and vendors, employees, and sometimes regulatory agencies. The ability to schedule and manage resources, time, and people will help bring the project to a successful conclusion.[3]

---

[3] Business continuity planning is an ongoing process of improvement, maintaining resource information and reexamining the basic strategies used to develop the plan. The plans must be "living" documents; the "project" is not a project but is a process and is never "concluded."

In broad terms, the following steps are followed to produce an effective plan:[4]

1. Identify the planning coordinator.
2. Obtain management support and resources.
3. Define the scope and planning methodology.
4. Conduct risk identification and mitigation inspections.
5. Conduct a business impact analysis (BIA).
6. Identify critical functions.
7. Develop recovery strategies.
8. Set up recovery teams.
9. Develop team recovery instructions.
10. Collect resource information.
11. Document the plan.
12. Train recovery teams.
13. Exercise the plan.
14. Maintain the plan.

1.   Identify the Planning Coordinator

A person within the organization is designated as the planning manager, coordinator, leader, or other appropriate title such as Business Continuity Planning Manager. This person is responsible for the management of the project (that is, the completion of the plan) and possibly for coordinating or leading the recovery effort subsequent to a disaster. The coordinator may also have major responsibilities for plan activation. Ideally, this person should be a management-level employee who has good people and project management skills, has a good understanding of the organization, and is detail oriented. A vast technical knowledge of risk and computer and telecommunications systems is not necessary, but it can be helpful. If a planning committee or a steering committee is established, this person should lead the committee. In a company with a dedicated business continuity planning department, this person is most likely the department manager, but individual planning coordinators may be assigned on a regional or national basis.

This position (Program Coordinator) is mandated in NFPA 1600 and suggests that a written position description is published. The ASIS standard approaches this position through the appointment of a management representative who is responsible for the direction of the Organizational Resilience system, irrespective of other responsibilities, and has, among other responsibilities to the program, the authority to implement the program.

[4]The Disaster Recovery Institute International (DRII), in an early attempt to standardize planning methodologies, has issued a common body of knowledge, which may list steps different from those listed here. Additionally, these steps are not completely inclusive of those found in NFPA 1600 or in the ASIS Standard.

2.   Obtain Management Support and Resources

No planning effort will be successful without the support of upper management. This support must be communicated to all levels of management and program participants and is one of the first steps, if not already established, that the Planning Coordinator should address. This support is often demonstrated through corporate policy and procedure that mandates the implementation of all elements of Comprehensive Emergency Management. The policy and procedure should reaffirm the organization's commitment to the protection of life, the environment, and of its property and other assets. In particular to Business Continuity Planning, it should, depending on the organization's approach to policy development, require the commitment of resources to identify and mitigate hazards, to implement the program and its continuity strategies, outline responsibilities that may include a planning or steering committee, and to provide for program maintenance and continuous improvement. It should completely, but briefly, define the program. The ASIS standard lists a number of other items they believe a policy should include.

Most agree that the development of a business continuity plan is a noble project, but all too often other priorities take precedence if participants are not held accountable to time lines and milestones. The timely completion of these milestones should be included in the goals and objectives for all expected participants. Although the best results are obtained by motivating participants in a positive manner to complete their development tasks, it always helps to carry the big stick of upper management support behind your back. Absent this support, it is often necessary to get the attention of and support from at least one member of senior or executive management such as the CFO. This person becomes the program "sponsor" or "champion."

In situations in which the driving force behind the project is not senior management, the results of a BIA should help demonstrate to senior management the potential impact to the organization if all or a portion of the company could not operate. Their support is often obtained as a result of this demonstration. Before conducting the BIA, you may need to define the scope of the project.

3.   Define the Scope and Planning Methodology

It can be a daunting, if not impossible, task to produce a program and plans for a large, worldwide corporate structure unless the job is accomplished in small pieces. If this is the case, narrow the scope of the project to a single division, geographical location, building, or from a functional perspective, concentrate planning to include only revenue producing functions—something small enough to allow a positive outcome. A successful project will add momentum for the completion of subsequent portions of the program throughout the remainder of the organization.

The selection of a starting point may be dependent purely on need or on the degree of risk. The risk or BIA will show where the planner's energy is best directed. Be prepared to go beyond the established scope when identifying interdependencies and strategies. The scope of the project may be to develop continuity plans for a manufacturing site in California, but the best strategies may involve the transfer of administrative functions and financial computing to the company's site in Illinois, and the transfer of manufacturing to Nevada, for example.

Once the scope of the project is agreed upon, decide how the planning effort is best managed. Many methods exist to manage the project. These include the following:

- Facilitation of internal development
- Consultants
- Use of template plans
- Single developer
- Planning standards for the business units or divisions
- Steering or planning committee

In most cases, the best planning method places the planner—that is, the program coordinator—in the role of an internal facilitator. The planner must bring his or her project and people management skills together to assist each individual business unit to develop its own plans. Each unit that may later become a recovery or continuity team must "own" its portion of the plan and become familiar with its contents. Working as a facilitator, the planner can tap the managers' or business unit leaders' technical knowledge and experience to guide the creation of their plans, while keeping them consistent in direction and format with the overall planning structure.

Considering the scope of the project, select the managers who will participate in the process. This may be the same group that participated in the BIA. In most cases, they will be the functional or department heads. Schedule interviews of at least 1 hour's duration to discuss the following:

- Business impacts
- Critical functions
- Recovery strategies
- Expectations and needs of the manager
- Description of the project
- Questionnaires or resource forms
- How the function operates under normal conditions and how the product or service is delivered or produced
- Interdependencies—for instance, where the input comes from and who receives the function's output. (For example, a manufacturing line might get its input from raw materials stores. Its output goes to Quality Assurance. If raw materials stores are not recovered, or not recovered first, manufacturing may have difficulty with its recovery. If a particular software application stops processing, other applications that depend on its output may stop or return inaccurate results.) These are sometimes referred to as upstream and downstream dependencies.

Schedule meetings over a 2- or 3-day period with participants from the following departments:

- Information Technology (IT)
- Human resources
- Legal
- Facilities
- Management or site manager
- Manufacturing and operations
- Finance (accounts receivable and payable, cash management, and payroll)
- Telecommunications
- Risk and insurance manager
- Security and safety
- Other pertinent functions

Before the meetings, issue a letter or memo to the participants to introduce yourself and let them know what you expect. Briefly outline the process so they can begin to think about their responses. Consider distributing, if used, a list of questions or software tool kits prior to the meeting. These meetings enable many of the remaining steps of the business continuity planning process.

Many business continuity planning consultants promise to come in and develop a plan for your organization within 3 months or sell you a template plan in which you just fill in the blanks, but your organization will gain very little from their product. The individuals responsible for its implementation will have little familiarity with, or commitment to, its contents. It becomes the consultant's plan, not your plan. Consultants should act as facilitators as described above and as advisors about the most effective strategies, if they are to deliver effective plans.

Use great caution when using templates (for instance, external or internal plans from other organizations). They are meaningless if not adapted to local conditions and needs. They do not require the "thought process" necessary to provide the training needed to execute the plan effectively after a disaster.

Likewise, Planning Coordinators and consultants are often asked to simply develop the plan themselves with little participation from the functional units. While this approach uses less of the organization's time and resources, it does not permit those responsible for the plans' execution the advantage of thinking though beforehand what will or won't work best for their units. Plans developed in this manner are usually the result of an immediate need, such as to fulfill the requirement of a customer, regulation or to make the organization more attractive for a pending Initial Public Offering (IPO). For these plans to be effective, a higher degree of training and testing is required, a necessity that, in these cases, is rarely accomplished. An astute auditor, however, will recognize such a planning process, especially due to the usual absence of prevention, preparedness, and mitigation.

In large multinational organizations, especially if the Business Continuity department is small, planning standards can be successfully developed when the scope of the project involves many diverse locations or business units. When developing these standards (planning guidelines), be aware of local requirements adopted or recommended and adjust yours for compatibility and simplicity. Standards referenced in this text may be used or developed internally as appropriate. A person(s) responsible for planning at each location or division is chosen. Planning guidelines, methods, and focused templates are developed and distributed. The corporate planner then trains these people in the expectations of the project, monitors their progress, and helps to implement the plans. When this method is used, plans for the entire organization are developed quickly (usually a 2-year process). Greatest success is achieved when the corporate planner develops the "basic plan" (see Chapter 17, Plan Documentation) and allows the divisions to concentrate on planning for only themselves.

Many businesses form a planning or steering committee composed of the company's or division's top managers to discuss business impacts and select strategies, and manage or audit the process. Both ASIS and NFPA 1600 mandate the use of a planning committee. The business continuity planner will coordinate the efforts of the committee, business units, and outside consultants. A variation of this method is often applied at the business-unit level, where a committee is formed to identify critical functions and strategies. Steering committees work best when the organization or planner has little knowledge of continuity planning principles and strategies or cannot gain an understanding of the operation or needs of the business unit.

Many planners insist that committees are the best way to produce effective continuity plans and can be used to resolve issues. Fundamental problems, however, can arise with the committee approach and are counterproductive under some corporate cultures. Committees can delay decisions, and corporate policies can dilute the effectiveness of the process. Often, much time is wasted discussing administrative issues. The Planning Coordinator must be a strong leader to prevent the committee from taking a direction the Coordinator's experience recognizes is not desirable.

4.   Conduct Risk Identification and Mitigation Inspections

The more hazards and risks you can identify and prevent or mitigate beforehand, the more you will minimize the effects of the disaster, allowing for a faster recovery and better enabling the continuity of operations.

Inspect the buildings, grounds, and community for any hazard that may injure employees, damage equipment or facilities, or cut off the supply of materials, resources, or services. Conduct the risk analysis at this point, which includes an inspection of buildings, grounds, and the surrounding community for any hazards that may harm employees or guests, or damage facilities or infrastructure. Look at current conditions and those that may be expected postdisaster, in particular any that affect support systems, access, supply chain, resources, or services. When searching for these hazards, the techniques learned from scenario planning are useful. Think through the causes and effects of likely scenarios and offer recommendations to prevent, prepare for, or mitigate their effects.

For example, some of the typical effects following a major earthquake may include the following:

- Structural damage and displacement
- Posttraumatic stress
- Loss of utilities (gas, water, electrical power) and other infrastructure
- Disruption of communications
- Transportation difficulties
- Inflated prices for goods and services (in a cash economy)
- Human resource problems
- Overloaded and nonresponsive governmental services
- Victims trapped under structures and debris
- Mass casualties, shortage of hospital beds and medical assistance
- Disruption of routines, loss of housing
- Difficulty obtaining food, water, and other basic needs
- Uncontrolled fires
- Increased illnesses
- Damaged or destroyed product and raw materials
- Canceled orders
- Loss of vital records

Scenario planning will help you foresee the postdisaster conditions and assumptions that must be considered.[5]

A review of the literature on the effects of a flood would make clear the need to control mold, mildew, and snakebites, but such benefits of experience may not be available in print on the results of a failure of a proprietary process or a hazardous substance spill.

Question the general manager, facilities manager, and other appropriate people on what could prevent emergencies, outages, and disasters. Does the company have an evacuation procedure? Are first-aid, food, and water supplies stockpiled? Are critical systems or equipment connected to uninterruptible power supplies? Are computer files backed up on a regular basis and stored off-site?

5.   Conduct a Business Impact Analysis (BIA)

When relevant risks and hazards have been identified, submit a report or a plan to the steering committee, senior management, or the sponsor of the process outlining

---

[5] Scenario planning is useful (some would say necessary) to understand, as previously stated, the conditions under which strategies are developed and implemented as well as constraints to, or the need for, resources postdisaster. In spite of the Department of Homeland Security's National Planning Scenarios, which number over 100, devising a continuity plan based on scenarios is not recommended. Plans based on scenarios can be duplicative in their tasks, cumbersome, and may not provide sufficient flexibility or detail if the scenario does not unfold exactly as planned. They may be insufficient to address the loss of critical functions without a degree of complexity. The ASIS Organizational Resilience standard, however, does require organizations to identify potential disruptive events and document how it will restore critical operational continuity and recover normal operations based on these events.

recommendations to prevent, prepare, or mitigate the hazards. This report can be combined with the results of the BIA, especially if the analysis has been completed informally, as is too often the case. The BIA identifies the financial and operational loss of functions over time and establishes the recovery time objectives (RTOs), critical functions, and processes. Business impact must be defined in light of business objectives. See Chapter 15, Business Impact Analysis. Step number 4 above is also often included as part of the BIA.

6.   Identify Critical Functions

The identification of critical functions is a major result of the BIA. Many planners believe it is a waste of time, effort, and resources to include in the plan functions that are not critical to the organization. Equipment and space at a hot site or other alternative location are expensive and limited; therefore, priority is given to the most important functions and employees. Remember that recovery operations are time sensitive. In many cases, there must be a logical sequence (order) of recovery actions, especially on the information technology side. Others argue that if a function is not critical, it should not be a part of the organization in the first place. I believe that every function should have a plan, but not necessarily a seat at the alternate site.

Generally speaking, a critical function can be a process, service, equipment, or duty that would have one of the following impacts on the company if the function were lost or if access to it were denied:

- Affect the financial position of the company
- Have a regulatory impact
- Reduce or destroy public or customer image or confidence or sales

7.   Develop Recovery Strategies

Selecting the best continuity strategy, or group of strategies, is also key to an effective recovery. This is the heart of the continuity planning process. Despite their cost, continuity strategies are often chosen based on desire, what others do, or what seems expedient at the time. The method you use to continue the operation of critical functions must be based on cost versus benefit, technical feasibility, the results of the BIA, RTOs, and the strategic vision of the organization. This is often mixed with the function or process owner's personal resources (relationships with vendors, other business managers, property owners, and educational facilities). The strategy must be realistic and adhere to any assumptions contained in the plan. Strategies should not require employees to radically change their normal work habits, routines, or organizational structure (the requirements of planning under the Incident Command System notwithstanding). They should not require extensive training subsequent to the disaster. During the BIA interviews with managers, discuss the alternatives they believe can be used to continue their operations for the short term and what actions are necessary to reestablish full operations at the present, alternate, or a new permanent location.

Most large organizations use two high-level, organization-wide (or region-wide) continuity strategies: one for information technology (computer systems, software, data infrastructure) and one for alternate work space. These usually involve alternate processing capability and recovery space at a location outside the area subject to the effects of potential disasters (this is where an understanding of the effects of individual hazards becomes important). After a disaster, information technology operations are transferred to an alternate processing site, and selected recovery team members go to the alternate work space location, connect their computers to the alternate processing site, and continue operations until the damaged location is repaired or replaced.

Individual functions, business units, or data processes can have their own strategies in place that may differ from or complement the organization's overall plan. The organization may plan to relocate team members to an out-of-state convention center, but the technical support group may simply transfer their operations temporarily to a third-party call center or have other technical support groups in the company located in other regions take up the slack or send the disaster-affected technical support employees to these locations.

The BIA may indicate that certain business processes cannot be out of service for more than 1 hour (or less) with very little data loss (recovery point objective), but other computer applications can disappear for 1 or 2 days without serious consequence (we will call these applications *Tier 3*). In this case, a *high-availability solution* is devised for the processes with low outage tolerance—for example, the construction of a second data center at a remote location that receives a real-time copy of data as it is processed. If there is a disaster, the second data center becomes the primary. This helps ensure a minimum of data loss. Copies of the Tier 3 applications are stored at the second data center and loaded on the system. Backup tapes from the Tier 3 applications are obtained and shipped to the second data center (if not stored there) and reloaded within its RTO. Processes are assigned to a tier, and different recovery solutions are devised for the groups to align costs with the need. Data from a Tier 2 application could be sent to the alternate data center by a batch method, wherein the backup data are stored until transmitted all at once, at certain time intervals.

High-availability solutions are generally the most expensive. A short RTO and recovery point objective (RPO), or lower outage tolerance, will result in a higher cost of the continuity strategy. Second data centers, as mentioned previously, are generally the best solution if no data loss is an issue, but are usually the most expensive. Third-party hot sites, network storage solutions, load balancing or server clustering (connecting multiple servers that share the processing and storage over a distance), and other information technology strategies are available. If the planner does not have the technical knowledge or if the expertise is not available internally, it is best at this point to bring in consultants to discuss these and other strategies.

During the interviews with managers, discuss options they believe can reestablish temporary (short-term) operations or to enable uninterrupted operations. Also discuss how the managers expect to implement the options and how long it will take. Repeat this process for a long-term outage. Very often the function leader has been through some type of outage or knows someone who has. What did they do to reestablish operations?

Discuss the feasibility of other strategies, and pick the ones that will work best, based on recovery or continuity needs and requirements (for instance, RTO).

Complete a cost/benefit analysis for each strategy, using information from the BIA. If the loss of the function will cost the company $100,000 after 10 days, it makes little sense to spend $300,000 on a strategy that will put it back in operation in 1 day, especially if less expensive strategies exist to resume at least partial operations in, say, 8 days. Once the analysis is completed, select the best recovery strategy.

It is important, of course, that a strategy can be reliably implemented. If the recovery strategy for a West Coast technical support center is to transfer support calls to a center on the East Coast, be sure of the following:

- Its telephone equipment has the capacity to handle the extra volume of calls.
- The East Coast support staff is knowledgeable about the products supported by the West Coast center.
- Provisions are made for the difference in time zones.
- Extra staff is available on the East Coast.
- Manuals and documents are available.

The following list represents a small number of strategies the planner can select to recover critical functions, data, and equipment. It is by no means exhaustive, and each entry should be researched by the planner to determine whether it is the best strategy suited to the situation. During the hazard inspection or BIA, determine to what extent any of these strategies or redundancies are already implemented:

- Hot, cold, and warm sites
- Relocation
- Work at home
- Telecommunications
- Third-party manufacturing
- Purchase of material from competitors
- Data systems
- Revert to manual methods
- Virtual manufacturing
- Workforce management
- Reciprocal agreements
- Equipment rental
- Rescheduling production
- Reallocation of resources
- Service-level or quick-ship agreements

Hot, Cold, and Warm Sites

A hot site is an alternative recovery location prepared ahead of time, in this case with desktops, servers, or a mainframe, and related equipment such as telecommunications. Hot-site vendors exist to provide this service on a first-come, first-served basis. The hot sites typically include a limited number of workstations and both data and voice communications infrastructure, enabling the organization to relocate employees temporarily. Organizations pay a subscription fee to the vendor and, when the hot site is needed, pay an additional "declaration fee"—to declare a disaster and reserve a system and space ahead of other possible claimants. The company brings its latest backup data tapes (has them shipped or electronically transferred to the site), loads its programs, and resumes operations at the hot site. Daily-use fees are usually payable as long as the hot site is occupied.

Because each site is limited in its capacity and as a necessity located a distance away from the area affected by the disaster, employees may be reluctant to travel great distances from their homes to occupy the hot site, especially after a large disaster. Depending on the potential impact of an outage, the use of an internal hot site may be the most practical solution for the rapid recovery or continuity of data systems. Duplicate data centers are maintained in different company locations, with all transactions of the main center immediately mirrored (duplicated or replicated) on the alternate system.

Plans should detail the step-by-step instructions required to transfer operations to the hot site and list the employees who are to occupy the site. Most hot-site vendors have personnel on staff to assist with developing the transfer plan.

A cold site consists of an empty facility or leased space where computer hardware, telecommunications, and furniture would be delivered to construct a temporary processing capability. At a cold site, nothing is prewired or ready for immediate operation. Obviously, this is a less expensive strategy, but because of the time required for setup, it may not be a practical solution. Something in between a hot and cold site is a warm site.

Relocation

Another common strategy is to simply relocate from one part of a damaged building or site to another. Executive suites, hotel rooms, convention centers, client and vendor offices, empty warehouses, or mobile home trailers are other options to consider to relocate some or all of your business functions. The use of circus-type tents is generally not a good strategy.

Work at Home

Many employees, given the proper resources ahead of time, can work effectively at home. This may free office or work space for those who can't.

Telecommunications

Even without widespread damage, phone systems, including the cellular system, can become overloaded and inoperable. Many of the strategies used to recover data systems are also used for telecommunications. These include emergency service and replacement agreements, divergent routing, radio systems (radio frequency and microwave), mobile switches, third-party call centers, and hot sites. Most switch vendors offer emergency service and replacement agreements. Divergent routing should be examined closely with a carrier representative. Cellular telephones add some degree of redundancy for low volume or emergency calls. Mobile satellite transmission can be used as a backup or as a form of diverse routing. Microwave transmission is a method to add redundancy to connections between buildings of a campus or across town. The simplest way to ensure the continuity of inbound communications is to transfer all calls to another company location if the equipment can accommodate the extra volume and if there are a sufficient number of knowledgeable operators to answer the calls. Commercial call centers are available to handle overflow traffic or to act as a substitute for your operators. Most call centers operate like data center hot sites, with similar fee structures. Their operators can take messages, forward calls, explain the situation, or if qualified, take orders and answer technical questions.

Third-Party Manufacturing

Many firms in the United States and Japan were directly affected by the Kobe earthquake because their only sources of raw materials or parts were from that region. Inability to get parts required these companies to reduce production, find and qualify alternative suppliers, order new parts at inflated prices, and suffer through delayed delivery schedules. The solution: Identify sole-source suppliers and take action to find alternatives far ahead of such problems. If the operation uses "just in time" manufacturing, arrange to warehouse a sufficient quantity of material to allow for delay caused by a disaster or contingent interruption. Some distributors will warehouse materials at your location, retaining ownership until the material is removed and used.

Another concern is the loss of manufacturing equipment, facilities, or personnel due to labor action, inclement weather conditions, or natural disasters. If it is not feasible to transfer operations to other locations within the company, make arrangements with a contract manufacturing firm to produce or assemble your product. As with reciprocal agreements, make these arrangements ahead of time and forward all production change diagrams to the vendor as they occur. Consider using these vendors to supply a small portion of your regular production to check quality, reduce ramp-up time, and familiarize the vendor with your operations and expectations.

Purchase of Materials from Competitors

A manufacturing plant recently destroyed by a tornado had no redundant processes or viable alternate sites. Their expected time frame to rebuild was 6 months. The company was certain to lose long-standing customers if delivery schedules were not met. Their solution was simple. They purchased their competitor's product at a slightly higher price, relabeled it, subjected it to stringent quality tests, and shipped it as their product with a note explaining the circumstances. Not one customer was lost.

Data Systems

Data recovery strategies include hot sites, spare or underutilized servers, the use of noncritical servers, duplicate data centers, replacement agreements, and transferring operations to other locations. Data policies and procedures will help to prevent "disasters" caused by users. To recover data systems, identify the critical applications and prioritize the order in which they are restored. If applications or operating systems are dependent on others, restore them first. Once the applications are prioritized, identify where these applications reside. This will tell you which server or system to recover first.

Servers that are on the same network (or can be easily connected) and that have excess capacity can be pressed into service to rescue a server that has failed. Some organizations keep spare, preconfigured servers in storage for immediate replacement if a primary fails. Unfortunately, this is a very costly strategy.

Duplicate systems—capable of processing normal operations, installed within the organization, and used to run test programs or other noncritical processes—can be pressed into service if a main system fails. Few managers, however, can justify the expense of such duplicate systems. The supply of commercial hot sites is limited and could easily be saturated in a regional disaster, leaving the organization without a recovery system or location.

Revert to Manual Methods

More and more functions rely on automated systems to perform their work. When the automated systems fail, businesses can revert to the manual methods used before the system was automated. For example, a mail-order electronics distributor types an order into a form that resides on a server. The server sends the "pick list" to the warehouse, deducts the item from inventory, and sends a report to accounting after it has billed the customer. If the server fails, the person who takes the order fills out a three-page NCR ("no carbon required") form and physically sends a copy to the warehouse, inventory control, and accounting. When the server is repaired, automated methods resume, and temporary employees are brought in to input the NCR forms into the system.

Unfortunately, with high turnover in many organizations, few employees remember how the job was done before automated methods were used. Often "new" manual methods are developed, and recovery teams are trained in their use. Many high-technology companies, however, cannot use manual methods to manufacture their products, and many processes in use today cannot effectively use manual means.

Virtual Manufacturing

Some companies have the ability to use virtual manufacturing if a production line or facility fails. This strategy worked so well in one case that the company didn't bother to restore its assembly line. Agreements, assembly diagrams, and data connections must be established before the disaster.

Workforce Management

Working extra shifts with the existing workforce or with temporary personnel is a simple strategy to recover from a short-term outage, especially when employees are cross-trained to perform a variety of functions. Decide what functions can be suspended and if employees from those functions can be borrowed. Every continuity plan must consider human resources issues. During recovery operations, ensure there is plenty of food, water, comforts, and rest for the recovery teams. Schedule all employees so that they do not work more than a 12-hour shift. Bring in masseuses for the management staff and other team members. Keep psychological counselors who specialize in posttraumatic stress on call and arrange brief 10- to 15-minute individual meetings with all employees. Those in need of additional counseling can be best identified in this manner.

Reciprocal Agreements

Excess capacity at other sites, similar industries, or even competitors can be used to remain in production until damaged facilities are repaired or replaced. The protection of proprietary information, disruption of the host's operations, and fluctuations in the amount of excess capacity can make this a difficult strategy.

Equipment Rental

If equipment is damaged or destroyed, many plans call for their temporary replacement with rentals. List this equipment and its sources in the plan. Whenever possible, have the rental company preconfigure the equipment to your specifications. Remember that other firms may be after the same equipment, so have alternate or out-of-town sources available. Arrange for priority agreements when possible.

Rescheduling Production

A priority task for many companies after a disaster is to determine the expected length of the outage and compare this to remaining capacity, current production schedules, critical deadlines, and pending product releases. Decide whether production schedules should be changed to concentrate on the most critical products or to eliminate others.

Reallocation of Resources

Similar to rescheduling production, firms should reexamine the assumptions, strategies, and critical time frames and compare them to the extent of the disaster. As necessary, reallocate resources among teams, functions, or sites.

Service-Level or Quick-Ship Agreements

The destruction of a building full of desktop computers would represent not only a monetary loss of equipment, work in progress, and possibly the data residing in the computers but also a major delay in recovery—because of the need to purchase, deliver, set up, reconfigure, and reload each computer. Once computers are installed and connected to the network server, reinstallation of the applications can be accomplished somewhat automatically. But even this, if it is possible, could require a lot of time. If the loss of the equipment is the result of an area-wide disaster, you will, as stated before, be competing with other large companies for replacement equipment. To avoid these delays, you can enter into agreements with computer manufacturers or third-party suppliers to deliver large numbers of preconfigured computers within 24 hours to your primary or alternative location. Your applications, configured to your environment, are installed by the vendor before shipment, saving you valuable time and resources.

8.   Set Up Recovery Teams

Some organizations rely on a single team of executives and key employees to direct recovery operations after a disaster. Sometimes referred to as the crisis management team, it decides what individuals or departments within the organization will do to effect the recovery. Their decisions may be based on detailed preplanning or on a loose set of recovery or continuity strategies.

A more effective method involves the formation of individual recovery and continuity teams arranged along departmental lines or drawn from several departments with similar functions (and therefore with similar recovery strategies). Large departments or teams may contain support teams, or subteams, that focus on particular functions or resources.

Each team is composed of a leader, an alternate leader, and essential personnel. The reporting hierarchy extends to a management team through the continuity coordinator (Figure 16-1).

This structure allows for a response that is selective (not all teams need to be activated in every recovery situation), coordinated (information flows efficiently up, down, and across the recovery organization), and focused.

A corporate team may exist in organizations with many divisions or multiple geographical locations. The corporate team is responsible for making strategic business decisions and will direct the recovery process on a regional basis. Team members will include top management and the business continuity coordinator. Other responsibilities may include the following:

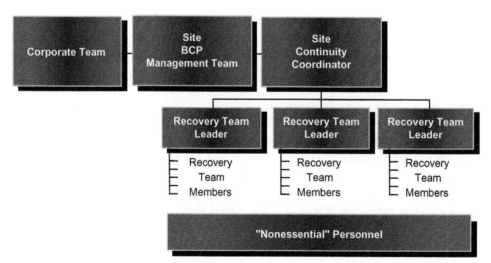

FIGURE 16-1 Continuity team structure.

- The safety of all personnel
- Assisting the site disaster recovery management team to decide whether an alternative work site is required
- Projecting and tracking the financial impact of the disaster
- Determining the need to review the strategic position of the company based on any change or expected change in financial position, production capacity, corporate image, or sales
- Working with the public relations director to develop messages and positions, and communicating the necessary management decisions to the public relations team— that is, activating the crisis management plan
- Resolving conflicts with the allocation of resource requirements among multiple sites affected by the disaster
- Ensuring that insurance claims are filed in a timely manner
- Monitoring the recovery operations and recovery expenses
- Keeping the board of directors updated on the position of the company and on the progress of the recovery operations
- Monitoring and assisting the site disaster recovery management teams, local recovery coordinators, or facilities teams with building restoration, relocation, and the acquisition of temporary or permanent replacement facilities

The site business continuity management team is responsible for the coordination of the continuity efforts of the local teams. In smaller organizations, it may assume many of the duties of the corporate team.

One member of the disaster recovery management team is designated as the continuity coordinator. The continuity coordinator is responsible for the overall operation of the recovery. The continuity coordinator activates teams as necessary if they have not self-activated and acts as the liaison among team leaders, the disaster recovery management team, and the corporate team.

The team leaders activate their plans and notify their team members. They are responsible for overseeing the implementation of their teams' recovery instructions. The qualities of a good team leader include the ability to take charge in an emergency situation, familiarity with the operations of the functions to be restored, and freedom from other significant recovery duties that may interrupt focus. For political reasons, the department manager is most often selected as the team leader.

Only employees "critical" to an operation are selected for the recovery team. "Nonessential" personnel are assigned to other duties or to other teams as needed, or are temporarily furloughed. In today's business environment, where staffing is lean, fewer employees (and business functions) are considered nonessential. The duties imposed on the team members by the recovery instructions must closely match the members' normal skills and scopes of responsibility. If members are to perform special functions outside their normal duties, they should receive continual training in these new skills beforehand.

Steps 9 to 11

Guidelines to develop team recovery instructions, collect resource information, and document the plan are included in Chapter 17, Plan Documentation. After these tasks are completed, it is time to train team members, validate the plan, and keep it up-to-date.

12.   Train Recovery Teams

All employees are trained in some aspect of the plan, even if it is to simply make them aware of its existence. The planning process should accomplish most of the orientation and training required to implement the plan. Those with an active role in the recovery should understand all aspects of their duties and all components of the plan. This includes the methods the organization and employees will use to communicate with each other and their responsibilities after a disaster. The importance of record keeping, lines of authority, and team structure must be emphasized. It bears repeating that if the duties of team members differ from their normal responsibilities or organizational structure, they should be well versed in these new skills beforehand, since the stress produced by postdisaster conditions may make even their normal tasks difficult. This is not the time to get to know a new "boss" or leader. An orientation is a good way to introduce the plan to the general employee population. This can be combined with home disaster preparedness training and disaster fairs. Workshops, videotape presentations, interactive CD-ROM, and intranet pages are effective for training targeted to specific levels of plan responsibility.

13.   Exercise the Plan

No plan is complete until every element has been subjected to some type of testing, exercise, or simulation. Simulating the plan will validate the effectiveness of strategies, ensure the accuracy of information, and increase the preparedness of the individuals who will execute the plan. It will pinpoint areas that need attention or improvement and reveal gaps in instructions, misplaced or absent assumptions, or the need for better strategies and tasks.

The term *testing* is no longer used because it connotes a pass/fail mentality; most planners believe their overall efforts are best served by promoting a positive outcome, and therefore they offer better motivation for the participants. Simulation is probably a better term because to many of us, exercise is not something done willingly. Still, some old-timers like the term *testing* because it places some degree of stress on the participants, thus creating a more realistic situation.

Many publications exist that guide the reader through the steps of a tabletop, departmental, functional, and full-scale simulation. A complete treatment of exercise planning is beyond the scope of this book because the planning for a full-scale exercise involves (or should involve) a significant effort. All portions of the plan must be simulated and contact information verified. A simulation must be planned in detail if it is to be effective and not disrupt normal operations. Begin simply by simulating individual business unit plans. Simulating the entire plan at once is a major event that can take 6 months to set up. Starting out with a small piece maximizes the chance for its success. Later, as your exercise program matures, include city, county, or state agencies in a full-scale simulation. At a minimum, plans should be simulated annually. Three levels of exercises are tabletop simulations, functional exercises, and full-scale exercises.

A basic or first-level exercise is a "walk-through" or tabletop simulation. A tabletop simulation is discussion with the continuity team members while executing their plan based on a scenario. The team is given the scenario and asked to apply or walk through their plan tasks. The continuity coordinator or manager can act as the controller who both runs the exercise and keeps it moving. The controller can change the parameters of the scenario as needed but should not do so unless he or she sees the need or advantage to go in a different direction to stimulate discussion or revise assumptions or resources. The controller, at the beginning of the exercise, makes it clear to the team members their purpose, the rules of the exercise (usually none), and the fact they are not there to find fault but rather to validate their strategies and tasks. Someone should be assigned responsibility to record the minutes of the exercise, noting what went correctly, what did not, and what action items result. Ideally, the scribe should not be part of the team. As the controller, be prepared if this is the first time the team has actually read the plan.

Scenarios must be believable, realistic, and relevant. An asteroid striking the Houston Astrodome will put many to sleep (unless you are a San Francisco Giants baseball fan). The scenario must present a situation that causes an outage that tests tasks and resources and that satisfies the objectives of the test. Present the scenario and let the team leader execute their plan.

A higher-level or next-step tabletop simulation can involve an exercise with one or more dependent teams to test their ability to coordinate tasks, provide input or output to each other, or uphold communications. Tabletop exercises are also used to test security plans and procedures and are more appropriately (usually) threat or event based.

Another simple but useful exercise is a "call-tree" simulation, whereby team members are contacted by the various means outlined in the plan. If an automated method is used, each team member is contacted and told to report a special code number at a convenient location the next morning to receive coffee and pastries or the like. Those who did not receive notification and those who don't report are queried about the reason.

Yet another exercise involves traveling to the alternate work site, especially if it is within driving distance, to familiarize team members with its location, main and alternate routes, and check-in procedures. Surprisingly, some team members may have difficulty finding the location even if just 30 miles away!

The relocation exercise can be combined with a "functional exercise" that actually tests the ability to physically accomplish a task. Many team members are issued laptop computers to log onto a different data center from the alternate work site. After the team relocates, members are required to log onto the site. A functional test can also include switching data processes to an alternate data site, testing communications, or similar activity.

A full-scale exercise can include actually switching data systems or restoring data at a hot site and alternate work site. Other types of full-scale exercises can include drills with city, state, or county emergency services or the activation of the Emergency Operations Center and the management team. The scripts are more detailed and complicated, with multiple simulators, controllers and observers involved. These positions involve the development of timed events lists, guidance documents, and training. Whenever the management team is asked to participate, it must be well conceived and executed.

To conduct a basic exercise:

1. Decide what to simulate and the scope.
2. Select objectives.
3. Develop a realistic scenario to test objectives.
4. Decide how to introduce the scenario to the participants. This can include a simulated newspaper article, prerecorded televised evening news broadcasts, or a tanker truck with a simulated spill of its contents.
5. Select test controllers and observers, if applicable.
6. Compose a timeline of events with their associated expected response objectives. This can be used to audit and evaluate the exercise. Write scripts for simulation volunteers and messages to inject events or new situations into the simulation. These steps apply mostly to a full-scale exercise.
7. Distribute memos to participants describing who, what, when, and where of the exercise.
8. Conduct the simulation.
9. Discuss the results and prepare a report.
10. Revise the plan based on the lessons learned during the exercise.

Obviously this is an abridged list of all steps necessary to conduct an exercise, especially as they increase in complexity (it does not speak to arranging necessary resources, for example). Limit the number of major objectives to no more than six. This will help keep the goals focused and uncomplicated. A secondary goal of a simulation is to enhance training. This is more effective when the objectives are limited.

14. Maintain the Plan

These plans must be "living documents." Employee and vendor contact numbers change often and must be kept current in the plan. This information must be examined quarterly. The plans must be reviewed annually to determine whether they still match the overall strategic direction of the organization, and be changed accordingly.

In large organizations, the team leaders are generally required to keep their plans up-to-date, with their efforts audited by the recovery coordinator or by internal audit. The results of the audit—that is, the freshness of their plan—can be reported in score card format to upper management. Many organizations try to interface their business continuity employee database with that of Human Resources so that any changes made in one update the other. Termination of an employee on a continuity team causes a flag that a team member needs to be replaced. Others are allowed to update their own information on line, if the planning software or database allows.

Keeping the information in the plans up to date is important, but secondary to the need to establish an ongoing business continuity planning management system. We repeat: Business continuity planning, as well as comprehensive emergency management, is not a "project" but a continuous progression of continual improvement and refinement, program maturity, and the application of sound management practices, led by a competent manager. A consultant who is retained to develop a plan and not a program, as is often the case, is underutilized, paid for by funds that are likely misdirected.

Summary

Business continuity planning is a management process that identifies a company's critical functions, develops cost-effective strategies to maintain or to recover those functions if they are lost or if access to them is denied, and lists the instructions and resources necessary to implement the strategies. Systems, applications, products, and processes are prioritized and recovered in a logical manner that will allow the firm to remain in business and to retain or gain market share over competitors that don't have a continuity capability.

It is in the planning process itself that the true value of continuity planning is realized. Although some planning methodologies are more effective than others, each must be adapted to fit the corporate environment and culture. In most companies, the planner will not be successful without the committed support of top management.

Business continuity consultants can be used to facilitate the project and advise the company on the best recovery strategies. The more common strategies include data backup; prioritization of systems, applications, functions, and equipment; and transfer of these operations to an alternate location. Strategies and plans must then be devised to continue operations and to repair, rebuild, or relocate the business. Do not forget to devise plans to move out of the alternate location and back into a permanent facility. Information and strategies must be validated and this information maintained through a properly managed program that focuses on constant improvement and program maturity.

17 ▪▪▪

Plan Documentation

The planning process delivers value to the organization by forcing its members to identify critical functions, place a value on the loss of these functions, and to think through the most effective recovery strategies. The importance of documenting these efforts is fundamental to the Business Continuity process. The lack of a plan will add delays to the organization's fight to recover, cause the misuse of scarce resources, and may ultimately set the stage for the failure of the business.

—**Eugene Tucker, CPP, CFE, CBCP, CHST**

A business continuity plan completes the definition of continuity planning by scripting the instructions to continuity team members so they may implement strategies and list the people and resources necessary for the team to accomplish their tasks.

Required Elements of the Plan

Business continuity planning can take many forms, depending on the type of organization, the applicability of regulations, standards, the organization's culture and document control procedures, and its level of sophistication (or limitations) in the use of technology (such as having interactive plans on smart phones, or web-based plans). Absent regulatory issues, the best format to use is one that the planner is comfortable with or the organization is used to and that satisfies the following conditions:

- The plan must be organized in a logical sequence.
- It must be "clean," simple, and easy to follow. Information that is difficult to find under stressful circumstances is useless.
- It must be complete but not overly detailed. It must contain sufficient information to allow someone who did not participate in the planning process to understand what is required to recover the business or to continue its operations with little or no disruption. It should not contain information irrelevant to the task at hand, such as hazard identification, mitigation recommendations, or justifications for the program, although this admonition may be contrary to the requirements listed in NFPA 1600 and the ANSI/ASIS SPC.1-2009 Organizational Resilience Standard.
- It must contain a glossary defining any terms used.
- It must assign responsibility for planning to individuals within the organization and describe the emergency lines of authority.
- It must outline specific resources and tasks required to carry out recovery or continuity operations.

- It must be flexible enough to address unforeseen events. No two disasters are exactly the same; neither are their response and recovery. According to FEMA's CPG 101 Comprehensive Preparedness Guide: Developing and Maintaining State, Territorial, Tribal, and Local Government Emergency Plans (March 2009), "No planner can anticipate every scenario or foresee every outcome. Planners measure a plan's quality by its effectiveness when used to address unforeseen events, not by the fact that responders executed it as scripted."
- The plan must allow for midcourse correction and adaptation.
- It must contain references to other plans or documents. Although all the information necessary to the continuity effort must reside in the plan, the plan should not contain a restatement of detailed operating procedures or instructions, such as a multivolume operating system installation set. Necessary documents, references, and other plans that are not immediately required and cannot be included in the plan should be listed in the vital records section. These must be duplicated and stored off-site for later retrieval. Note: In electronic-based plans, these documents, if not too large, can be included as document attachments.
- It must state assumptions upon which the plan is based or by which it is constrained.

Multihazard Functional Planning

Multihazard functional planning is a format the Federal Emergency Management Agency (FEMA) suggested that governmental agencies use to develop their emergency operations plans. An effective business continuity plan follows the multihazard functional format, especially if a recovery team methodology is used. Known as "all-hazard emergency operations planning,"[1] it is based on the premise that although the causes of emergencies and disasters vary, almost three-fourths of them produce common response requirements. The jurisdictions can then develop task-based plans around these requirements or functions rather than around each anticipated hazard. Some hazards do produce unique needs; these needs, requirements, and responses are appended to the "basic plan," under this methodology.

The main part of the plan—known as the *basic plan*—outlines the overall emergency organization and its policies, assumptions, activation, and lines of authority. Its primary audience is the jurisdiction's executive- and management-level staff.

Functional annexes are subplans that focus on specific functions the jurisdiction will perform in response to the disaster. Shelter management, evacuation, and search and rescue are examples. Each annex may contain its own appendix of tasks and

---

[1] In March 2009, FEMA revised and replaced its *Guide for All Hazard Emergency Operations Planning, SLG 101* with *CPG 101*. This revised guide, "while maintain[ing] its link to the past, it also reflects the changed reality of the current operational planning environment."

requirements for dealing with the specifics of a particular hazard. Standard operating procedures (SOPs) and checklists may be included in the annex.

Plan Organization and Structure

Like the multihazard functional plan, the business continuity plan consists of a basic plan and departmental or team plans (similar to annexes, but not necessarily organized in the same manner). The basic plan contains administrative and descriptive details required to implement the plan, as well as information that is common to the recovery or continuity effort of two or more (or all) teams.

The basic plan may contain the following sections:

- Table of contents
- Policy
- Scope
- Objectives
- Assumptions
- Activation procedures and authority
- Emergency telephone numbers
- Alternate locations and space allocations
- Recovery priorities or recovery time objectives (RTOs)
- Pertinent information for most or all teams
- Plan distribution
- Training
- Exercising
- Plan maintenance
- Confidentiality
- Appendix
- Team recovery plans

ASIS and NFPA mandate a number of procedures for response and recovery that include communication, short- and long-term performance objectives, crisis and incident management, and many others from both a high level and a detailed perspective. The organization must establish and maintain a procedure to comply with applicable legislation, policies, regulatory requirements, and directives along with a strategy to keep the procedures up to date. NFPA also requires the establishment of financial and administrative procedures (financial framework) to support the program before, during, and after an incident that is uniquely linked to response, continuity, and recovery operations. This includes procedures for the request and treatment of funds and for the capturing of financial data for subsequent cost recovery (a wise procedure to follow). How these are documented are not exactly specified, but most should not reside in the business continuity plan itself. NFPA's records and document management program lists a number of elements that may not match the organization's program.

Table of Contents

The plan should contain a table of contents after its title page. If the plan is electronic, it is useful to include document links to each subject heading.

Policy

The plan can briefly outline or reference the existence of management's policy to develop, exercise, and maintain the plan. Some planners include a description of the responsibilities of separate divisions, sites, or departments for developing their own plans. The policy should mandate the planning process and business continuity program management, including important elements like an annual business impact analysis, exercises, and maintenance. It must assign responsibilities and allocate budget. It should tie business continuity planning performance to performance goals and objectives and bonuses. See Chapter 16 for additional information.

Scope

The dimension of the recovery process encompassed by the plan is briefly but completely discussed. If the plan pertains to a single building or site, refer to it by building numbers, site name, and exact street address. Inform the reader of any pertinent functions, locations, or contingencies not included in the plan. If the plan includes crisis management and response issues, bring them to the attention of the reader.

Objectives

What, in general terms, will the plan accomplish? Objectives can include the following:

- Ensuring the safety of employees, the environment, and assets
- Minimizing economic losses resulting from interruptions to business functions
- Providing a plan of action for an orderly recovery or continuity of business operations

Assumptions

List any assumptions upon which the plan is based or by which it is limited. It is nearly impossible to plan for the absolute worst-case scenario, in which everything ceases to exist. Most of us now know that California will not fall into the ocean (or, worse, see all its residents move east). Common assumptions include the following:

- Buildings will be either partially or totally damaged or inaccessible.
- Most key personnel identified in the plan are available following a disaster. (This is becoming less of a valid assumption as was demonstrated when air traffic was shut down post–September 11. Pandemic planning will carry just the opposite assumption.)
- Alternate facilities identified in the plan are available for use in a disaster.
- Backup data and valuable papers located in off-site storage will be readily available.

- Critical resources will be available.
- Most employees are trained in facility evacuation and relocation procedures.

Including the above assumptions in the plan does not mean that strategies to mitigate their occurrence are not designed and implemented.

Activation Procedures and Authority

List the people or circumstances who have the authority to activate the entire plan, individual team plans, or multiple team plans. Typically, any member of the management team or the recovery coordinator can activate the entire or any portion of the plan, whereas team leaders can independently activate their individual teams, with notification to the recovery coordinator.

Some plans allow for a graduated activation (level I, II, III); some, in an effort to get to the head of the line when competing for space at a hot site, will let the type of disaster dictate how the plan is activated. For example, in a localized disaster, such as a fire in the data center, the team will first assess damage before declaring a disaster. However, if the disaster is regional, such as an earthquake, and the data center is damaged, the Activation Authority will immediately declare a disaster to the hot site, pay the declaration fee, and then assess the damage and only then decide whether there is a need to relocate to the hot site.

Disaster levels can be defined as follows:

Level I disaster. A level I disaster is one resulting in facility inaccessibility or loss of power or other critical services for an expected period of up to 48 hours (this time factor can change according to the recovery time objective). Damage from a level I disaster is not large in scale. It may consist of minor damage to one or more buildings, lack of access due to weather or city infrastructure conditions, or hardware and software problems.

Level II disaster. A level II disaster exists when the outage is expected to last 2 to 5 business days. Damage from a level II disaster is more serious than level I damage and may result in heavier losses to equipment and documentation (files, reports, contracts) due to a prolonged event, such as a fire or flooding.

Level III disaster. A level III disaster is one in which the disaster results in outage anticipated to last in excess of 5 days. A level III disaster is severe and could include the total destruction of one or more buildings, or service within the buildings, requiring significant facility restoration or replacement.

Data-centric or information systems team plans may use a different activation sequence, each of which may involve a specific response by a different group of people. If a disruption is expected to last less than two hours, a "stage 1" situation is declared, the team leader or shift supervisor is notified, and instructions are implemented by the on-duty personnel. If the outage is not resolved within this time frame, or if it is apparent that more time is required, the response will escalate to stage 2 or 3. This will invoke an additional set of instructions, notifications, or full recovery team activation. Organizations must develop activation sequences that make sense for their situation.

Emergency Telephone Numbers

Include a section in the basic plan for emergency telephone numbers that may be common to all teams. This will normally include police, fire, and ambulance (paramedics). List both the emergency number (i.e., 911 in the United States, 999 in the United Kingdom and many other countries, or 000 in Australia) and a secondary number that connects directly to their dispatch center. Radio and television station numbers and call signs, poison control, local medical clinics and hospitals, utilities, and transportation contact numbers can also reside here.

Alternate Locations and Allocations

Virtually all recovery plans include an alternate location to which the team can transfer its operations to maintain its critical functions or resume the delivery of service. Common alternate locations include the following:

- Hot or cold site, hosted alternate site
- Vacant or shared space in another portion of the building, site, or corporation
- Hotel conference rooms or convention centers
- Mobile trailers
- Third-party contract manufacturers
- Training centers
- Vendors and supplier's facilities
- Competitors
- Work from home

Although the company may have an overall relocation strategy, certain teams may choose to relocate to different locations. All alternative locations should be listed in this section.

Allocation refers to the number of team members that will relocate. Not all employees may be needed during the initial phases of a recovery. The allocation will indicate how many employees or team members are needed so resources and space can be anticipated. Some planners include the total head count and the short-term and long-term space requirements (square footage) in this section, but this may be best left in the facilities, real estate, or other team plan.

Hot-site or relocation floor plans to expedite the setup of workstations, communications, and systems should be located elsewhere in the plan.

Recovery Priorities or Recovery Time Objectives

The recovery priorities section is a listing, usually in tabular form, of the organization's departments, teams, or critical functions in order of importance as determined by the business impact analysis. The RTOs, as well as any times of the year (quarter, month, or other business cycle) during which the loss of these functions is especially critical, are also listed. This information is useful to the management team if a decision to reallocate resources becomes necessary.

Pertinent Information

This section may contain other information that most, if not all, teams need to know that is specific to the recovery or continuity operation. Describe emergency financial/purchasing procedures. In a postdisaster situation, it may be necessary to increase signature authorities, streamline purchasing procedures and approval levels, or other strategies to expedite the replacement of equipment, materials, or services. These changes should be agreed to beforehand and clearly explained in the plan.

Plan Distribution

List the plan recipients in this section. If the organization maintains document control procedures, they should apply to the business continuity plan as well. Initial publication and revision dates are important to avoid confusion and to help ensure that plan recipients have the latest version. Large organizations will not list in the plan every person who is entitled to receive a copy. Senior members may be listed by name, and the remainder by title or generic description such as "team leaders" and "team members."

As mentioned in other locations of this publication, the entire plan is not distributed to every continuity team member. This is appropriate only for the continuity coordinator, plan administrator, off-site storage, and the Emergency Operations Center. Of course, there may be others who need the "phone book" version of the plan, but team members must receive only the information they will need in a continuity or recovery situation. This will help them to remain focused in their duties, locate important information in a stressful situation, and maintain security of the information contained in the plan. This is accomplished by distributing only a copy of the basic plan and the individual team plan to the team members. Some planners argue that the team members should receive only their team plan, excluding the basic plan.

Paper plans are useful and necessary; when all else fails, the ultimate business continuity planning tool, the pencil and eraser, becomes the base element. However, paper plans can become cumbersome, especially because as updates are issued, the outdated plan should be collected and destroyed. In today's electronic age, plans can (and should) be converted to some type of portable data file for use on a laptop computer, smart phone, notebook, Internet, or intranet. Many of the business continuity plan development software programs have the capability to publish individual plans on the World Wide Web or intranet, or can convert their output to a Word file, Adobe Acrobat file, or other type of format. Large organizations will control plan distribution, especially if an enterprise-wide database or BCP software program exists, through team leaders. The possession of the latest version of the plan can become an audit item.

Team members will maintain paper and electronic copies of the plan with at least one copy off-site at all times. All team members with laptop systems should ensure that their team plans are regularly replicated to their system.

Training

The training and orientation requirements should be briefly referenced in the basic plan.

Exercising

The exercise program requirements should also be referenced in the plan and should include generic schedules such as "all Tier 1 functions (functions with an RTO of less than 24 hours) will conduct at least one tabletop exercise and one full-scale exercise per year." Exercise scenarios and other related documents should reside elsewhere.

Plan Maintenance

Like training and exercising, the people (generally) responsible for maintaining the plans are identified, along with the method and frequency of updates.

Confidentiality

The basic plan should contain a statement outlining the confidential nature of the information contained in the document. Training and orientations will also mention the confidential nature of the business continuity process and its related documents. Security of the plan should be set up in such a manner that information is restricted to only those persons with a need to know.

Appendix

The appendix will contain the glossary, forms common to most teams, supporting information, and common documents. Team continuity plans can also contain their own appendices.

Team Recovery Plans

Business continuity plans, especially those for large organizations, can be many pages long, rivaling the size of most large metropolitan area phone books. The mere size of these plans will cause most people to place them forever on their bookshelves, to be read only by an auditor. It is best to distribute to each individual recovery team only the portion of the plan that is meaningful to it.

The recovery team plan should contain at least the following information:

- Review of pertinent information from the basic plan
- Overview of the team plan
- Team member contact list
- Scripted continuity instructions
- Resource listings
- Blank forms, contracts, and other documents

Information from the Basic Plan

Activation procedures, alternate workspace locations, communications, organization, and structure may be summarized in the team plan. The duties and responsibilities of the team leader and how the team relates to the recovery or continuity operations are discussed. This will allow the team plans to function independently of the basic plan, keeping the information focused and the size of the plan manageable. Some planners believe this is not necessary or desirable.

Overview of the Team Plan

A review of the general business functions of the department or team, along with its critical functions and continuity or relocation strategies, are listed here. Briefly include other relevant information that would be useful for a team member or outsider to gain an understanding of the continuity or recovery objectives.

Team Member Contact List

A listing of the team members is given, detailing their titles or positions. Home phone numbers, office phone numbers, personal or company cellular phone numbers, pagers, and other alternate phone numbers of team members must be completely listed in the team plan. Ideally, home addresses are also listed because this can give the Emergency Operations Center the ability to make personal contact if necessary, conduct zip code sorts to understand the impact on employee populations, and so on. These numbers should be treated as confidential. Call trees, if used, are placed in this section. It is possible that all team members will not be needed during the early stages of the recovery. Indicate at what phase or day they should be called. Regulations in some countries severely restrict the publication of home addresses and telephone numbers. Restricting this information to only team members may help, but as much contact information as possible should be included. Many companies establish a toll-free number and team voicemail boxes at off-site (distant) locations for team members to call to receive information or instructions. The ability to notify and communicate with team members (and with the Emergency Operations Center) should have as much divergence as possible.

Scripted Continuity Instructions

Include the instructions or tasks the team must follow to implement its continuity (recovery) strategies. When scripting continuity instructions, bear in mind the fundamental difference between a city, county, or state preparing for the response to a regional disaster and a business organization's response to the same situation. The governmental jurisdictions will tend to approach their response from a very centralized point of view, controlling or making decisions in the Emergency Operations Center (EOC). This is necessitated in part because, in a large community or geographical region, it is almost impossible to anticipate response needs and resources at a micro level. A business organization, however, presents a more controlled, foreseeable environment, allowing a more decentralized, scripted approach. This allows the organization to make decisions beforehand based on predicted situations or potential conditions that streamline its response and allow decision makers in the organization's EOC to concentrate their efforts on situations that were not anticipated. In other words, decentralize tasks to the continuity or recovery teams. Plans should permit the management team or team leaders to adapt recovery instructions as required to meet unique conditions as they occur. Although the instructions should not be changed without justification and knowledge of why they were created, assumptions and strategies may change sufficiently, since the plan was developed to necessitate adaptation during the recovery.

Separate instructions can exist for the team leader and team members; they can be further separated by time (initial 24 hours, initial 48 hours, more than 5 days), by phase (response, restoration, recovery), or by whatever period or method is most practical. As stated before, these instructions should be brief but complete. They should be understood by someone reasonably familiar with the type of operation the team is attempting to recover; avoid, however, technical terms, abbreviations, and acronyms. The instructions should minimize the need for decision making after the disaster.

Resource Listings

Resources listed in the plan must be sufficient to implement each team's continuity strategies. This is usually validated through plan exercises and scenario planning. Unless provided at a relocation site or taken off-site each day, assume nothing is available to the team to continue their operations. In other words, if the building burned down while you and your team were at lunch, what resources (equipment, forms and supplies, contact phone numbers, software, and vital records) would be needed? If the team members are required to travel to alternate locations and purchase equipment, are they issued corporate credit cards, laptop systems, and communication devices? Each team must list resources that include equipment requirements with a complete description (i.e., make, model, etc.); forms, supplies, softest requirements, and vital record needs. Answer the question: What minimal critical equipment will each team require to function after the disaster? This should include, but not be limited to, PCs (if possible state configuration or hardware requirements), terminals, printers, fax machines, calculators, phones, modems, special equipment, laboratory equipment, production machines, and so on. All forms and supplies regularly used that would be required to conduct business for 5 days (i.e., letterhead, vouchers, diskettes, special forms, etc.) should also be listed in the team plans. Regular office supplies such as paper, pens, staplers, and so forth are not usually included because they should be readily available. List those critical records or reports on which your unit depends or those records that must be available to send to regulators and others. Model numbers, configurations, floor plans, cabling diagrams, tax tables, milestones, and other due dates, along with both the short-term and long-term quantities of items needed, are also listed.

All contact phone numbers (home, cellular, pager, other) of critical vendors are listed in each team section of the plan. List as many contact phone numbers for critical vendors as practical. If possible, one number should be a corporate 800 number located outside the state. Include purchase order numbers, license numbers, and passwords as necessary.

A complete customer list, or only critical customers who need or should be notified of the outage, along with their contact numbers can be included in the plan. Consider calling customers to inform them of a new temporary address or of the firm's continued ability to deliver its services. Consider completing these notifications even if the company is not directly affected by the disaster. Firms in San Francisco lost business after a major Los Angeles earthquake (more than 500 miles away) because the customers did not understand the geographical separation or the nature of earthquakes.

Blank Forms

Forms that are needed by the team and included in the appendix are also found here. Copies of contracts or agreements with vendors may appear in the team plan.

Summary

The business continuity plan should be brief, to the point, but encompass enough information to implement the strategic objectives and resource requirements of the organization and of the individual continuity or recovery teams at a time when stress and physical and mental exhaustion are elevated.

Emerging standards contain many references to the documentation and maintenance of plans, but only two (or perhaps one) documents should be used as tactical, operational documents (with other necessary documents appended or referenced): the "basic" plan and individual recovery or continuity "team" plans. The basic plan contains information necessary and common to all teams. The individual team plans list the tasks and resources team members will use to carry out their continuity strategies. Plans are maintained in a manner where they can be accessed off-site, are flexible, and their security and the confidentiality of their contents respected.

18

Crisis Management Planning for Kidnap, Ransom, and Extortion

Kidnapping, abductions, and hostage situations have been a human practice since the beginning of time. These practices, however, have changed radically in the last 15 years.

—Charles Seoane Norona, CPP, ASIS International Seminar on Kidnapping, Dallas, Texas

Many business executives, continuity planners, and security directors incorrectly believe the term *crisis management* is synonymous with business continuity planning. There are important distinctions between the two, although the methods used to arrive at plans for each are basically the same (see Chapter 17). Planners who miss these distinctions believe their plans will allow them to respond and recover from all major threats to the organization. This mistake can cost the firm millions of dollars in lost market position, diminished reputation, and devaluation of the company. Union Carbide illustrated this point after the value of their stock declined 27 percent immediately after the toxic gas leak in Bhopal, India. In 1989, Perrier was the market leader in bottled mineral water, with sales over 1 billion bottles per year. Their bottling plant in Vergezem, France, was tooled for a projected market growth of 20 percent. One year later, the news media reported that benzene, a chemical linked to cancer, was found in the water. Sales plummeted to 500 million bottles, and market growth was essentially nil. The Vergezem plant could no longer operate profitably, resulting in its acquisition by Nestlé.

The security profession uses the term *crisis management* to include terrorist acts, kidnapping, labor violence, civil disorders, industrial disasters, and natural catastrophes.[1] The insurance industry uses the term to mean the response to (but not the recovery from) an emergency or disaster such as a fire or tornado.[2] To further complicate our understanding of this concept, some business continuity vendors use the term to mean a hot-site recovery operation.

Both a crisis and a disaster will affect the financial position of the company. The goals of their response are essentially the same, but continuity planning is generally focused inward toward mission critical business processes (functions), whereas crisis management is focused outward to manage the image or public perception caused by an incident or disaster. Crisis management planning ventures slightly outside the traditional

---

[1] Fay, J.J., 1987. *Butterworth's Security Dictionary—Terms and Concepts*. Butterworth Heinemann, Woburn, MA.
[2] Head, G.L., editor, 1989. *Essentials of Risk Control, Vol. II*. Insurance Institute of America, Malvern, PA.

role of the security professional by the inclusion of speculative risks in both the analysis of threats and the design and implementation of response and recovery strategies.

Maxialo Ckonjevic defines a crisis as "extreme threats to an organization which have the potential for significant negative results to important organizational values, functions and services and which could result in major damage to the organization, its employees, products, services and reputation."[3]

It is possible to experience a disaster without a crisis. The reverse is also true. The temporary loss of the *Exxon Valdez* after it spilled 250,000 barrels of oil into Prince Edward Sound was not initially a disaster. Oil wells continued to pump, transportation and billing systems functioned, and no refineries went off-line. But because of the manner in which the company responded to the incident, Exxon lost many uninsured dollars in goodwill and future sales and gained closer regulatory scrutiny. A disaster can escalate to a crisis if a business does not recover within its outage tolerance (recovery time objective).

Business continuity focuses on the loss of critical business functions. Crisis management considers loss to its image and goodwill. Major components of crisis management include communication and public information plans. Management is most likely thinking along these terms when defining crisis management.

Crisis planning can conform to the comprehensive emergency management model. Once the threats are identified, preventative measures are developed and implemented. Examples of these measures include tamper-proof product containers, corporate procedures, and controls. Preparedness can take the form of predeveloped statements or press releases, the selection of a company spokesperson, and media training for the management team. Response includes the rapid collection and dissemination of information designed to minimize the negative impact of the incident or to prevent the escalation of a situation to a crisis. Recovery includes strategies to restore public, employee, or shareholder confidence.

This chapter examines the framework of crisis management planning and places it in the perspective of planning for kidnap, extortion, and ransom. The odds of any business organization being the victim of a kidnap, extortion, or ransom demand are slim. Nevertheless, we outline here the steps to be taken by a crisis management team (CMT) because a kidnapping could nonetheless occur at any time. More important, however, the CMT approach can be used to resolve any crisis of magnitude sufficient to threaten the financial existence of a corporation. We go into specific details regarding the threat of kidnap and extortion because we have worked with organizations that, not having plans, had to develop them on an ad hoc basis to deal with these problems. This real-world experience is an example to the reader of how to develop, before it occurs and not after, a planned approach for dealing with such a crisis.

---

[3] Ckonjevic, M., 1997. FBIC, CGCP, Presentation to the "Survive" conference, San Francisco.

Threat Identification

Some companies spend billions of dollars investing in their corporate image through trade and brand-name identification. The value of this identification could represent the company's largest asset. Threats against it must be identified. Threats outlined in the crisis management plan include the following:

- Accusations against, or the arrest of, a company official
- Boycotts
- Demonstrations
- Errors and omissions
- Fines and penalties imposed by regulatory agencies
- Missed production milestones
- Good neighbor policies and community concern
- Hazmat and environmental issues
- Health of key executives
- Hostile takeover attempts
- Human error
- Insider trading
- Kidnap and ransom
- Labor disputes
- Major injuries
- Mismanagement
- Negative research or a death from clinical trials
- Patent infringement
- Product liability
- Product recalls and defects
- Product tampering
- Reorganization
- Rumors
- Terrorism
- Third-party crime (assault, rape, robbery)
- White collar crime
- Workplace violence

These risks are best identified through a risk analysis and crisis scenario planning than with a business impact analysis. Use scenarios to uncover the sources of a crisis. When developing scenarios, look at situations the competition has experienced or is likely to face. Select each anticipated event and subject it to as many "what if" questions as practical, such as, What if an executive is kidnapped domestically, overseas, for ransom, by what group? and so forth. See Chapter 13, Mitigation and Preparedness for a brief explanation of scenario planning.

Government officials and business executives have been attractive targets for kidnappers and extortionists since long before the 1960s. Domestic terrorist activity directed against executives, corporations, and political leaders in Asia, Europe, South America, and recently Mexico and the United States has made the general populace justifiably apprehensive. As in the case of the Hearst family, corporations, their executives, and members of the executives' families have been, and are expected to continue to be, victims of kidnappings and extortion, which could inflict heavy losses on corporations or on the personal well-being and wealth of the victim's family.

There are two basic reasons for executive kidnappings: personal gain and political objectives. These kidnap (extortion) victims generally possess one or more of the common elements of money, power, or high public visibility. Individual chances of being kidnapped are extremely low, but the odds increase rapidly if the potential victim is wealthy; controls large amounts of money; is associated with such cash-driven industries as banking, savings and loans, gambling casinos, or food marketing; or works in industries such as airlines or public utilities. An executive's chances of being kidnapped are further increased if his or her industry is often victimized by terrorists or extortionists, or if his or her company has a history of paying ransom demands.

Plan Documentation

To mount a successful response to kidnappings, extortion, and other threats, a plan to deal with such crises must be formulated in advance. The major responsibility for advance planning belongs to the organization. Organizational planning is more effective than individual effort, and it is more likely to be implemented and thus successful. Therefore, organizations must develop crisis management skills that are adaptable to any demand made on them.

Although a crisis management plan can stand on its own, a business continuity plan that does not include crisis management will generally fall short of the objectives of the organization and emerging standards. Crisis management can exist as a subsection of the business continuity management team plan, as part of the public relations team plan, or as a combination of both, depending on the construction of the document and the size of the organization. Some planners believe the development of separate documents for continuity planning and crisis management is the most effective approach. Others point out that separate plans are cumbersome and require extensive duplication of information and upkeep. The complexity of the risks will guide the planner toward the best document format.

Because extortionists can often inflict heavy losses on organizations, it is imperative that the CMT prepare a readiness plan that will minimize these losses. This plan must fix corporate objectives and limitations, and it must be designed to be effective when the CMT is operating under the emotional strain of responsibility for human life, often with limited data and time for making decisions.

The plan must resolve the fixed elements of a crisis, so as to require the CMT to make only those decisions during a crisis that are affected by immediate variables. Also, it must have sufficient flexibility to enable the CMT to develop alternative strategies after gathering information and analyzing threats under rapidly changing crisis conditions.

In the event of a kidnapping, provisions for gathering personal data, such as employee and family biographical sketches, as well as medical and other requirements of the employee and his or her family, must be incorporated into the plan along with methods to make these data readily available during the crisis period.

The resource section of the plan should include phone numbers and addresses of the team members, major customers, media contacts, brokers, local officials, and regulatory agencies. The plan should include instructions on how to best contact these resources and officials. Ensure that equipment and supplies such as projection machines, sound amplification equipment, battery-driven bullhorns, sign-in sheets, and name tags are available in preparation for any news conferences.

The crisis management plan, like the continuity plan, is a confidential document. It contains important strategic information and phone numbers of key executives. Its distribution must be limited. Both are living documents that must be maintained and simulated. Testing of the plan should include role playing and on-camera interviews. Rehearse responses well in advance of the need.

Plan Activation

Activation of the crisis management plan is a bit more difficult than the continuity plan. An earthquake, fire, or system crash is an obvious sign of trouble, but the signals of an impending crisis are often less obvious. Subtle signals can appear in trade journals, a rise in terrorist threats, or an increase in customer service calls. It is important to distinguish among an issue (a matter of dispute and differing points of view), an incident (a sudden, unexpected occurrence that requires prompt action), and a crisis. A crisis is far more crucial than most issues and incidents—it is a major turning point, resulting in permanent and drastic change to the organization. These events can have an internal or external impact (fire or earthquake), can be sudden or gradual (product sabotage or community pressure after a toxic release), can be short-term or long-term in nature (bomb threat or months-long strike), and can be human in origin (rumors about product safety). The failure to respond at the appropriate level can lead to serious errors, spread alarm, and create a siege mentality. The analysis of the threats should include a list of indicators or symptoms of a potential problem so that the team can evaluate and respond to them as quickly as possible. This response should be proactive; a nonreactive position may place the organization in a defensive position.

Institute a system to monitor emergencies, problems, and controversies. Clipping services that capture news reports about the company from broadcast media, newspapers, and the Internet are a good source of information that can warn of an impending crisis or provide feedback on how the company is perceived after a crisis. Summaries of complaints from customer service or technical support should be tracked for this purpose.

Crisis Management Team

Crisis management can be addressed most advantageously by a CMT. The CMT should consist of a group of senior management personnel who have the authority to make decisions for the entire corporation during a crisis. Because a small unit is generally capable of reaching decisions more quickly, the CMT should consist of the least number of individuals possible. It should also be dynamic in nature and should consist of a team leader, a secretary, and a senior member from the following departments:

- Executive management (chief executive officer)
- Finance (chief financial officer)
- Operations (chief operations officer)
- Risk manager
- Sales and marketing
- Legal and/or regulatory affairs
- Public relations
- Human resources
- Telecommunications
- Security

On a CMT, the chief executive officer can act as the team leader. A secretary or administrative assistant is included to log events, actions, times, and expenditures. As with continuity planning, a complete and accurate record of events will help the company wade through the inevitable legal, regulatory, and insurance problems once the crisis is stabilized. Technical advisors can substitute for the chief operations officer to assist the team with the development of consistent and accurate assessments, projections, and press releases. Technical advisors can include hazardous materials experts, the Federal Bureau of Investigation, and rescue or hostage negotiation teams. A representative from the telecommunications department is included to help the team decide if any extra load on the telephone switchboard can be accommodated by the installation of extra trunk lines, embargoing nonemergency calls, and arranging for additional switchboard operators.

The CMT will collect and evaluate information on the scope of the incident and attempt to prevent its escalation to a crisis. Team members will review strategies designed to mitigate the adverse effects of the incident or crisis and attempt to benefit from any opportunity the situation may present. The repurchase of stock or the suspension of trading, increased advertising, or public service announcements (or the suspension of advertising) and the recall of products are some strategies the team may select. They will look at how organizational dynamics, financial position, and public opinion have changed since any strategies in the plan were first developed.

The CMT is not a substitute for law enforcement or the organization's security department; rather, it is a complementary support organization. Decisions that affect the company directly and require corporate decisions or responses should be handled by the CMT in partnership with law enforcement. For example, a trained hostage negotiator working in conjunction with law enforcement personnel should be designated to act as the intermediary between the extortionist and the victim company. The extortionist must clearly understand that the negotiator has neither the authority nor the capacity to make decisions or commitments on behalf of the company. Used in the proper context, the trained negotiator often provides time for the CMT decision process to work effectively.

Most firms have capable, qualified, and responsible people in their employ who can and should conduct all negotiations in these situations. These people, many of whom are schooled and experienced in industrial relations or similar types of business negotiations, need only receive training in how to negotiate in criminal situations in order to become effective.

The use of outside consultant resources for negotiations is impractical, for several reasons. Most extortion situations are resolved rather quickly. The one exception is a protracted terrorist kidnap, for which publicity is a key ingredient of the case. By the time outside consultants could arrive at the scene and receive an update on the progress of the case, local law enforcement, combined with company executives, would probably have the situation stabilized and in some cases resolved. Coming on the scene 6 to 12 hours after a kidnap-extortion situation begins, the outside consultant will likely have no impact whatsoever. In addition, remember that the basic tenet of risk-control consulting is to make the client as self-sufficient as possible in matters pertaining to security; this includes crisis management programs. The best use of consultants is to have them help the corporation to develop the CMT, and then to have them available for telephone consultation during the crisis.

There are several basic areas of concern to be addressed by the CMT as it provides corporate leadership during a crisis. The protection of assets, which in this case includes personnel, is of primary concern. Experience with a systems approach to assets protection, as well as a knowledge of the types of adversaries encountered, should be provided to the CMT by the organization's security director (consultant), who can also assist by acting as liaison with law enforcement agencies.

The CMT will require the assistance of its legal counsel to examine such issues as employee and stockholder rights in relation to the legal standing of the company regarding various strategies and monetary payments to extortionists. Information from the financial arm of the corporation is needed to develop the monetary base for CMT operations, and its assistance is needed to set the corporate strategy and limitations regarding ransom of any particular corporate employee. The CMT must also consider the long-term effect of crisis decisions on employees of the company.

The CMT must be given autonomous control over decisions the corporation must make during a crisis, consistent with an advance plan approved by the board of directors. Every action to be taken by the company not dependent on the specific nature of the crisis should be rehearsed, much like a fire drill; only the variable decisions will then have to be handled. Even those decisions will be addressed from a perspective of preset goals, limitations, and strategies.

A corporate crisis-management capability will enable professional law enforcement personnel to respond to the crisis with better initial information and a clearcut base from which to operate. Also, this capability will lessen considerably the probability of loss through matters growing out of the original crisis, such as stockholders suits, employee negligence suits, wrongful-death suits, insurance cancellations, and expropriations of assets by foreign governments irritated by the way in which a corporation handled a problem.

Handling the Initial Contact

When an extortionate demand is received, the CMT, the organization's security department, and law enforcement should be advised immediately and the crisis management program put into action. The actions taken during the first crucial moments after an extortionate demand is received may well determine the eventual outcome. Because most threats are transmitted by telephone, recording devices and tracing capabilities should be discussed with the local telephone company. Recording an extortionate call will not only preserve its details for later analysis in decision making but also may provide investigators with background noise and voice-print characteristics, leading to the place of origin of the call and the identity of the extortionist.

People who may receive initial extortionate communications are, in many respects, vital sources of information for the CMT. As such, a training program should be developed to ensure proper implementation of the procedure regarding handling this type of call. Individuals handling such calls should be instructed to remain calm, record or write down all data given by the extortionist, express cooperation, and ask questions to lengthen the time of the call. An attempt should be made to calm the extortionist and secure proof that the hostage is being held and is unharmed. The recipient of the call should attempt to talk with the hostage and to give the hostage the opportunity to relay critical information through a prearranged code. Above all, people receiving the initial call should bargain for time; if possible, they should end the conversation in such a manner that additional contacts with the extortionist will be necessary before a ransom is paid. This allows for the opportunity to trace and record a second call from the extortionist, as well as providing time to implement the crisis management program and setting the stage for a controlled negotiation response.

Many extortion demands are transmitted in the form of a written threat. The letter and its envelope should be protected from unnecessary handling and preserved for fingerprints, handwriting, and printing and typewriting examinations, as appropriate.

Following receipt of a written threat, steps should be taken to identify the source of the document if it was not mailed. It may be necessary to interview all employees immediately in order to develop information leading to the identity of the person who delivered the communication.

During the initial phase of the crisis, it is imperative to determine whether the demand is a hoax. In a kidnap case, the whereabouts of the alleged victim must be established immediately. Employee family biographical fact sheets can be of critical importance at this time. Several notable kidnap hoaxes have involved calls to the executive's family in which the caller pretended to be a telephone company representative. The caller would state that the family telephone was being serviced and request cooperation in not answering the telephone for the next hour. The executive would then be called at work and told his family has been kidnapped. Naturally, when he called home and got no answer, he would panic and comply with the extortion demands, believing that his family has been taken hostage. A family fact sheet containing the telephone number of friends and neighbors who can confirm the whereabouts of the family can be one means of thwarting such a scheme.

Ransom Considerations

Payment of ransom is a decision to be made solely by the corporation or the victim's family. Law enforcement officials will discuss the pros and cons of ransom payment with the top officials of the organization and with the family of the victim. They will not, however, make the final decision as to paying or not.

Policy and the limitations on payment of ransom should be developed by the corporation and approved by the board of directors. This way, directors or executives insulate themselves from civil liability. For example, a shareholder could allege that the executive approving or making the ransom payment had not acted legally, did not have corporate authorization, and therefore was personally liable to the corporation for the amount diverted. If the payment of ransom or any other action taken in response to the extortion demands was itself a violation of local criminal law, the civil liability position could be aggravated. Finally, if the executive approving or making the payment failed to consult other executives or directors but was nonetheless able to obtain the cash or other assets and complete the transaction, shareholders could allege that the other executives and the directors were negligent in failing to consider the possibility of such extortion and in failing to require appropriate controls. In such a case, liability might be alleged against all involved.

Inadequate action or improper action by the corporation leading to death or injury of an employee might result in claims against the corporation for damages by the employee or his family. This is particularly true if there were no contingency plans and the injured employee was exposed chiefly because of his corporate employment.

The above examples are not all-inclusive, and corporate counsel should be consulted in all such matters. This is not to suggest that no action be taken to free a kidnap victim merely because of the potential of civil liability. Instead, it emphasizes that the way to minimize or avoid such liability involves preplanning and prior authority.

It is not possible to fix any categorical limits on the amount that should be paid for the release of a kidnap victim. In most companies, the only likely financial gauge of the impact of the death of an executive or official is the amount of "key man" life insurance contracted by the company. This type of insurance is intended to cover the cost to the firm of replacing a deceased official and the interim expenses or losses likely to result from his or her sudden absence, and it does not address the sensitive area of public and employee attitude toward the company; nonetheless, the amount of such insurance is at least a rough standard that can be used as a first step in considering ransom payment amounts.

Another alternative in this situation is to refuse to pay ransom altogether. It has been suggested that paying a ransom of any type may induce others to try again, that the possible individual loss of life is a necessary cost. This is the position taken by many governments in regard to the kidnap and ransom of government officials, but it may be untenable when applied to private business enterprises. At least within the United States and Canada, the business community and the general public may not accept the position and its potential cost.

If a decision is made to pay a ransom, the net impact on the enterprise may ultimately be much smaller than the amount paid, for two reasons. First, active cooperation with law enforcement from the very beginning will improve the chances of capturing the kidnappers and recovering all or part of the money. Second, commercial insurance can be purchased to cover a portion of the ransom payment actually made.

If the decision has been made to meet ransom demands, law enforcement authorities will assist in preparing the ransom package. Plans must be made for availability of funds in appropriate denominations. It takes considerable time and effort to record currencies used in ransom payments, and this step should be completed as much in advance of the payoff as possible. Large amounts of money, in small denominations, produce surprisingly heavy, bulky packages; $1 million in $10 bills weighs about 225 pounds, for example. This should be kept in mind during negotiations.

Where nonmonetary demands are made, such as supplies, publicity, or chartered aircraft, the responsibility must be the result of thoughtful decision on the part of the CMT. In formulating a policy, the possibilities for ransom should not be limited to money. However, political demands, such as the release of prisoners and the provision of arms, generally cannot be influenced by officials of the kidnapped victim's enterprise.

Preventive Security

Unfortunately, experience has shown that a dedicated group of terrorists (extortionists) can penetrate all but the most sophisticated personal and corporate security systems. However, a company's demonstrated crisis response capability and awareness by executives of personal and corporate security practices will likely decrease the chances that a corporation or its executives will become victims of a kidnap or extortion attempt. The key is to alert the corporation and its executives to the level of danger where the particular executives reside, the need to avoid patterns or routines in personal behavior, and the increased vulnerability of wives and children. These are all important factors in forming an appropriate preventive security plan.

Some of the more obvious suggestions include steps toward ensuring the physical security of executive residences, instructing children on appropriate precautions, limiting the dissemination of personal information to only those deemed in need of such information, and securing automobiles. Additional and more detailed security precautions can be developed through consultation and internal planning.

Suggestions for Kidnapped Individuals

Based on information developed in past cases, it is clear that kidnapped individuals should control their fear and realize that professionals are working for their safe release. Problems should be analyzed and decisions made based on the individual's present condition; a display of anxiety could be contagious and counterproductive. In many instances, a victim can actually control an abductor's actions through his or her dominating personality, leadership qualities, and calm approach to the situation.

If the victim is troublesome or appears to jeopardize the abductors' plan, serious harm to the victim may result. The victim should attempt to convince his captors that his or her well-being is essential to their success. Those working for release will be simultaneously making every effort to convince the abductors that their goals will not succeed under any circumstances unless the victim is set free, alive and unharmed. An attempt should also be made to develop a relationship with the abductors so as to change their perception of the victim from that of an "object" to that of a "person," similar to them. Attempts to cooperate with the abductors should be made with this in mind.

If given a chance to communicate with people working for their release, victims should attempt to give maximum information through prearranged code words, phrases, or verbal mannerisms that have been developed by the CMT. If the victim recognizes the captors or any detail of the kidnapping, it is imperative that this knowledge be kept from his captors because it may cost the victim his or her life.

There are almost no circumstances in which an escape attempt is recommended. The key word, however, is "almost." The possibility of escape should not be considered if the victim is goaded by impatience. Escape attempts should be viewed as a last resort, not a time-saving device. By considering the abduction a long-term venture, the victim will be less tempted by impatience. Escaped victims could find themselves lost in a remote, inaccessible, alien region, without transportation, money, food, water, or shelter, and perhaps unable to speak the local language. If recaptured (and not killed or seriously injured in the process), the victim will likely be treated more harshly. Thus, escape should only be considered as a life-saving effort when success is reasonably certain and the likely alternative is death.

Media Control

The CMT will also review methods to best communicate the position of the company to the public. The public may include the news media, stockholders, employees, customers, potential customers, regulators, and potential victims. The company must present a consistent, coordinated response. Establish media policy and communicate the policy to all employees. Employees should be instructed not to speculate on conditions and to refer all media inquiries to the appropriate spokesperson. The team will also decide what information not to release. Extortion or the kidnapping of an executive is best kept confidential.

The premature release of information could have an overwhelming and demoralizing effect on public or stockholder confidence. A premature "do not use" directive by a water department or a suspected food contamination warning will not only satisfy the goals of a terrorist but also can cause economic impacts that are well documented. A delayed disclosure could cause severe illness or injury, increase civil and regulatory liability, and diminish the credibility of the entity's members. The failure to release any information in some cases could be the best course of action (such as with a bomb threat) but could also have negative consequences that the firm should be prepared to address.

The most important resource for the CMT is accurate information on the scope of the problem. Effective crisis management often means effective information management from the standpoint of both rapid access to data and conditions for the dissemination of information to the public. The manner in which information is controlled, and who is in control of the information, will ultimately shape public perception. Use information to enhance the firm's credibility. Train the management and public relations team members (if your organization is large enough to have a separate public relations team) in on-camera media relations as mentioned earlier in this chapter. Role play as much as possible as a part of this training.

A media relations expert should be available to the CMT for a positive, controlled response to media inquiries during the crisis—because one asset to be protected is the public image of the corporation. In this regard, it is imperative that responses to the media be coordinated with law enforcement officials to avoid premature release of information, which may jeopardize a victim's life.

The Emergency Operations Center or Command Center where the management team meets should be off limits and out of sight of the media. Any sign of stress or confusion should not be shown to the outside. Assign a person to direct visitors and relatives or friends of victims to a quiet room away from the press. Keep press, visitors, and victims apart.

Establish a secure room near the entrance for reporters and equip it with the latest technology used by the press. This can include extra phones, facsimile machines, data ports, satellite hookups, and so forth. Obtain equipment in advance to set up and support logistical needs for the largest number of reporters that can be expected. Assign a public relations representative to the press room to answer questions (within limits) and to make other decisions and arrangements. Establish written ground rules for press briefings. Arrange to have the facilities and telecommunications team set up live and taped information, and provide sign-in sheets and name tags with affiliations. Maintain a log of media contacts, and videotape all media briefings.

Establish a compendia of information before an incident that include the following:

- Binders containing all recent press information, background on the company (its general description, employees, products, income, donations and charities, awards, inspection certificates), media advisories for the event, and special information phone numbers. Distribute these as necessary.
- Produce background ("B-roll") raw video footage about the company and its facilities and products. Make copies available to broadcast media in the event of a disaster. This will help the company control information, help reporters meet deadlines, and possibly prevent news crews from filming events or locations the company may wish not to broadcast.

Depending on the type of the crisis, consider establishing a public fact center, separate from the press office, to disseminate information to customers and the general public. Use this center for rumor control with live and taped telephone response. Update taped information on a separate line.

Summary

Business continuity and crisis management are similar in that they both involve preplanning that minimizes the need for decision making after the plan is activated. Because much confusion exists about what the term *crisis management* implies, the security manager is often responsible for developing this capability simply because management does not know where it belongs in the organization. The risk analysis process is used to identify and categorize the exposures. Operational and media strategies are developed, implemented, tested, and maintained through the use of scenario planning. A primary focus of the CMT is the identification of an impending situation and the control of public perception to prevent an incident from escalating to a crisis. A crisis management plan is developed and formatted in the same manner as a business continuity plan. The plan will allow for a rapid, proactive response. These plans can be combined or can exist as separate documents.

To prevent or minimize the harm that might result from executive kidnappings and other forms of extortion, the business community should recognize the need for and take the necessary steps to develop crisis management plans. The responsibility for developing such plans lies with the corporation itself. It is only through such planning, both internal and in consultation with experts, that the tragedies inherent in such crises may be avoided or minimized. For more detailed information, see Appendix B, Sample Kidnap and Ransom Contingency Plan.

Bibliography

Barton, L., 1993. *Crisis in Organizations: Managing and Communicating in the Heat of Chaos.* Southwestern, Mason, OH.

Ckonjevic, M., 1997. FBIC, CGCP. Presentation to the "Survive" Conference, San Francisco.

19

Monitoring Safeguards

In God we trust; everyone else we monitor.
—Anonymous FBI inspector

The security professional may be asked to review or design the security for an entire facility. If a facility has a security system in place, the security professional may be asked to perform a risk analysis survey to ensure the system is performing adequately or is in need of being updated. The system, upon review, will prove to be either adequate or inadequate in meeting the needs or objectives of the company, which generally can be broadly defined as protecting of the company's assets.

Monitoring or Testing the Existing System

One technique for making an adequacy determination is to monitor or test the existing security program and systems to determine whether they are still doing the job for which they were initially designed. This is especially true for systems that were put in place 15 or 20 years ago and never updated.

We were once assigned the task of evaluating the security system for a research and development (R&D) division of a large computer manufacturing firm. The R&D division, in existence at the same location for 5 years, had taken over facilities previously occupied by the consumer products division, which had been dissolved during their reorganization 5 years earlier.

All the administrative support systems, including security and safety, already at the facility had remained in place to serve the newly expanded R&D effort. In fact, the only significant change at the facility was that instead of manufacturing digital wristwatches and calculators, the facility was now developing artificial intelligence data and advanced computer-aided design information for the U.S. Department of Defense (DOD). Five years later, however, at the time of our survey, the security system was found to be totally inadequate for the company's new mission.

We were asked simply to design a test to answer the question, "How well is the R&D facility protected against industrial espionage?" The answer was obvious even without testing: the R&D programs were completely unprotected! In 5 years of existence at this location, no one had bothered to review or to test the security program and systems to see whether they were still meeting the client's objectives. The client was spending in excess of $750,000 a year to protect products from theft that the company had stopped producing at this location 5 years earlier!

One might question why a multinational corporation with vast financial resources at its command would ever allow a situation such as this to exist. The answer, sad though it may be, is that such situations are not the exception; they are generally the rule. Situations like this are seen many times at many locations. Absent a serious security problem, productivity is generally the most important issue and the one consideration upon which management's attention remains focused. It has been defined as the "knee-jerk reaction" to security problems.

From the standpoint of achieving a functional security system, one that can be counted on to work when it is needed, testing of safeguards (countermeasures) in a production environment is probably one area most likely to be overlooked. This is in sharp contrast to the scientific and engineering fields, in which the need for periodic testing is usually accepted as an article of faith.

The Scientific Method

Most high school students have had an introduction to the "scientific method," the basis for all modern science and technology. The scientific method is, simply stated, a very basic problem-solving approach—namely, the gathering of data to be used to confirm or reject a developed hypothesis.

Few of us will object to the statement, "People and procedures must be tested in a number of scenarios, and testing, to be effective, must be an ongoing process." This same principle, however, is seldom applied in the real world of security. Yet, in no other way than by periodic, programmed testing can the integrity of any system or procedure be ensured or, conversely, can system flaws be detected before the system fails and catastrophe strikes.

Depending on the type of security in question (procedures, hardware, electronics, or manpower), testing can take many forms and have many objectives. Here we are mostly concerned with tests that evaluate performance and reveal weakness, failures, or potential flaws in the design of the system—testing that will uncover problems that otherwise might remain hidden from view. Periodic tests, from a security perspective, are invaluable and should be included in every designed security program and system, no matter how large or small.

Five Basic Types of Testing

There are five basic types of testing, which can be summarized as follows:

- *Functional testing*: determines whether hardware, such as a closed-circuit television camera or an electronics access control system, will do what it was designed to do
- *Safety testing*: determines whether the object or a procedure can be used without causing injury, loss, or harm
- *Performance testing*: normally concerned with conformance to timing, resource usage, or environmental constraints (an example is an anti-intrusion alarm)

- *Stress testing*: checks a person's or an object's tolerance to abuse or misuse under deliberately introduced stress techniques
- *Regression testing*: usually applies to an object, system, or procedure that has been altered to perform a new function and must still perform some of the functions for which it was originally designed

It is well to remember that testing can apply equally to people, systems, procedures, methodologies, and objects. Also, regardless of the application, testing must have a specified objective. From a security viewpoint, it is wise to question, before initiation, what the test objective is and why a particular test is deemed important. Other questions that should be answered include the following: "Are the tests adequate? Are the results valid? Can this type of test uncover a weakness or flaw that might otherwise remain hidden? Is this the best test to use?"

The best time to perform a hardware or electronics equipment test is during design, or at the latest, during installation, before the product or system is accepted and put on line for operational purposes. This facilitates changes, enhancements, and deletions. Efforts should be made to facilitate testing. Many electronic circuits are designed to include self-test circuits, diagnostic lights, fault detectors, and other built-in test aids.

Tests can be broken down into component segments. This makes it possible to test various sections at different times. It is sometimes more desirable to conduct functional tests this way. Performance testing is another area in which it may be desirable to test only parts or segments of the whole system. The more complex the system, the more difficult, time-consuming, and expensive it is to test the complete system at one time. The modular or segment approach can be designed to be reliable and less time-consuming, and from management's standpoint, to have less impact on production schedules.

Depending on the facility involved, tests can be relatively easy, such as verifying and updating biographical data. Tests can also range from the use of "tiger teams" and mock penetrations to that of a complex security environment, such as a DOD facility. Most tests fall somewhere between the simplest testing and complex penetration efforts, but each type is an essential part of the security program and should not be overlooked. The results of all tests should be studied thoroughly and subjected to interpretation by objective analysis.

Avoid Predictable Failure

It is possible to reduce one's identified risk by testing. Murphy's Law states, "Anything that can go wrong will." A relevant corollary could be, "Any system that is not periodically tested will eventually fail." Systems that are routinely tested also occasionally fail! The idea, however, is to avoid predictable failure. The testing concept is consistent with the objective of reducing risk by diminishing uncertainty, which is, after all, one of the fundamental principles of an effective security program.

With all the obvious benefits to be gained by testing, one might logically inquire, "Why don't security professionals do more of it?" The answer is simple—time and money! Even routine testing is both time-consuming and costly. Testing complex systems, on the other hand, can be enormously expensive in terms of time and money. A routine fire evacuation drill in a manufacturing plant, for example, can cause thousands of hours lost to production.

Some Audit Guidelines

In some cases, test expense can be reduced by the use of some basic audit techniques, such as the following:

- Statistical sampling: limiting the number of test cases to a statistical (representative) sample of the universe being tested
- Restricting the value of input parameters or limiting the scope or field of inquiry
- Scheduled testing: breaking the audit or test into halves, quarters, or eighths and scheduling it over a period of months or even years, instead of doing the entire audit or test at one time

A word of caution: every test shortcut has its price in terms of potential risk. Management must weigh the concerns about potential (not actual) risks against actual cost. This is particularly true if the potential risk is insured—that is, perceived to be someone else's problem should anything drastic occur.

There is much that one can learn from testing techniques used by other disciplines, such as safety. As an example, Herbert H. Jacobs lists the attributes of an effective measurement (test) system, as follows:[1]

- Administratively feasible
- Adaptable to the range of characteristics to be evaluated
- Constant
- Quantifiable
- Sensitive to change
- Valid in relation to what it is supposed to represent
- Capable of duplication with the same results from the same items measured
- Objective, efficient, and free from error

In our research, we have been especially concerned at the almost total absence of a generally accepted set of practices for testing security programs and systems. This being the case, the security professional can selectively borrow from related disciplines and adapt their principles or practices to help solve some security problems. One such field—auditing—sets forth the following guidelines, which can be borrowed and used with little or no modification:

- An audit is part of management's control.
- Management is the planning, organization, direction, and control of activities to achieve desired goals.

---

[1] Jacobs, H.H., 1980. Toward More Effective Safety Measurement System. In: Tarrants, W. (Ed.), *Measurement of Safety Performance*. Garland, New York.

- It is necessary in a successful business process to set policy, establish procedure, assign responsibility, institute an accountability system, and measure performance.
- Exceptionally good levels of security performance are achieved when risk control is perceived as an important and integral part of planning, organization, and direction.
- Risk control management systems must be integrated into the mainstream of all management functions.
- There is usually a noticeable difference between published policy and procedure and what actually occurs in most organizations.
- Seldom is an activity as effectively managed as those responsible for it say it is.

Auditing risk control (security) programs can serve as an appraisal of management's performance in relation to established company policy and procedure. The basic objective of the audit (test) is, however, the qualitative analysis of the existing security system to determine whether performance is effective and acceptable.

As stated by a senior executive of M & M Protection Consultants, "It has been our experience in conducting audits of the effectiveness of hazard (risk) control programs that there are usually two such programs in place at every location—the one management thinks it has and the one it really has!" He concluded that a high degree of failure is implicit if the hazard (risk) control program that management really has is a great deal less effective than the one management thinks it has.

Develop a Plan of Action

It is no longer acceptable conduct for security practitioners to ask themselves, "Are we testing the right things, or are we testing things right?" What is necessary is to develop a plan of action for submission to management to make sure the job is accomplished. Some suggestions are as follows:

- Review the existing test procedures, if any.
- What kinds of tests are being conducted, by whom, when, and where?
- Are tests cost-effective and proven to reduce (eliminate) identifiable risks?
- Are records of past tests conducted being maintained for future use?
- Are there better, less expensive tests available that can be adapted for use at this facility?
- Are tests being conducted that can be eliminated as no longer functional or effective?
- Can you identify high-risk areas within the organization that are not being tested (audited)? Can you develop a suggested test program for management's review and approval?
- For complicated tests (audits and surveys), would it be in the best interest of the company to invite an outside expert or consultant in to conduct a review? To give a second opinion?
- Is it within your capability to develop and implement formal testing policy and guidelines in those areas where you have operational responsibility?

Risk control specialists must seek out existing testing systems and promote development of new ones through which the effectiveness of security programs can be measured. Performance examination is a necessary element of the security professional's job description. One should always strive to do the best job possible, recognizing that under the best of circumstances, any measurement or test that one may develop, adopt, or adapt will have shortcomings. Any test, however, is better than none. The axiom "Test it, don't trust it" is a safe course of action. Ignoring the problem in hope that a failure, which may cause a serious incident, may never occur, is unacceptable.

In conclusion it is well to remember the 2011 disaster in Japan. The lesson to be learned here is even with built-in redundancy and periodic testing, an unforeseen event or circumstance may occur. It is from such events that we learn how to plan for the future. Probably every nuclear power plant in the world was subjected to immediate inspection with the question uppermost in management's mind, "Can It happen here?" The problem, however, remains the same: complacency. We seem to need a disaster of large magnitude, such as the "Japanese Trifecta," to wake us up. And the answer of course is, "Yes, it can happen here."

20

The Security Consultant

There are four broad areas of competence that a consultant must achieve in order to be considered qualified. The four areas are experience, education, professional credentials, and personal and interpersonal skills.
—Charles A. Sennewald, CPP, *Security Consulting*, 3rd edition,
Butterworth-Heinemann, 2004

In-House versus Outside Advice

Many companies call on outside consultants to perform studies, make evaluations, and offer recommendations for implementing or improving their security programs. Some companies have benefited from the experience and knowledge that consultants can bring to bear on problems encountered during surveys. Other companies have not been so fortunate. Disappointments are a result of a number of factors. For one thing, employees sometimes regard an outsider as an interloper, a stranger, one who has no real feeling for the company or its employees. Rank-and-file employees as well as supervisors and line managers may be resentful and secretive, thus preventing the "outsider" from obtaining a full understanding of problems as they presently exist within the company. No matter how experienced the consultant may be, his or her first task, and it is often a difficult and time-consuming one, is to learn the intricacies of the company, its ingrained processes, procedures, and methods of operation. This is often referred to as the "corporate culture." Absent a full understanding of the corporate culture, the consultant's recommendations, usually seen first by line managers in the form of a written report, may produce a less than positive reaction. Some line managers may spend more time defending the status quo than implementing what may be valid recommendations for improving their operations. Employees generally know that outside consultants charge large fees for work that may well be done, at substantially less cost, using inside resources. Additionally, some so-called security consultants represent manpower or hardware firms and are salespeople in the truest sense of the word and not consultants at all. The title of "consultant" has been misused perhaps more in the security field than in any other profession. The result of using a salesperson who is not a qualified consultant is that the client often ends up paying for more "security" (manpower or hardware) than is actually needed.

One example frequently encountered during surveys will suffice to make this point. A financial corporation dealing in wholesaling precious metals—gold, silver, and platinum—was found to be utilizing 16 closed-circuit television (CCTV) cameras in a 3,000-square-foot office area. One fixed CCTV camera was mounted on the ceiling of an interior corridor located about 25 feet from the security console. This CCTV camera was targeted on the security console area, which, among other things, contained 16 CCTV display monitors. The security officer at the console was monitoring one CCTV display screen that presented him with a video image of himself at work! This can hardly be considered a cost-effective use of CCTV. Another technique that we like to cite is the oversubscribed contract security guard service. Recommending the elimination of just one guard post (coverage 24 hours per day, 7 days per week) can save a client about $95,000 per year. This is more than the usual cost of hiring a consultant. Thus, the client can receive all the other benefits a security consultant brings to the table while saving money on contract manpower costs.

Security consultants can and do provide valuable services to their clients, provided the client does a reasonably good job of selecting the right consultant in the first place. As a onetime professional security consultant, one who earned his living by plying this trade, we would caution prospective users of consulting services to use the same solid business judgment and standards in selecting a security consultant that one would in selecting any other type of consultant. In order to do that, perhaps a brief look into the historical development of the security consultant will be a worthwhile journey.

The field of protection consulting is now relatively established. This was not true 30 years ago when we wrote the first edition of this text. Protection consulting had its origins in the insurance industry, principally with regard to property (fire) protection. The field then grew, as a natural extension, into accident prevention (casualty) and safety consulting. Last, but certainly not least, came security (crime prevention) consulting.

Security consulting probably got its start just before the United States entered World War II, with the development of the defense industry and its secret and top secret projects. Originally the emphasis was on perimeter protection, access control, and document classification as the principal means to protect defense secrets. The requirements for a security program of some sort were contractual in nature; that is to say, adequate protection was deemed to be necessary before the facility would be considered safe for secret or top secret defense projects. It was only when an obvious flaw or hole in the security program was detected that an outside inspector came in and analyzed the situation. The inspector's job was to make recommendations to improve the security sufficiently for the facility to retain its clearance for secret or top secret work.

It is probably safe to say that most security consulting assignments then were based on problems that had already occurred. Little, if any, thought was given to prevention. It was during the 1960–70s that professional security consulting came into existence. It was also about this time that enlightened developers, owners, and managers began to recognize that to increase efficiency and reduce cost, security had to be built into facility design and not added onto a building project as an afterthought. Today, it is not uncommon for architects to seek the services of qualified protection consultants to ensure that their final designs take into account the security requirements for the buildings or project under consideration. As such, we now see some security consultants specializing in the business of design engineering.

Working with architects and engineers on complex design and construction projects is not a task to be assigned to an apprentice security consultant. Clearly, a combination of education and experience leading to professional maturity is needed here. It is said that a wise man knows his limitations. In the consulting field, mistakes can be costly. One's professional reputation can suffer if one takes on a project for which one is not fully qualified, and fails. Huge industrial complexes, such as nuclear power generating facilities or a large hospital complex, will probably require the services of a team of consultants, because of the multifaceted and varied disciplines required to survey such complex environments. In the team approach, consultants are selected because of their expertise in the particular fields or specialty for which their talents will be utilized, recognizing that no one consultant can be expert in all fields of endeavor having to do with security or any other discipline.

Using the team approach to consulting assignments can reduce time and expenses for most large projects. Often it is the only way some large projects can be managed, because of the many specialty areas encountered in these environments. No one security consultant should be expected to be an expert in all phases of security management, procedure, hardware, and electronics—though most, by necessity, have a general idea of the proper application of the various security systems that may be used under specific conditions.

Why Use Outside Security Consultants?

We recall a telephone conversation with a security professional who asked to be referred to a text or written guide to help him design an electronic access control system for a building under development that was to house a financial institution's data processing center. At that time we knew of no such textbook (there have since been published a number of excellent texts on the subject of design security). We asked him, "Why don't you contract the job out to a security consultant with design engineering experience?" He stated, "I can't do that! My boss expects me to be able to handle every security problem that comes up, regardless of how complicated. I would be putting my professional reputation on the line if I ever admitted I didn't have the skills necessary to design an access control system." The simple answer to this situation is—nonsense! No professional from any discipline should be expected to be able to solve every problem that arises. This situation would be analogous to a general practitioner in the field of medicine calling a surgeon, stating that he or she had a patient who needed brain surgery for the removal of a tumor, then asking the surgeon to recommend a textbook so that general practitioner could read up on the subject before performing the operation himself or herself.

The above example is by no means uncommon. The question concerning when to use the services of an outside consultant does frequently arise. Some of the more common questions regarding the use of outside versus inside resources to do a security survey or consulting job are as follows.

Why Do I Need Outside Advice?

An independent consultant can furnish objective opinions without prejudice and without regard to internal pressures or corporate politics. The consultant can, in effect, "let the chips fall where they may."

More often than not, a competent security director or manager knows what his problems are and has even identified the solutions. In these cases, the outside consultant can furnish a "second opinion" to reinforce the initial opinion, especially regarding cost-effective solutions to complicated problems.

When one seeks outside advice and assistance, one will surely seek help from a professional with a high degree of experience in dealing with the problems at hand. As mentioned earlier, the second opinion technique is common practice among other professions and disciplines. Yet in the security field, we find a great reluctance on the part of some professionals to admit to their obvious limitations. Unlike manpower or hardware salesmen, the truly independent security professional has only one concern—the best interest of his client.

Manpower and hardware consultants (read salespeople) are generally limited in scope and are understandably biased toward their own products or services. Their first loyalty is to the company that employs them, and rightfully so. Nevertheless, security professionals have little reluctance in accepting proposals for service from contract security salespeople. The very same security professional will agonize over the prospect of recommending an outside security consultant to do a comprehensive security survey of his entire operation, including procedures, manpower, and hardware. So the next question often asked is the following.

How Can I Justify the Cost of a Consultant on a Limited Budget?

One must not lose sight of the fact that most security surveys are full-time propositions. Assuming that the in-house professionals are fully employed at their day-to-day occupations (and who in this business will admit that they are not?), where will they find the time to conduct a meaningful audit or survey?

Professional consultants usually have available to them library and research assistance unavailable to the average security practitioner. The library resources have been collected, catalogued, and indexed over a period of many years. Admittedly, with the advent of the Internet, this is less critical today than in the past.

Few security professionals, however, have developed the depth of knowledge necessary to do risk assessment in a multidisciplined environment. Most professionals tend to become specialists in certain fields—government, finance, utilities, hospitals, and retail, to name a few. It is not that most professionals are not capable of broadening their scope; it is just a fact of life that few of us do, preferring the "comfort" of our own field of expertise or practice.

An outside consultant can investigate the financial aspects of the necessary manpower and hardware solutions and then negotiate these cost factors with corporate management. Not every in-house security professional is schooled in the financial and negotiating techniques necessary to sell program changes. Most consultants, however, are.

Will an Outside Consultant Provide Assistance in Setting Up the Recommended Program?

This touches on a very common fear—that the consultant will make broad-brush recommendations and then walk off into the sunset, counting his or her excessive fee, leaving a difficult job for those who have to implement the consultant's recommendations. In actuality, consultants can continue to be employed to the extent that they and management feel is necessary to achieve the level of protection required to solve the problems identified during the survey. Risk assessment is at best a matter of opinion, with much uncertainty. The continued presence of the consultant with input at the implementation or installation stages can materially contribute to the final success of the project. And physical presence is not always necessary. As we tell all our clients, "Night or day, we are only a telephone call away."

Most consultants do not provide contract services. Instead, they usually recommend several reliable firms in the immediate geographical vicinity that have reputations for providing quality service. The consultant then assists the client by drawing up minimum specifications and requirements, which the client furnishes to several firms, requesting that each submit a written bid. After the bids are returned to the client, the consultant can assist the client in reviewing the bids and selecting the service that meets the client's requirements at the best (not necessarily the lowest) cost. Once the service is accepted, the consultant can inspect, guide, provide administrative oversight, and critique the implementation or installation of the service.

This same procedure is applicable whether the product is security manpower, hardware, or electronics. But, as with all other phases of the survey, the consultant's key role is to function as the client's representative. Successful consultants function in the best interest of their clients at all times. This means scrupulously avoiding even the mere impression of a conflict of interest.

Security Proposals (Writing and Costing)

A security survey can range from a simple telephone call, to a 1-day on-site review with verbal conclusions and recommendations, to a full field study. The last would encompass a comprehensive review of all risks, complete with a fully documented report detailing the entire security effort. Consulting assignments may also include plan development and review of blueprints and purchase specifications for access control and anti-intrusion alarm systems and other sophisticated security hardware and equipment.

To avoid misunderstanding the parameters of the task to be performed, both client and consultant should establish at the outset the specifications of the tasks to be performed. Probably the best way to accomplish this is with a written proposal.

Before a client asks for, or a consultant begins to prepare, a proposal, it is important that each have a basic understanding of the results expected to be achieved. This can be tricky. Often clients have only a limited idea of their problems and may not be able to articulate their needs. Some clients have not made a realistic appraisal of their problems and thus may not have realistic expectations regarding the solutions. The only way to ensure that both parties understand exactly what is to be accomplished is by outlining the issues in a written proposal.

Written proposals can take many forms, but five basic elements are common to most. They are the introduction, proposal, management, cost, and summary.

Introduction

The introduction identifies the client, geographical location, and problem in very broad terms. It also identifies the consultants and the firm that is submitting the proposal.

Proposal

The proposal must clearly state the need to be fulfilled, most often expressing it as a statement of work or scope. It sets forth in very specific terms both the problem and the proposed review or study that will be undertaken to gather the data necessary to solve the problem and meet the client's needs. It will also later serve as a general planning outline for the consultant doing the field work.

Outlined below are some basic subject areas that may be considered in developing this part of the proposal. These areas are not all-inclusive, and your proposal must be tailored or modified to fit the specifics of the task involved. Some are presented here as examples.

Security Objectives

There are usually four prime objectives that need to be developed during the evaluation of a facility: the risk assessment, vulnerability assessment, criticality assessment, and security functions.

Losses

An in-depth assessment will be made of all incidents leading to losses at the site. Crime experience in the local area, investigation of existing shortages, and any incidences of fire, malicious damage, and vandalism will all be addressed in the assessment.

Security Organization

A review will be made of the security organization and structure as it pertains to vested authority, policy, assignment of responsibility, and cost-effectiveness.

Security Regulations and Procedures

A comprehensive review will be made of the security program in effect. This would include access control, personnel identification, package inspection, after-hours security procedures, liaison with police and other law enforcement agencies, and security indoctrination of employees.

Guard Force

A review of the present guard force or protective service will be made to cover organization, cost-effectiveness, training, report writing, and cost effective utilization of manpower.

Personnel Security

This phase will include a review of background screening of employees, use of badges and passes, and termination procedures.

Physical Security Conditions

The survey team will evaluate the physical conditions, including all aspects of peripheral and interior security and security of objects that are protected or may need protection. This evaluation will consider the present facility, temporary conditions during construction, and proposed expansion plans.

Utilities

A security examination and evaluation will be conducted of critical utility and power sources, including backup power and generators for standby power. Examples are gas, telephone, computer, sewerage, water, and electricity. Storage practices and related security provisions will also be included.

Construction of Security Facilities

Detailed advisory information will be provided to the architect and engineer concerning the methods of construction and the installation of equipment that affects security. Examples are guardhouses, vaults, computer rooms, anti-intrusion devices, electronic card-access systems, and CCTV.

Security Hardware

A locksmith will evaluate the existing security hardware, such as physical deterrents, locks, key scheduling, associated hardware, and installations. Recommendations will cover repair and replacement of existing equipment and suggested material for new construction.

Alarm Systems

Evaluation will be made of the existing alarm system and subsystems, to include expansion and improvements that may require substantial systems additions or complete replacement. State-of-the-art system conformity and performance will be considered. Interior and exterior intrusion detection systems, fire-detection and fire-suppression systems, and building evacuation plans may be part of this task. Note: Much of this can and should be delegated to a fire protection specialist.

Communications

A security evaluation will be made of the existing and proposed communications networks. These will include wired interior systems, telephone, cell phones, Global Positioning System (GPS) and computer systems, radio facilities, and Internet and satellite networks.

Surveillance

Security monitoring by CCTV, still and motion camera photography will be evaluated as applied or considered for future applications.

Security and Fire Safety Hardware

Security containers, security hardware, locks, and products employed for life safety, fire control, and fire extinguishments, unless delegated to a fire protection specialist, will be evaluated.

Procurement
Methodology for procurement, including sourcing, cost estimates, and scheduling, will be provided. Successful implementation of any security program hinges largely on a well-defined and executed procurement contract.

Management

The management section of the proposal will identify and fully describe the consulting organization, its experience, its personnel, and if necessary, a sampling of client companies that may be used as references. In any event, management, administration resources, logistics involved, and capabilities should be spelled out in some detail and should fully qualify the consultant and his firm for the task at hand. Usually included in this part of the proposal are biographical sketches of the consultants who will actually be performing the survey.

Cost

Cost figures are the best-guess estimate of the consultant doing the job. They are only a yardstick and are subject to change if the scope of the inquiry changes when the job is underway. Nevertheless, the client is entitled to a reasonably accurate estimate of the cost and to prompt notification when the job is underway if the scope (and thus the cost) is going to change. Some clients specifically outline the task to be accomplished and send the outline out for several firms to bid on and then accept the return proposal with the lowest figure. This technique, found most often in government entities, is called a request for proposal (RFP). It is also used by large multinational corporations with well-structured purchasing departments. The cost proposal will generally include the following factors:

- Direct labor (manpower) cost
- Travel and expenses
- Miscellaneous cost, if any
- Overhead rate (usually in percentage)
- General and administration (includes reports)
- Total estimated cost
- Profit
- Total proposal cost

The wise consultant will also include a 10 percent contingency fee, based on the total cost figure, to take care of such unforeseen problems as the following:

- Potential delays on site
- Meetings before, during, and after the on-site work commences
- Responses to follow-up inquiries after the final report is submitted
- Other unanticipated cost connected with the project

Table 20-1　Proposal Pricing Worksheet

Name & Title of Contact: _____

PROSPECT NAME AND ADDRESS _____

  Street _____ Phone No._____

  City & State _____ Zip Code _____

DESCRIBE SPECIFIC SERVICE TO BE PROVIDED: _____

_____

_____

_____

_____

_____

BILLING INSTRUCTIONS: _____

  Describe _____

_____

_____

_____

_____

  Type Report Desired _____

  Date Report Desired _____

QUOTATION GOOD UNTIL _____

TIME	TRAVEL & EXPENSES (T&E)
No. Locations Involved:_____	Travel Hours: _____
On-Site Hours*: (Add 25–50% for Foreign) _____	Travel Costs: _____
	Air _____ Cabs_____
Report Preparation Hours*: _____	Rail _____ Car Rental _____
Ratio Guide for Report Prep. to On-Site	Bus _____ Personal Car _____
	Lodging: _____
Simple Reviews　1:1	Food: _____
Complex Reviews　2:1	Misc.: _____
*Above Hrs. Converted to Cost: _____ Support (typist, etc.) Cost: _____	Note: Normal domestic travel will range $100 to $130/day
TOTAL TIME COST: _____	TOTAL T&E: _____

INCIDENTALS

Films _____ Printing _____	Slides _____ Binders _____
Projector _____ Pictures _____	Tapes _____ Other_____

TOTAL

TIME COST _____　　**Estimated by:** _____

T&E COST _____

INCIDENTALS COST _____　　**DATE:** _____

10% CONTINGENCY _____

GRAND TOTAL _____　　**APPROVED BY:** _____

Summary

The summary is used to highlight the details of the proposal, as set forth in the previous four sections. It also contains the total cost of the proposed project, as obtained from section four. This section should identify the benefits the survey hopes to accomplish in terms that even the most recalcitrant, bottom line-oriented, bean-counting executive can understand. It must leave the reader with the positive feeling of having just read a proposal prepared in a timely, efficient, and professional manner. A late, poorly prepared, and disjointed proposal is a reflection of what the future holds regarding the primary task. Don't expect more or less from a consultant's proposal than you would expect to receive for the principal task. A proposal pricing worksheet (Table 20-1) is included to assist both consultants and clients in developing cost figures for submission with proposals.

Evaluation of Proposals and Reports

Charles Hayden, PE, CPP, retired, formerly of the San Francisco office of Marsh & McLennan Protection Consultants, developed a list of minimum criteria to be applied in evaluating proposals and reports prepared by consultants. The following criteria were submitted to and adopted by the client. They are reproduced here with the approval and permission of Mr. Hayden.

The proposal should set forth a reporting procedure. Will the reports be periodic or final? When (date) can the client expect the report to be submitted, how, by whom, and in what form? Remember—keep the language and the format of reports simple.

The report (proposal) should fully satisfy the purpose for which the evaluation was made. The objective(s) of the evaluation should be identified and achieved. The scope of the evaluation must be consistent with the purpose and objective(s). The methodology used must be stated and must ensure that all significant information is collected, collated, and analyzed. The documented qualifications of the consultant must be adequate to perform the task.

Conclusions drawn in the report must include:

a. Application of appropriate standards, acceptable practice, and/or experience.
b. Credible estimates of comparative risk (probability/time) and potential damage/loss.

Recommendations for abatement of risk must be appropriate and effective in regard to:

a. Priority.
b. Cost.
c. Estimated reduction of risk and potential damage.

Appendices

Appendix A
Security Survey Work Sheets

This is a basic guide that may be used to assist in performing physical security surveys in most industrial settings. Questions have been prepared for the purpose of reducing the possibility of neglecting to review certain areas of importance and to assist in the gathering of material for the survey report. Although the list is comprehensive, it is not all-inclusive. Individual adaptation will almost always be necessary to fit a specific environment or special circumstance.

Attached as Annex A and B are some specific questions that pertain to hospitals, universities, and colleges.

General Questions before Starting a Survey

- Date of survey.
- Interview with (name of decision maker).
- Number of copies of the survey report desired by client, to be sent to:
- Obtain plot plan. Plot the production flow on plot plan and establish direction of north.
- Position and title of all people to be interviewed.
- Correct name and address of the facility.
- Type of business or manufacturer.
- Square footage of production or manufacturing space.
- Property other than main facility to be surveyed is located at:
- Property known as:
- Property consists of:
- What activity is conducted here?
- Is there other local property that will not be surveyed? Why?
- If plot plan is not complete, sketch remainder of property to be surveyed.

Number of Employees

- Administrative—total number all shifts.
- Skilled and unskilled—total number on each shift:
  - 1st shift
  - 2d shift
  - 3d shift
- Maintenance/cleanup crew
- Normal shift schedule and break times

- Salaried:
  - 1st shift
  - 2d shift
  - 3d shift
- What days of the week is manufacturing in process?
- Are employees authorized to leave plant during breaks?
- Are hourly employees in a bargaining unit?
- Are company guards in a bargaining unit?

Cafeteria

- Where is the cafeteria located?
- What are the hours of operation?
- Is it company or concession operated?
- What is security of proceeds from sales?
- What is security of foodstuffs?
- What is method of supply of foodstuffs?
- How are garbage and trash removed?
- Where is location of vending machines?
- Where is change maker, if any?

Credit Union

- Where is the credit union located?
- How is money secured?
- How are records secured?
- How is office secured?
- What are the hours of operation?
- How much cash is kept on hand during day and overnight?

Custodial Service

- Staffed by outside contractor or company employees?
- What hours do janitors actually start and complete work?
- Do they have keys in their possession?
- How is trash removed by them?
- Who, if anyone, controls removal?
- Who controls their entrance and exit?
- Are they supervised by any company employee?

Company Store

- Where is the company store located?
- What are hours of operation?

- What method is used to control stock?
- How is stock supplied from plant?
- Number of clerks working in store?
- How is cash handled?
- When and who performs inventories?
- How are proceeds from sales secured?
- How is the store secured?

Petty Cash or Funds on Hand

- In what office are funds kept?
- What is the normal amount?
- How are these funds secured?
- What is process for the control and security of containers?
- Who has general knowledge of funds on hand?

Classified Operations

- Is government classified work performed?
- What is the degree of classification?
- How are classified documents secured?
- What is security during manufacture?
- What is classification of finished product?
- Are government inspectors on premises?
- Is company classified research and development (R&D) performed?
- Is company classified work sensitive to industry?
- What degree of security is it given?
- What degree of security does it require?
- Where are the locations of the various processing and storage areas?

Theft Experience

- Office machines or records.
- Locker room incidents.
- Pilferage from employees' autos.
- Pilferage from vending machines.
- Pilferage from money changer.
- Thefts of company-owned safety equipment.
- Theft of tools.
- Theft of raw material and finished product.
- Are thefts systematic or casual?
- Have any definite patterns been established?
- Are background investigations conducted before employment of personnel?
- What category of personnel is investigated?
- What is the extent of investigations?

The foregoing questions, answered properly, will assist you in developing the degree of control required for various areas. This information can be secured through interviews. You should also have a working knowledge of the organizational and operational plan of the facility. Before starting your detailed examination and study, you should take a guided (orientation) tour of the facility to acquaint yourself with the physical setting. Take notes during this tour, as necessary.

I. Physical Description of the Facility
  - Is the facility subject to natural disaster phenomena?
  - Describe in detail the above if applicable.
  - What major vehicular and railroad arteries serve this facility?
  - How many wood-frame buildings? Describe and identify them.
  - How many load-bearing brick buildings? Describe and identify them.
  - How many light or heavy steel-frame buildings? Describe and identify them.
  - How many reinforced concrete buildings? Describe and identify them.
  - Are all buildings within one perimeter? If not, describe.

II. Perimeter Security
  - Describe type of fence, walls, buildings, and physical perimeter barriers.
  - Is fencing of acceptable height, design, and construction?
  - What is present condition of all fencing?
  - Is material stored near fencing?
  - Are poles or trees near fencing? If so, is height of fence increased?
  - Are there any small buildings near fencing? If so, is the height of fence increased?
  - Does undergrowth exist along the fencing?
  - Is there an adequate clear zone on both sides along fencing?
  - Can vehicles drive up to fencing?
  - Are windows of buildings on the perimeter properly secured?
  - Is wire mesh on windows adequate for its purpose?
  - Are there any sidewalk elevators at this facility? If so, are they properly secured when not in operation?
  - How are sidewalk elevators secured during operation?
  - Do storm sewers or utility tunnels breach the barrier?
  - Are these sewers or tunnels adequately secured?
  - Is the perimeter barrier regularly maintained and inspected?
  - How many gates and doors are there on the perimeter?
  - Number used by personnel (visitors, employees)?
  - Number used by vehicles?
  - Number used by railroad?
  - How is each gate controlled?
  - Are all gates adequately secured and operating properly?
  - Are railroad gates supervised by the guard force during operations?
  - How are the railroad gates controlled?
  - Do swinging gates close without leaving a gap?
  - Are gates not used secured and sealed properly?
  - What is security control of opened gates?
  - Are chains and locks of adequate strength used to secure gates?

- Are alarm devices installed at the gates?
- Is CCTV used to observe gates or other part of the perimeter?
- How many doors from buildings open onto the perimeter?
- What type are they—personnel or vehicular?
- How are they secured when not in use?
- What is security control when in use?
- How many emergency doors breach the perimeter?
- How are emergency doors secured to prevent unauthorized use?
- Are there any unprotected areas on the perimeter?
- What portion of the perimeter do guards observe while making rounds?

III. Building Security

  A. Offices
  - Where are the various administrative offices located?
  - When are offices locked?
  - Who is responsible to check security at end of day?
  - How and where are company records stored?
  - How are they secured?
  - Are vaults equipped with temperature thermostats (rate-of-rise, Pyro-Larm)?
  - Are offices equipped with sprinklers?
  - Fire extinguishers?
  - Are central station and local alarms installed to protect safes, cabinets, etc.?
  - Are file cabinets locked?
  - Are individual offices locked?
  - Does the company have a secure computer file server or communications room?
  - What type of fire protection does the facility have installed?

  B. Plant
  - When and how are exterior doors locked?
  - When and how are dock doors locked?
  - Are individual plant offices locked?
  - Are warehouses apart from production area secured?
  - Are certain critical and vulnerable areas protected by alarms? What type?
  - What are these areas? What do they contain?

  C. Tool Room
  - Is one or more established?
  - Departmental or central tool room?
  - What is the method of control and receipt?
  - How is tool room secured?

  D. Locker Rooms
  - What is basis of locker issue to individual?
  - What is type of locker—wall or elevated-basket type?
  - How are individual lockers secured?
  - Does company furnish keys/locks?
  - Who or what department controls keys/locks?
  - What control methods are used?
  - How and when are keys and locks issued and returned?

- - Are issued uniforms kept in lockers?
- - Are unannounced locker inspections made?
- - Who conducts inspections and how often?

E. Special Areas That May Require Additional Attention

If the facility houses the following types of activities, they may require special individual inspection. Recommendations can be based on any or all of the applicable portions of the checklist. After the initial inspection tour, you should design a checklist applicable to any special areas encountered.

- - Research and development areas.
- - Laboratories.
- - Storage areas for valuable, critical, or sensitive items.
- - Finished-product test areas.
- - Finished-product display areas.
- - Vehicle parking garages apart from the facility.
- - Vacant or used lofts, attics, etc.
- - Mezzanines or sub-basements.
- - Aircraft hangars, maintenance shops, and crew quarters.

IV. Security of Shipping and Receiving Areas

- Locate all shipping docks, vehicle and railroad.
- What are the hours of operation of docks?
- What is the method of transportation?
- What is the method of inventory control at docks?
- What is the method of control of classified items?
- What is the security of classified or "hot" items?
- What supervision is exercised at the docks?
- Are loaded and unloaded trucks sealed?
- Who is responsible for sealing vehicles?
- What type of seals are being used?
- How are truck drivers controlled?
- Is there a designated waiting room for truck drivers?
- Is it separated from company employees?
- Are areas open to people other than dock employees?
- Do guards presently supervise these areas? Is this necessary?
- What is the method of accounting for material received?
- Is shipping done by parcel post, UPS, or FedEx?
- What is the control at point of packaging?
- Who controls stamps or stamp machines?
- Who transports packages to post office?
- What is the method of transport to post office?
- Where is pickup point at plant?
- What controls are exercised over the transport vehicle?
- Are inspections of operations made presently?
- Who conducts these inspections and how often?
- Does the facility have ship-loading wharves or docks?

- Are contract longshoremen used?
- How do longshoremen get to and from the docks?
- If they pass through the facility, how are they controlled?
- How are ships' company personnel controlled when given liberty?
- Are any specific routes through the facility designated for longshoremen and ship personnel?
- If so, how are they marked and are they used?
- Are these personnel escorted?
- If they are not escorted, what measures are taken to supervise them?
- Is there any way in which these personnel could be kept from passing through the facility?

V. Area Security
- Can guards observe outside areas from their patrol routes?
- Do guards expose themselves to attack?
- Are patrols staggered so patterns are not established?
- What products are stored in outside areas?
- Is parking allowed inside the perimeter?
- If so, are controls established and enforced?
- Where do employees, visitors, and officials park?
- What security and control are provided?
- Are parking lots adequately secured?
- Is there a trash dump on the premises?
- How is it secured from the public?
  - Is it manned by company employees?
  - Is its approach directly from the manufacturing facility?
  - Do roads within the perimeter present a traffic problem?
  - Do rivers, canals, public thoroughfares, or railroads pass through the plant?
  - Are loaded trucks left parked within the perimeter?
  - If so, what protection is given them?
  - Do the roads outside the facility present a traffic problem?
  - What are these problems, and how can they be remedied?
  - Is there any recreational activity within the perimeter, such as baseball?
  - Are these areas fenced off from the remainder of the property?
  - Could they logically be fenced off?

VI. Protective Lighting
- Is protective lighting adequate on perimeter?
- What type of lighting is it?
- Is lighting of open areas within perimeter adequate?
- Do shadowed areas exist?
- Are outside storage areas adequately lighted?
- Are inside areas adequately lighted?
- Is the guard protected or exposed by the lighting?
- Are gates adequately lighted?
- Do lights at gates illuminate the interior of vehicles?
- Are critical and vulnerable areas well illuminated?
- Is protective lighting operated manually or automatically?

- Do cones of light on perimeter overlap?
- Are perimeter lights wired in series?
- Is the lighting at shipping and receiving docks or piers adequate?
- Is lighting in the parking lots adequate?
- Is there an auxiliary power source available?
- Is the interior of buildings adequately lighted?
- Are top secret and secret activities adequately lighted?
- Are guards equipped with powerful flashlights?
- How much and what type of lighting is needed to provide adequate illumination? In what locations?
- Do security personnel report light outages?
- How soon are burned-out lights replaced?

VII. Key Control, Locking Devices, and Containers
- Is there a grandmaster, master, and submaster key system? Describe it.
- Are locks used throughout the facility of the same manufacturer?
- Is there a record of issuance of locks?
- Is there a record of issuance and inspection of keys?
- How many grandmaster and master keys are there in existence?
- What is the security of grandmaster and master keys?
- What is the security of the key cabinet or box?
- Who is charged with handling key control? Is the system adequate? Describe the control system.
- What is the frequency of record and key inspections?
- Are keys made at the plant?
- Do keys have a special design?
- What is the type of lock used in the facility? Are locks adequate in construction?
- Would keys be difficult to duplicate?
- Are lock cores changed periodically at critical locations?
- Are any "sesame" padlocks used for classified material storage areas or containers?
- If a key cutting machine is used, is it properly secured?
- Are key blanks adequately secured?
- Are investigations made when master keys are lost?
- Are locks immediately replaced when keys are lost?
- Do locks have interchangeable cores?
- Are extra cores properly safeguarded?
- Are combination locks three-position type?
- Are safes located where guards can observe them on rounds?
- How many people possess combinations to safes and containers?
- How often are combinations changed?
- What type of security containers are used for the protection of money? Securities? High-value metals? Company proprietary material? Government classified information?
- Are lazy-man combinations used?
- Are birth dates, marriage dates, etc., used as combinations?
- Are combinations recorded anywhere in the facility where they might be accessible to an intruder?

- Are combinations recorded and properly secured so that authorized individuals can get them in emergencies?
- Is the same or greater security afforded recorded combinations as that provided by the lock?
- Where government classified information is concerned, does each person in possession of a combination have the proper clearance and the "need to know"?
- Have all faces of the container locked with a combination lock been examined to see if combination is recorded?
- Are padlocks used on containers containing classified material chained to containers?

VIII. Control of Personnel and Vehicles
- Are passes or badges used? By whom?
- Type used? Describe in detail.
- Is color coding used?
- Are badges uniformly worn on outer clothing?
- Are special passes issued? To whom? When?
- Who is responsible for issue and receipt of passes and badges?
- Are badges and passes in stock adequately secured?
- How are outside contractors controlled?
- How are visitors controlled?
- How are vendors controlled?
- How many employee entrances are there?
- What type of physical control is there at each entrance and exit?
- Where are the time clocks located?
- Is it possible to consolidate clock locations to one or two main clock alleys?
- Is there any control at time clock locations?
- Are there special entrances for people other than employees?
- How are the special entrances controlled?
- Are fire stairwells used for operational purposes?
- Does the facility have elevators to access various floors?
- What control is exercised over their use?
- Are elevators used by operating employees?
- Do the elevators connect operational floors and office floors?
- Does this present a problem in personnel control?
- Are the elevators automatic or self-operated?
- If automatic, are floor directories posted in them?
- Do avenues within the buildings used for emergency egress present a problem of personnel control?
- Examine pedestrian flow from entrance, to locker room, and to work area.
- Can changes be made to shorten routes or improve control of personnel in transit?
- Are personnel using unauthorized entrances and exits?
- If government classified work is being performed, do controls in use comply with the Defense Contract Agency requirements for safeguarding classified information?
- Are groups authorized to visit and observe operations?
- How are these groups controlled?
- Are registers used to register visitors, vendors, and nonemployees? Do they contain adequate information?

- Are these registers regularly inspected? By whom?
- Are employees issued uniforms?
- Are different colors used for different departments?
- What control is exercised over employees during lunch and coffee breaks?
- Do guards or watchmen ever accompany trash trucks or vending machine servicemen?
- Is parking authorized on premises within the perimeter?
- Are parking lots fenced off from the production areas?
- What method of control of personnel and vehicles is there in the parking lots?
- Is vehicle identification used?
- What type of vehicle stickers or identification is used?
- How are issue and receipt of stickers controlled?
- If executives park within the perimeter, are their autos exposed to employees?
- If nurses and doctors park within the perimeter, are their autos exposed to employees?
- Where do vendor servicemen park?
- Do vendor servicemen use plant vehicles to make the service tours? Are small vehicles available?
- How are outside-contractor vehicles controlled?
- What method is used to control shipping and receiving trucks?
- Are the parking facilities adequate at the docks?
- Does parking present a problem in vehicle or personnel control?
- What is the problem encountered?
- During what hour does switching of railroad cars occur?
- Is it possible for people to enter the premises during switching?
- Are there adequate directional signs to direct people to specific activities?
- Are the various buildings and activities adequately marked to preclude people from becoming lost?
- Are safety helmets required?
- Are safety shoes required?
- Are safety glasses required?
- Are safety gloves required?
- Are safety aprons required?
- Are full-time nurses or doctors available?
- Is there a vehicle available for emergency evacuation? What type is it?

IX. Safety for Personnel
- How far away is the nearest hospital in time and distance?
- Are any company employees or guards trained in first aid?
- Is a safety director appointed?
- Is there a safety program? What does it consist of?
- How often does the safety committee meet?
- Is a first-aid or medical room available?
- How are medicine cabinets secured?
- Who controls these keys?
- How is the first-aid room secured?
- Are any narcotics on hand?
- If so, has narcotics security been established?

- Are the required safety equipment items worn by employees? By visitors?
- What is the safety record of this facility?
- How does it compare with the national record?
- Are areas around machinery well policed?
- Does machinery have installed guards where needed? Are they used?
- Are mirrors used where needed to allow forklift operators to observe "blind" turns?
- Could or would mechanical devices used for forklift control improve safety?
- What type of device could be used? Pneumatic alarm system? Signal light?

X. Organization for Emergency
- Are doors adequate in number for speedy evacuation?
- Are they kept clear of obstructions and well marked?
- Are exit aisles clear of obstructions and well marked?
- Are emergency shutdown procedures developed, and is the evacuation plan in writing and periodically updated?
- Do employees understand the plans?
- Are emergency evacuation drills conducted?
- Do guards have specific emergency duties? Do they know these duties?
- Are local police available to assist in emergencies?
- Are any areas of the building in this facility designated as public disaster shelters?
- If so, are controls established to isolate the area from the rest of the facility?
- Do the emergency plans provide for a designated repair crew? Is the crew adequately equipped and trained?
- Are shelters available and marked for use of employees?
- If the plant is subject to natural disaster phenomena, what are they? Floods? Tornados? Hurricanes? Earthquakes?
- What emergency plans have been formulated to cope with potential hazards?
- When and what was the latest incident involving a natural disaster?
- Did it result in loss of life or loss of property?
- Attach a copy of the emergency procedures.

XI. Theft Control
- Are lunchbox inspections conducted?
- Is a package-pass control system being used? Describe it.
- Is a company-employed supervisor assigned to check the package-pass system regularly?
- Is a company official occasionally present during lunchbox inspections?
- Are package passes serially numbered or otherwise containing control numbers?
- Is security of package passes in stock adequate?
- Are comparison signatures available for inspection?
- Is the list of signatures kept up-to-date?
- What action is taken when employees are caught stealing?
- What controls are established on tools loaned to employees?
- What controls are established on laundry being removed?
- What is the method of removal of scrap and salvage?
- What controls are exercised over removal of useable scrap?
- Is control of this removal adequate?

- Are vending and service vehicle inspections being conducted?
- Do employees carry lunchboxes to their work areas?
- Are railroad cars inspected entering and leaving the plant?
- Are company-owned delivery or passenger vehicles authorized to park inside buildings of the facility?
- Does this parking constitute a possible theft problem?
- Do guards check outside the perimeter area for property thrown over fences?
- Do guards occasionally inspect trash pickup? Does anyone?

XII. Security Guard Forces

- What is present guard coverage—hours per day and total hours per week?
- Describe in detail guard organization and composition.
- Number and times of shifts each 24-hour period during weekdays and weekends?
- Number of stationary posts? When are they manned?
- Number of patrol routes? When and where are they, and when are patrols made?
- Are tours supervised? How?
- How many tour stations? Locate them on your plot plan (use different colors or shapes or symbols for different floors and routes).
- What is length of time of each tour?
- Is there additional coverage on Saturdays, Sundays, and holidays?
- Do the patrol routes furnish adequate protection for guards?
- Are the guards required to be deputized?
- Are armed guards required?
- How do guards communicate while on tour?
- Are written guard instructions available? If so, secure a copy.
- If no written instructions are available, generally describe duties of each shift and post.
- What equipment does the guard force have issued? Need?
- Do they require security clearances? What degree?
- Do they require special training?
- Is there a training program in force?
- What communications are available to the guard force to call outside the facility?
- Is the number of guards, posts, and tour routes adequate?
- Are mechanical or electronic devices used by the guard force?
- Do the guards know how to operate, reset, and monitor the devices properly?
- Do the guards know how to respond when alarms are activated?
- Are guard duties included in the emergency plans?
- Do guards know their duties? Emergency duties?
- Do guards make written reports of incidents?
- Are adequate records of incidents maintained?
- Are the guards familiar with the use of firefighting equipment?

Recommendations for changes must indicate each post, tour, and so forth, by number of hours for weekdays, weekends, and holidays, as well as a brief description of the guard's duties. List the total hourly guard service coverage at present and the total coverage after recommendations. If your recommendations increase guard service coverage, you must justify the hours and the cost.

Some Reference Materials

- DOD Industrial Security Manual—Classified Information
- American Standard Practices for Protection Lighting
- General Electric Brochure, How to Select and Apply Floodlights
- Factory Mutual, Organizing Your Plant for Fire Safety
- Security Equipment Brochure Catalogue
- Alarm Installation Estimate Work Sheets
- NFPA—Quarterly Reports
- National Safety Council—Previous and Current Reports

Annex A: Hospital Surveys

(Use applicable portions of the basic industrial security survey work sheet.)

A. Pharmacy
   - Where is the pharmacy located?
   - What are operating hours of pharmacy?
   - Is pharmacist registered and licensed?
   - Is license displayed in pharmacy?
   - How many pharmacists are employed?
   - How are narcotics received and recorded?
   - How are narcotics issued and recorded?
   - How are narcotics secured in pharmacy?
   - How are narcotics secured by nurses on wards?
   - How are narcotics secured in emergency room?
   - Are medicines issued by prescription only?
   - What type of inventory control is used for accounting?
   - What type of personnel control is used at the pharmacy?
   - Can entrance be easily gained, or is "Dutch door" used?
   - How are keys to pharmacy secured?
   - Are keys carried away from hospital by pharmacists?
   - Are "reach-through" storage cupboards used?
   - Are "reach-through" refrigerators used? If so, how are outside doors secured? Who possesses the keys?
B. Morgue
   - Where is the morgue located?
   - Does morgue remain locked when not in use?
   - Who is responsible for morgue security?
   - Who inventories items found on cadavers?
   - How are they inventoried and secured?
   - Who is authorized to release cadavers to undertakers?
   - Are local police escorted when they enter morgue?
C. Linen Department
   - Where is the linen department located?
   - What type of inventory control is used?

- Is linen laundered on property?
- Where is the laundry located?
- How are various items marked for identification?
- Is soiled linen accounted for upon receipt?
- Is clean linen issued by receipt?
- Are both the laundry and linen departments adequately secured?

D. Security of Receipts
- Where are daily receipts paid and stored?
- How are receipts secured?
- How much cash is normally accrued on one business day?
- How often and how is it deposited?
- Are containers furnished to secure patients' valuables?
- How are these valuables inventoried and secured?

E. Emergency Room
- Are security guards present?
- Do local police remain with patients they bring in?
- How are patients under the influence of alcohol or narcotics controlled?
- Are emergency medicines and narcotics properly secured?
- Who inventories valuables of patients arriving unconscious?
- How is this inventory done and when?
- Is a female nurse assigned or on call?
- Is the emergency entrance clear of obstructions?
- Is emergency vehicular approach kept clear?
- Are emergency phone numbers posted at or near the telephone?

F. Security Furnished to Nurses
- Where are the nurses' quarters located?
- Are nurses escorted to their quarters?
- Are nurses escorted to parking lots or to local transportation? If so, what time does this occur, and who escorts them?
- Are nurses' quarters included in guard tour system?

G. Security of Resident Doctors' Quarters
- Do doctors reside in the hospital?
- What is location of their quarters?
- How are keys to quarters issued?
- Are visitors allowed in quarters?
- Are doctors' quarters included in guard tour system?
- What is theft experience at quarters, if any?
- Are visiting doctors furnished residential quarters?
- What security is provided to residential quarters?
- Do doctors normally leave their medical bags unattended?
- What is theft experience at this location, if any?
- What security is exercised over doctors' parking area?
- Are signs displayed reminding them to lock their cars?

H. Dietary Department
- Where is the dietary department (kitchen) located?
- What type of inventory system is used?

- When are "dry" or canned goods received?
- When are fresh meats received?
- Who inventories foods and supplies received?
- How is food issued for preparation?
- What are the hours of preparation?
- How is food for breakfast meal issued?
- What are the hours of operation of the cafeteria?
- How are personnel authorized to use cafeteria identified?
- What is percentage of turnover of dietary employees?
- Are these employees' backgrounds investigated?
- What system is used to issue food to bed patients?
- Do any floors have individual kitchens? If so, what system is used to issue food from stock?
- How are stock rooms secured?
- Who has possession of keys? Who controls keys?
- How is garbage removed?
- Is garbage ever inspected upon removal?
- How is combustible trash removed?
- Is this trash inspected upon removal?
- Are employees allowed to bring parcels, packages, or briefcases to work?

I. Identification and Control of Visitors
  - What are the authorized visiting hours?
  - How many and what entrances are used?
  - Are visitors issued passes? Do they register?
  - Who issues passes or registers visitors?
  - Are passes required to be returned?
  - Are passes color-coded by location within the hospital?
  - How many visitors are authorized per patient at one time?
  - Are visitors policed by the hospital staff?
  - What system is used to prompt visitors to leave?
  - What are the areas or locations where visitors are not authorized to go?

J. Emergency Evacuation Plan for Patients
  - Have emergency evacuation plans been formulated?
  - Are emergency evacuation drills conducted?
  - Are staff members familiar with the plans?
  - Are the procedures posted in strategic locations?
  - Are emergency exits plainly marked?
  - How are ambulatory patients evacuated?
  - How are bed patients evacuated?
  - Does the guard force participate in the evacuation of bed patients?
  - Where are patients housed after they are evacuated?
  - Are nurses, hourly employees, and staff members trained in emergency removal of bed patients?
  - Are all hospital employees included in the emergency evacuation plan?
  - Are local police and fire department personnel included in evacuation of operations?
  - Does the emergency evacuation plan include specific routes for specific cases?
  - Do the routes used for evacuation of bed patients conflict with the routes used by ambulatory patients?

- What system is used to sound the alert for fire?
- Is the method coded or disguised so as not to cause panic?

K. Parking Facilities

- Are the parking lots for doctors, staff, employees, and visitors separated? Posted?
- Is employee parking adjacent to the linen or dietary departments?
- If so, are fences erected so employees cannot gain entrance to their automobiles at will?
- Does the problem exist of visitors and/or doctors parking in fire and emergency lanes?
- Is emergency parking available for doctors?
- Does the administrator have a parking violation ticket system established? Describe it.
- Could visitors entering the hospital from the parking lot be canalized to pass a guard?
- Could this be accomplished for visitors entering from the street?
- Examine closely the traffic flow plan. Is it possible to organize the flow pattern and obtain a more efficient flow using one-way arteries or other devices?
- Is it possible to change the designation of the various parking areas (for instance, doctors' parking where visitors now park, and vice versa)?

Annex B: University and College Surveys

A. Locking, Key Control, and Security Containers

Note: All questions included in the basic industrial security survey work sheet will apply to universities and colleges, but the locking problems of such institutions are more complex and need additional scrutiny.

- Is there a system used for issuing keys to lockers to students at registration time?
- Is there a deposit charge made on the key?
- How much is the deposit?
- What type of lock is used on student lockers?
- How much difficulty is encountered if a lock must be changed when the student fails to return a key?
- Does the institution keep a record of keys issued to lockers?
- Are keys issued to students to any classrooms?
- Who keeps a record of such issue?
- Is there a deposit charge made on these keys?
- How much?
- Are locks changed between terms when such keys are not returned?
- Is any method provided to deny entrance to classrooms during hours when no one has authority to be in them?
- What protection is provided over examinations to be given?
- Is this material kept in combination or key-locked containers?
- If key locks are used, how securely is the room locked?
- Are windows to such rooms locked with key locks?
- Are windows to such rooms accessible from the ground or fire escapes?
- Is any special locking protection provided in areas where damage could be done to important experiments?

- What type of lock is used to protect areas where college funds are kept, bookstores, food storage areas, cafeteria offices, supply and equipment rooms, libraries, museum areas, valuable collections, dispensary narcotics, cabinets, and so forth?
- Are they adequate?
- How are doors to girls' dormitories locked and controlled?
- How are windows at first-floor level, basements, and those accessible by use of fire escapes protected?

B. Protective Lighting

- Are open areas of the campus sufficiently lighted to discourage illegal or criminal acts against pedestrians?
- Is the campus equipped with emergency call stations placed at strategic locations and in open areas?
- Are there any areas that are covered with high-growing shrubs or trees where the light is not sufficient?
- Are the outsides of buildings holding valuable or critical activities or materials lighted?
- Are interiors of hallways and entrances lighted when buildings are open at night?
- Are areas surrounding the dormitories and living quarters well lighted?
- Are lighting fixtures used for this purpose placed in a position to shine out from the building into the eyes of a person approaching?
- Are halls, building entrances, and dormitories well lighted?
- Are campus parking lots lighted sufficiently to discourage tampering with parked cars or other illegal activities?
- Are areas where materials of high value are stored well lighted? Safes, libraries, bookstores, food storage areas, and the like?

Appendix B
Sample Kidnap and Ransom Contingency Plan

This plan was designed to fit the needs of a particular company with a unique kidnap and ransom exposure. It is not a blueprint that can or should be followed by any other organization, although parts of it will apply to most kidnap and ransom situations.

I.   Introduction

Even the most carefully tailored security procedures sometimes are not enough. As evidenced by repeated bombings, kidnappings, extortion plots, and other acts of violence, security procedures are not infallible; they can be compromised or penetrated. For this reason, it is vitally important for this company to have a written contingency plan that outlines some of the steps to be taken in the event a kidnap or extortion plot against the company becomes an actuality. The threat/risk ratio for _____ has been assessed as low to medium. Therefore, some security precautions are deemed to be essential. Should the threat/risk ratio change, it will be necessary to review the existing security program to reassess the level of security and determine whether it is adequate in view of the changing circumstances.

Note: The contents of this report are confidential. The details contained herein should receive limited distribution. The plan is classified as "Business Confidential."

II.   Basic Plan

A.   Policy Regarding Ransom

Company policy is that a reasonable ransom or extortion will be paid in the event of a kidnap of one or more corporate officials or members of their immediate families. The same applies in the event of an extortionate plot against the company. The limitations of our insurance coverage are as follows:

1. Dollar limitations
2. People covered (general)

Note: A copy of the insurance policy covering kidnap and extortion should be attached hereto as an exhibit and remain a permanent part of this file. [The existence of this document is always confidential.]

B.   Crisis Management Team Composition

Members of the crisis management team should include those few individuals having authority to implement and carry out the policy as dictated by the board of directors and the procedures contained in this plan. The presence of more than five people on this team could easily lead to confusion at a time when confusion is least desirable. In addition, the team should be aware that all the resources, in terms of manpower and material, of the company are available for their use on an ad hoc basis.

Members of the team should include:

1. The coordinator, chairman of the board, or chief executive officer
2. President and chief operating officer
3. Executive vice president of finance
4. Executive vice president of operations
5. Executive vice president of administrative and technical services

Other members of the team may vary, depending on the nature of the threat or demand and whether the crisis occurs in [company's location], the United States, or a foreign country.

C.   The Coordinator (and Alternate)

The coordinator's function is to implement the plan and procedures and to coordinate the crisis operation. The coordinator should also be the person with the top decision-making authority. In the event that _____ becomes a victim, the next person in line of succession in the crisis management team would be the alternate coordinator. In the event he or she is unavailable, the next in line of succession would be _____. [The line of succession must be worked out in advance. You may wish to reduce this succession policy to writing. A copy of that policy should then be attached to this document as an exhibit.]

The coordinator, _____ (code name "Mr./Ms. Adams," when dealing with a kidnapper or extortionist), will not act as the negotiator. It is vitally important that the task of negotiator be assigned to only one person—a person who has had training as a negotiator in criminal situations. In this case, we strongly recommend that _____, Vice President, Industrial Relations, who has been trained in union bargaining and negotiations, be trained for criminal-type negotiations. In the event of a kidnap or ransom against the company, [he/she] would become an ad hoc member of the team, serving in the capacity as negotiator and advisor. Mr./Ms. _____ should also direct the preparation of the list of names and related information, which will become a permanent part of this plan (see Exhibit E.1).

In addition, the coordinator shall:

1. Formulate plans and procedures for handling crisis situations.
2. Gather an advisory staff (if deemed appropriate) to generate information and perform services to facilitate these procedures.
   Example: A member of the legal staff may be necessary to review the plans for compliance with established corporate policy.
3. Maintain in a secure place the current crisis management plan and procedures.
4. Communicate these plans and procedures to only authorized individuals and follow up to ensure that these individuals are fully cognizant of any changes in the plan or procedures.
5. Maintain current personal information and biographies of all corporate executives in a secure place. The personnel department maintains a very limited amount of biographical information pertaining to company executives. Enclosed with this document is a biographical inventory. We recommend all executives complete the document, to be placed in their individual personnel files where they can be quickly located in the event of an emergency.

■ ■ ■ ━━

Exhibit E-1 List of Executives and Personnel

1. Executives and Publicity Identified Personnel
   Name   Title

   _____   _____

   _____   _____

   _____   _____

2. Branch or Profit Center Executive Personnel
   Name   Title

   _____   _____

   _____   _____

   _____   _____

3. Personnel to Authorize Ransom Payment (2 needed)

   _____

   _____

4. Name of Financial Institution
   Contact: _____
   Telephone Number: _____

5. Personnel to Administer Payment and Plan/Draw Payment (1)

   _____

   _____

6. Personnel to Handle Police, Press Contact (1)
   Press: _____
   Police: _____
   Federal Bureau of Investigation: _____

7. Persons to Be Notified of Demand
   Name   Telephone (Office)   (Home)

   _____   _____   _____

   _____   _____   _____

   _____   _____   _____

━━ ■ ■ ■

6. Recruit and train the personnel necessary to carry out the crisis management program.
7. Exercise good judgment in determining the course of action in any crisis situation not covered by approved policy.
8. Implement plans and procedures according to the management plan.

D. The Crisis Management Center

The purpose of the crisis management center is to serve as the focal point for directing a coordinated and planned response during a crisis situation. It should be located within the organization's headquarters facility at or in the executive conference room. It should be furnished with all documents, supplies, and communications that may be needed during a crisis. At the minimum, items such as tape recorders, office equipment, computers, and a log to record all calls and actions taken will be necessary.

E.   Crisis Management Plan (CMP) Implementation

When an executive, employee, or family member becomes the victim of a kidnap, or the company becomes the victim of an extortion or terrorist plot, the organization will respond by implementing the CMP. The authority to implement the plan must be clearly spelled out. Implementation criteria should be defined:

1. Who has the authority to implement the crisis management plan?
    a. Chairman of the board or chief executive officer
    b. Alternatively, the president and chief operating officer
2. What are the minimum circumstances that must exist for this authority to become effective (example: if a threat of kidnapping is received)?
    a. Time period of duration that this authority will remain in effect is: (example: until the crisis is satisfactorily resolved)
    b. Succession of this authority should the holder be removed or incapacitated: (names)

F.   Crisis Management Program

When an extortion demand or threat is received, it should be immediately reported to the decision-making authority, as outlined in the crisis management plan. The decision as to when this crisis management program should be implemented will depend on the following factors:

1. Threat verification (true/false)
2. Threat analysis:
    a. How valid is the threat?
    b. Who is doing the threatening?
       - Terrorists (political). Like their counterparts in other countries, these people are the most dangerous. One of their principal goals is publicity, which can be accomplished most effectively by shock tactics. A large multinational company or utility represents a prime target.
       - Criminals. Many criminals in foreign countries have turned to kidnapping, extortion, and other terrorist tactics for criminal gain. Because the goals of these people are identifiable, they are usually open to bargaining, and their demands can be negotiated.
       - The Mentally Ill. This is the fanatic whose sense of values is at odds with those of society. This person is often prepared to die or go to jail for a cause. In this category fall the cunning, the clever, and the inept. They are always unpredictable and therefore hard to plan for and deal with. Also, as in the case of the "Unabomber," they are also difficult to identify, locate, and arrest.

         Note: Most kidnappings are carried out by criminals or the mentally ill. Although kidnappers seeking publicity or nonmonetary rewards usually select large companies, those wanting money frequently select a prominent official of a medium-sized company. Most such organizations have considerable amounts of money available and do not have blanket policies against paying ransom money.

         Note: Threat analysis in complicated cases is usually best left to trained security, risk management, or law enforcement personnel. The tools of threat analysis should be used only as aids in decision making. Total reliance on any one method or tool may cause serious error. In a situation as complicated and dangerous as a kidnapping, all factors must be weighed in order to arrive at effective resolutions.

G. Verification of the Validity of the Threat

1. Does a threat exist?

2. Is the threat as serious, more serious, or less serious than the creator of the threat would have us believe?

3. What is the present vulnerability of the intended victim of the threat? How will this vulnerability increase if we:
  a. Ignore the demands?
  b. Grant the demands?
  c. Engage in prolonged negotiations?

H. The Threateners

1. Can we identify the individual or group responsible for the threat?

2. Can we pinpoint the origin of the threat (physical location)?

3. What is the previous history or experience of this type of threat in this specific environment? (Law enforcement input here is usually necessary.)

4. What is the previous history of other organizations experiencing this type of threat in this specific environment? (In some foreign locations, the police may not be privy to these data.)

5. What type of threat are we faced with?
  a. A simple extortion?
  b. A simple threat, no demands?
  c. A demand without a threat?
  d. What does the type of threat indicate about the demanders' view of the company?
  e. What does the type of threat indicate about the group or person making the threat?

6. How was the threat delivered?
  a. Verbally, telephone (see Exhibit E.2, ransom demand telephone checklist)
  b. By messenger:
    i. Delivered
    ii. Mailed
    iii. Found at crime scene in a protected area
  c. What does the manner in which the threat was delivered indicate about:
    i. The individual or group making the threat?
    ii. The location of the individual or the group making the threat?
    iii. The nature of the group or individual making the threat?

7. What is the nature of the demand, if any?
  a. Who, or what, is the precise victim?
  b. Who is or will become the victim(s) if the demand is not met?
  c. Are demands:
    i. Within the realm of possibility? (Release of political prisoners, for example, is something most corporations cannot influence, much less accomplish.)
    ii. Of propaganda value to the threatener? (Usually a tactic of terrorists)

8. Who is making the demand?

Note: All the above information, plus any other data available that may be helpful, must be collected and analyzed to evaluate the validity of a threat. Time is a critical element in threat analysis. In situations in which time is very short, it must not be wasted on deciding what categories of data are most necessary to be collected before analysis.

All genuine threats will be manifestations of careful preplanning. Only a preplanned response will suffice to meet such a threat. Such preplanning in threat analysis must begin at the very inception of the threat. Remember, time is of the essence. Few sophisticated kidnappers (extortionists) will allow you much time for decision making. Much of the necessary planning can occur before the receipt of a threat.

I. Resources of Verification

Listed below are some resources one might consider using to verify the genuineness of a threat, providing, of course, that there is sufficient time:

1. Corporation-processed pre-event data (may indicate a kidnapper has inside information)
2. Prearranged codes and procedures—for example, "Mr./Ms. Adams" (the coordinator) to "Mr./Ms. Able" (the kidnapper-extortionist). This will preclude an opportunist from taking advantage of publicity to divert an extortion payment or otherwise interfere with the recovery.
3. Local and national law enforcement liaison
   a. The Federal Bureau of Investigation (FBI)
   b. The Department of Public Safety (state level)
4. The Office of Security, U.S. Department of State (if the victim is overseas)
5. Host government liaison (if overseas)
6. Propaganda analysis (for terrorists)
7. Psychological stress evaluator (PSE), if conversation tapes are available
8. Psychiatric analysis
9. Psycholinguistics (tapes only)
10. Graphology (document examination)
11. Forensic document examination
12. Voice-print analysis (tapes only)
13. Electronic tracing
14. Noise analysis
15. Previous case histories

> **Note:** In most cases, the initial threat is communicated by telephone so that demands can receive maximum attention with a minimum of delay. The information contained in the original threat messages is of vital importance.

Ideally, the initial threat message should be recorded. In practice, this is seldom accomplished, although, with preplanning, any secondary and beyond-threat messages can and should be recorded. (Usually the FBI or local law enforcement will arrange to do this.)

Alternatively, the threat call must be reduced to detailed notes. A form to assist in this task is attached and should be furnished to the central telephone operator or the person who usually handles incoming calls. It should be located at or near the telephone in an inconspicuous place, for ready reference (see Exhibit E.2, sample ransom demand telephone checklist).

All demands must be communicated to the crisis management team by oral or written message. Analysis of the communication itself can often reveal a great deal about the character of the individual making the demand and may also reveal whether the threat implicit in the demand is genuine.

■ ■ ■ ▬▬▬▬▬▬▬▬▬▬▬▬▬▬▬▬▬▬▬▬▬▬▬▬▬▬▬▬▬▬▬▬▬

Exhibit E-2 Ransom Demand Telephone Checklist

Time of Call _____

Make every attempt to gain as much information from the caller as he or she will furnish, but do not give the caller the impression you are reading questions from a checklist or are trying to keep him or her on the line so the call can be traced. Write down the responses of the caller word for word.

_____

Would you please repeat your statement?

_____

Who is making this demand?

_____

How do I know this is not a joke? We get many pranks here.

_____

IF A KIDNAP:
What is (he, she) wearing? _____
May I talk to (him, her)? _____
Could you explain what you want? _____
I will have to give your demands to my superior. We will want you to include the word (key word)* and the number (key number)** in all future communications with us.
If the caller gets into specifics on payment, ask, "What do you want?"

_____

If money: what currency and how do you want it?_____
Where and when should the ransom be delivered? _____
How should the payment be made? _____
End the call on a positive note by assuring the caller that his or her demand will be communicated to the proper person in the company as soon as possible. Leave the caller with the impression that his or her call has been understood and action will be taken. Make note of the following information.
Time call ended _____
Background noises _____
Sex of caller _____
Approximate age _____
Any accent_____
Any voice peculiarity such as lisp or stutter _____
What was the caller's attitude?_____
Was the caller sober? _____
Did the caller sound educated? _____
What did you notice about the call that you find unusual? _____
If the caller seemed familiar with the building or operation, indicate how_____
_____

* Recognition code has been established as: Mr. Adams.
** Private unlisted telephone number.

■ ■ ■

J.   Analysis

The purpose of threat analysis is to turn any form of threat into a manageable problem that can be analyzed and then eliminated, neutralized, or controlled by a crisis management team. A schematic of the crisis management process is as follows:

1. Preplanning
   a. Resource identification
   b. Crisis program operating on standby
2. Threat reception: competent reception of threat and the circumstances of its receipt
3. Threat verification: what must we know to be certain that the threat is real?
4. Threat analysis: what must we know to determine:
   a. Threat level?
   b. Identification of the threateners?
   c. Safety of personnel/assets?
   d. Validity of negotiations?
   e. Origin of threat and location of our personnel?
   f. Real goal of the threatener?
   g. Creation of a risk matrix (if deemed necessary)?
5. Threat response: what steps must be taken to eliminate, neutralize, or control the threat and guarantee the safety of our personnel and assets?

K.   Extortion Demands

The demands criminals, terrorists, and mentally ill individuals make in these cases usually fall into one or more of the following categories:

1. The amount of ransom money for the safe return of a kidnapped executive or a member of his or her family depends on:
   a. The wealth of the organization
   b. The criminal's (terrorist's) needs
   c. An intention to demand a "measured quota" from the organization
   d. "Value" of the kidnap victim in the eyes of the criminal (perception)
2. Medical supplies for hospitals, public works, and the like, in exchange for the hostage
3. Public recognition of their cause (terrorists)
4. Release of fellow terrorists or members of their organization jailed by authorities
5. Protest against national politics and policies, or those of the victim organization
6. To embarrass the organization, victim, or victim's family

   Note: Prenegotiation preparation and training should cover the possibility of more than one of the above demands being presented.

L.   Presentation of Demands

Characteristics of the demands of some criminals and most terrorist organizations are:

1. The demands are nonnegotiable (at least in the beginning).
2. All demands (if more than one) must be met in full.

3. Time periods are usually short and are often established at the outset. The initial demands and time frames are rarely realistic.
4. The consequences may be the prompt carrying out of the threat if the demands are not met.

M.   Response

The response of the organization to the demand will probably be determined by the policy of agreeing to negotiate for "reasonable" ransom demands, as set forth above (paragraph II.A).

N.   Counterdemands and Proposals

Depending on the information available, the crisis management team may respond as follows:

1. By asking for (actually, demanding) proof that the executive (victim) is still alive and unharmed. (Captors can be required to supply an item of personal identification; however, a handwritten letter—containing a key phrase or code that we dictate—would be the preferred proof. The letter should contain the date and time it was written.)
2. By asking for the exact time and place of the executive's (victim's) release if an agreement on demands can be reached.
3. By asking for time to study the demands and raise the currency. (You may not get it, but ask for it anyway.)

   Note: It is extremely important that the crisis management team signal that all reasonable demands can be negotiated and will be met, provided that the safety of the executive (victim) can be assured. The reverse should also be emphasized—that without a firm guarantee that the victim will be released unharmed, neither the money nor any other demand will be delivered. The remaining part of the plan can then be accomplished by negotiation. The best policy regarding negotiation is to play for time, total agreement, and guarantees.

O.   Insurance

The insurance policy should be given close scrutiny at this point. Look closely at the coverage and restrictions to ensure that you are in full compliance. The following items should be reviewed:

1. Publicity regarding the policy
2. Genuineness of the extortion demand
3. The specific names or titles of the people covered in the policy
4. Coverage of executives in particular job categories
5. Whether payment must be made under duress
6. Cooperation by the insured with law enforcement

■ ■ ■ ━━━

Exhibit E-3 Negotiator with the Family

1. KEEP THE FAMILY AS CALM AND AS UNWORRIED AS POSSIBLE. Assure the family that the company is doing, and will do, everything possible for the safe release and return of the victim. Assist the family by doing small chores. Try to have the family resume normal activities as far as is possible. Act cheerful and confident.

2. If possible, the family members should not be interviewed by the news media. Younger family members, especially teenagers, might reveal detailed information that might jeopardize the success of your operation.

3. If necessary for its safety, comfort, and security, move the family to a safe haven—for instance, a motel in a secluded location—until the situation is resolved. The police will usually cooperate by assigning protection to the family. If not, hire a private bodyguard.

4. COOPERATE WITH OTHER NEGOTIATORS SO THAT EVERYONE CONCERNED IS AWARE OF THE SITUATION. Do not discuss the situation with anyone other than your fellow negotiators.

━━ ■ ■ ■

■ ■ ■ ━━━

Exhibit E-4 Negotiator with Law Enforcement

1. AS SOON AS PRACTICAL, NOTIFY LAW ENFORCEMENT OF THE SITUATION. Most terrorist situations are under the joint jurisdiction of federal and local authorities. The Federal Bureau of Investigation (FBI) and the U.S. Postal Service are concerned with possible federal law violations; local police are concerned with possible local law violations.

2. BE HONEST AND FRANK WITH LAW ENFORCEMENT. In all situations, law enforcement will cooperate with you, and with one another, for the safe return of the victim. Law enforcement agencies will do nothing to jeopardize the safe return of the victim and will conduct investigations in a covert manner until the victim is returned.

3. HONESTLY ASSESS THE CAPABILITIES OF LAW ENFORCEMENT. In some communities, local police have limited capabilities and experience. In such cases, it will be better to notify the FBI first for their primary investigative activity and to make sure that a capable investigation is conducted. Local law enforcement, in this case, would handle secondary investigation upon notification by the FBI.

4. COOPERATE WITH OTHER NEGOTIATORS. Cooperate so that everyone concerned is aware of the situation. Do not discuss the situation with anyone other than your fellow negotiators.

5. LOCAL POLICE ARE CLOSELY CONNECTED TO THE PRESS. A telephone call to the local police switchboard or emergency number will usually be monitored by the press and television news reporters. If you are to exercise any control over the press, let the FBI notify the local police in every instance.

━━ ■ ■ ■

Exhibit E-5 Negotiator with the Media

1. REMAIN IN CONTROL OF ALL INFORMATION RELEASED TO THE MEDIA. It is better for the safe return of the victim that as little detailed information as possible be released to the press, television, and radio. You can admit that a situation exists, but do not reveal any details about the family situation, the amount of ransom demanded, or the details of the ransom delivery. In a terrorist situation where the safety of personnel is at stake, it is better to release too little rather than too much information. Law enforcement advice should be sought about the release of specific details. Criminals and terrorists read the papers and listen to news broadcasts. If too many details are furnished:

   • Sick, antisocial, or greedy people may enter the picture and complicate things by fraudulent attempts to get money (this is why code names are necessary in all negotiations with the kidnappers).

   • Overly aggressive news reporters might complicate funds delivery by close surveillance.

   • The efforts of law enforcement to secure the return of the victim, to apprehend the kidnappers, or to recover the ransom funds might be jeopardized.

2. A SINGLE PERSON SHOULD HANDLE ALL CONTACTS WITH THE MEDIA. All other people should refer any contacts from the media to that one person. This will prevent the media from playing one official against another to obtain more information. It will also permit the controlled release of nonvital information. It is standard media procedure to induce a person by feeding his or her self-importance, to release small details that are then used to confront a second person in an attempt to extract more substantial information.

3. DO RELEASE TO THE MEDIA ALL DETAILS ABOUT SPECIAL MEDICATION OR TREATMENT NEEDED BY THE VICTIM.

4. REMEMBER AT ALL TIMES THAT YOU DO NOT HAVE TO ANSWER ANY QUESTIONS by the media or anyone else, except in a court under subpoena. Many people feel that they should, or have to, answer questions from the media or well-intentioned citizens. Learn to say, "I'll have to get back to you later with the answer to that question."

5. COOPERATE WITH OTHER NEGOTIATORS so that everyone concerned is aware of the situation. Do not discuss the situation with anyone other than your fellow negotiators.

Exhibit E-6 Negotiator with Terrorists

1. TRY TO MAKE SURE THE VICTIM IS ALIVE. If you can talk to the terrorist, tell him honestly:

   • That you will do everything for the release of the victim.

   • That you just want to make sure that the victim is unharmed.

   • That you would like the victim to write a note to you or say something to you so that you know positively that he is alive and unharmed. (Dictate a key phrase to be included in the note.)

2. **ASK THE TERRORIST TO REFER TO HIMSELF OR HERSELF BY A CODE NAME** that you agree upon so that you will know that you are talking to the same person each time. Do not reveal the code name to anybody until after the victim is released. It is common for several different people to try to collect a ransom in any publicized kidnapping. Give each person calling a different code name. Use a neutral or even a favorable code name, not a derogatory one—"Robin Hood" rather than "Dirty Tom."

3. **OBTAIN AND REPEAT INSTRUCTIONS FOR FUNDS DELIVERY.** If possible, work out alternate instructions in case you cannot comply fully with the original instructions.

4. **ENDEAVOR TO LESSEN THE AMOUNT OF FUNDS BEING DELIVERED.** Since the safety of the victim is most important, do not haggle over the amount of funds. However, if the opportunity arises, tell the terrorist that you can obtain one-half or one-third for immediate, same-day delivery to any spot he or she wants, but delivery of the full amount might take longer since you need higher authorization. If the terrorist is intransigent, drop the subject immediately.

5. **ASSURE THE TERRORIST THAT THE TELEPHONE IS NOT BEING TAPPED, THAT LAW ENFORCEMENT HAS NOT BEEN INVOLVED, AND THAT THE NEWS MEDIA ARE BEING KEPT OUT** (but only if he or she brings up these subjects first).

6. **COOPERATE WITH OTHER NEGOTIATORS** so that everyone concerned is aware of the situation. Do not discuss the situation with anyone other than your fellow negotiators.

Exhibit E-7 Sample Notification of Company Policy

The personal safety and well-being of all of your employees and their families is very important to the company. This cannot be overemphasized. While your company does not believe that the company, or the employees, will be the object of any criminal actions, the following procedures for action are being set out for your guidance.

In any situation involving a hostage, ransom, or extortion, the only important consideration is the safety of our personnel, of their family members, and their safe return.

In any criminal situation or in any questionable situation, notify a company official as promptly and as completely as is possible. Do not delay action to investigate the matter fully just so you can give complete details to a company official.

The company official, and his alternate, are:

Notify _____

_____ Office Phone

_____ Home Phone

_____ Alternative

_____ Office Phone

_____ Home Phone

Distribution: All company offices, managers, and supervisory personnel.

Exhibit E-8 Executive Biographical Inventory (to be retained in the individual's personnel file)

NAME: _____

NICKNAME (FOR IDENTIFICATION): _____

ADDRESS: _____

ALTERNATE ADDRESS (SUMMER, ETC.): _____

DESCRIPTION: AGE _____ BIRTH DATE _____

PLACE OF BIRTH _____ HEIGHT _____

WEIGHT _____ COLOR OF HAIR _____

SEX _____ NOTICEABLE PHYSICAL TRAITS _____

HOME PHONE: _____

WIFE: _____

ADDRESS OF WIFE: _____

DESCRIPTION OF WIFE: _____

VEHICLES: _____

LICENSE	STATE	DESCRIPTION
1. _____	_____	_____
2. _____	_____	_____
3. _____	_____	_____

CODE: FAVORITE BOOK _____ FAVORITE SPORT _____

MEDICAL EMERGENCY INFORMATION: _____

DOCTOR'S NAME: _____ PHONE: _____

SPECIAL MEDICATION: _____

(BACK OF CARD)

HOUSEHOLD MEMBERS:

1. NAME: _____ RELATIONSHIP: _____
   ADDRESS: _____
   DESCRIPTION: _____
   MEDICAL EMERGENCY INFORMATION: _____

2. NAME: _____ RELATIONSHIP: _____
   ADDRESS: _____
   DESCRIPTION: _____
   MEDICAL EMERGENCY INFORMATION: _____

3. NAME: _____ RELATIONSHIP: _____
   ADDRESS: _____
   DESCRIPTION: _____
   MEDICAL EMERGENCY INFORMATION: _____
   FINGERPRINTED? _____
   BLOOD TYPE: _____
   LOCATION IF NOT IN FILE: _____
   RECENT PHOTOGRAPH: _____
   LOCATION IF NOT IN FILE: _____

Appendix C
Security Systems Specifications

[Company Letterhead]

Date _____
Bidder's name and address

_____

_____

Dear Mr. _____,

The XYZ Company invites you to participate in the bidding process to provide an integrated intrusion detection/fire detection, access control, and closed-circuit television system at the facilities located at 2727 Sepulveda Street, Torrance, California.

Attached to this request for proposal (RFP) is the specification that provides the requirements for the system integration. A bidders' conference and a job walk will be held at [time] on [date] at 2727 Sepulveda Street, Torrance, California. Responses to the RFP are due by [time] on [date]. Contract award will be approximately 30 days following receipt of the RFP. Should you decline to participate in the bidding process, please advise me as soon as possible. If you have questions regarding the specification before the bidders' conference and job walk, please call John Doe, Security Manager, XYZ Company, at (310) 555-4005.

Sincerely,
Richard Murphy
(Title)
Enclosure

Introduction

A *specification* is "a detailed, exact statement of particulars, especially a statement prescribing materials, dimensions, and quality of work for something to be built, installed, or manufactured."[1]

A specification for a security system is a part of a request for proposal (RFP) and should provide the bidders with as many details as possible. The need for a selection/evaluation team to prepare the RFP is mandatory and should consist of security, finance, procurement, facilities, operational personnel, and other functions deemed appropriate. The selection/evaluation team is normally

---

[1] *The American Heritage Dictionary of the English Language*, Third Edition.

chaired by security or procurement. The selection/evaluation team jointly prepares the specification, jointly reviews the responses to the RFP, and jointly makes the selection of the successful bidder.

Before issuing the RFP, a bidder's questionnaire should be sent to a select group of suppliers who perform the type of work being requested. The questionnaire should ask questions that are designed to qualify bidders who can perform the desired work while eliminating those bidders who cannot perform the desired work. Information sought may include the size of the company, the length of time in the business, organization chart, financial statement, and references (past and present) where similar work was performed. To aid in the evaluation process, a form should be developed that assigns weighted values to the questions asked in the questionnaire. Another form should be prepared that also assigns weighted values to various portions of the proposal when evaluating the responses to the RFPs. An evaluation criterion includes pricing as well as the overall responsiveness to all elements of the RFP. These forms will aid the selection/evaluation team in the selection of the successful bidder.

After transmission of the RFP, but before the receipt of the responses, a bidders' conference and job walk should be conducted. The bidders' conference provides the bidders with the equal opportunity to ask questions about the project. The job walk provides the bidders with a familiarization of the facility, including the location of the system and devices. A record of the bidders' conference should be made to include attendees, questions and answers, and additional information as deemed appropriate.

A transmittal letter is required to accompany the RFP. The transmittal letter identifies the project schedule, including (a) the date, time, and location of a bidders' conference and job walk, (b) the date and time that the responses to the RFP are due, and (c) the expected date of contract award. The format of the specification will be tailored to the facility and should include at least the following subjects.

Introduction

This section defines the overall system to be procured, including access control, closed-circuit television (CCTV), intrusion detection, or an integration of a number of systems.

Scope of Work

The scope of work identifies the physical location of where the work will be performed and the work to be completed by the contractor, such as construction, electrical, conduit, systems hardware, software, training, and the supplies to be provided.

System Requirements

System requirements identify how the buyer expects the system to perform as well as any specific requirements that are unique to the project.

User Requirements

User requirements are those operational provisions that the buyer desires incorporated into the system, such as software particulars, specific format of input and output data, hardware system capabilities, and hardware type.

Example: Requirements Specification for an Integrated Electronic Security System

Introduction

This specification contains the requirements for an integrated intrusion detection, access control, fire detection, and CCTV monitoring system for the facility identified in the RFP letter. All quotations must ensure that any inability to comply with these requirements is clearly stated in the quotation submission.

Scope of Work

The quotation is to include (a) the integration of access control, intrusion detection, fire detection, and CCTV; (b) all necessary hardware; (c) photo-identification badges; (d) installation of all hardware; (e) ergonomic console; (f) training of systems operators pertaining to hardware and software; and (g) commissioning of the system as specified including all wiring and equipment. Fire detection will be in accordance with all applicable regulations. The physical location and the number of intrusion detection devices, access control devices, and cameras are contained in the attached drawings. The location and size of the proprietary central monitoring station is also contained in the attached drawings. The quotation should include the contractor's recommendations pertaining to the above requirements.

System Requirements
Intrusion Detection/Fire Detection
The present intrusion detection and fire detection systems and devices will be integrated into a console located at the security control center. The access control, intrusion detection, and fire detection systems will be fully integrated.

Access Control/CCTV
A. The access control system will be a Microsoft Windows–based personal computer (PC) system.
B. The supplier will provide a console in sufficient size to house all state-of-the-art equipment (access control, intrusion detection, fire detection, and CCTV monitors), as well as provide for future expansion. The attached drawings indicate the size and location of the control center.
C. Current light requirements will be taken into consideration when specifying the type and location of cameras. Where required, the supplier will provide the light-level requirements.
D. A digital imaging system will be used to create a photo-identification badge. The system should be compatible with generating badges for employees, nonemployess, and visitors in the log-in and log-out processes.
E. The photo-identification badge will incorporate the following characteristics:
  1. Colored-coded to designate area access visually—for example, color background and/or colored bar
  2. Company logo on the front
  3. Color photograph on the front (sized to be visible from a short distance)
  4. Employee's name on the front (sized to be visible from a short distance)
  5. Signature block on the rear

6. Name and address of the company on the rear
7. Statement on the rear indicating that if the badge is found, it should be returned to the address indicated
8. Clips utilized to affix the badge to the outermost garment, chest high, on the person. Necklaces, or other such devices, will be provided for use by those who do not wish to use the clip.
9. The badge will include sufficient fields to incorporate personal details—for example, name, Social Security number/employee number, department number, card number, access authorizations, and expiration date.

F. The badge will incorporate the following two technologies:
1. Magnetic stripe
2. Proximity

G. The badge technology will be compatible with the existing time and attendance system.
H. Initially, about 1,000 photo-identification/access control badges will be required.
I. For comparison purposes, cameras, monitors, recorders, multiplexers, switchers, quads, or other peripheral equipment will be Brand X or Brand Y. Cameras and monitors will be quoted in color except when physical conditions may require the use of black and white.
J. The supplier will specify all building requirements—for example, conduit, door hardware, and electrical, as needed, that are not included in this specification.
K. The system will provide for modular expansion of the hardware as well as cardholder capacity for future needs.
L. Access monitoring will be able to detect and report the following conditions: (a) valid request, (b) lost card, (c) wrong time, (d) wrong door, (e) invalid card, and (f) unknown card.
M. The system should include battery backup support.
N. The system database will encrypt the operator passwords to prevent unauthorized viewing.
O. The system will provide for anti-passback.
P. The system will be compatible with other industry-standard office equipment and software programs, in particular the latest version of Microsoft Office.
Q. Door monitoring will include the ability to report door forced open and door held conditions.

User Requirements
Access Control
A. The requirement is for a Microsoft Windows–based compatible user presentation.
B. The system will include comprehensive online help screens that relate to the currently active window.
C. A "print screen" command will be required for all screens.
D. System operators will be associated with a log-on password and user ID.
E. The system will have the capability of restricting cardholder by dates, times of day, and reader locations.
F. The system will provide an audible alarm at the console for all unauthorized ingress/egress from any door.
G. The system will provide a detailed audit of the arrival and departure times at any of the card readers. The report will include the ability to sort by any of the personal detail fields.

H. History reporting will be incorporated to provide the ability to review all system alarms, access control activity, and operator actions. Report capability will be through operator's display, printer, or magnetic media. Sort capability will include any of the personal detail fields. In addition, the data in the system will be archived from the system to a digital media to ensure it is preserved.

I. To provide for short-term usage, each card record will have a start and end date validity period. Upon expiration of the valid period, cards will become automatically inactive without operator action.

J. The system will provide a means to back up the system's database.

Closed-Circuit Television

A. Cameras will be pan/tilt/zoom as well as fixed; however, they will be able to provide clear and recognizable images.

B. Cameras and monitors will be color except when physical conditions require the use of black and white.

C. External cameras will be able to view the entire perimeter of the buildings.

D. Cameras are to be installed in weatherproof housings where the elements dictate.

E. The system will provide an audible alarm at the console for all unauthorized ingress/egress from any door.

F. Nine-inch monitors at the control center will be used for each access control point indicated on the attached drawings.

G. A 19-inch monitor will be used for pull-down images from any of the 9-inch monitors.

H. All images will be recorded on a 24-hour digital media and retained for a minimum of 30 days.

I. The number of cameras and monitors is shown in the attached drawings. The location and size of the monitoring control center are also shown in the attached drawings.

Conclusion

The specification example contained in this chapter pertains to an upgrade of an existing system and is provided only as a guide. The example is not wholly designed to be applied in every situation. Each facility and each system is unique and should be addressed accordingly. Security systems and devices, as well as related software, are complex, and the state-of-the-art changes rapidly. If the knowledge of state-of-the-art systems and devices and related software is not available within the buyer's organization, assistance should be sought from an independent and objective outside source. Reliance solely on the input from suppliers is not recommended. Procurement specifications also apply to security services, such as contract security personnel, alarm monitoring, consulting, and investigations, to name a few. Although the specification for hardware procurements is different from that of service procurements, the process is similar.

Index

1 percent flood, 175–176
100-year flood, 175–176
3D policy, crime insurance, 93

A
acceptable risk, 29
accidents
  hazards, 6
  nuclear, 7–8
accuracy, survey report, 72
ACFE (Association of Certified Fraud
    Examiners), 45–46
acquisition cost, 34
active shooter, workplace violence, 199
administrative controls, mitigation, 121–122
adverse events, ascertaining, 21
AED (Automated External Defibrillators), 189
after-the-fact analysis of notice, 88
AIG (American International Group), 94–95
Al Qaeda, 119–120
alarm system
  proprietary, 36
  security checklist, 17–18
ALE (annual loss expectancy), 22–23, 23t
All Hazards Planning approach, 101
allaying fear, 54
alternate communications, 123–125
alternate power sources, 123
American International Group (AIG), 94–95
American National Standards Institute (ANSI),
    103
anecdotal evidence, use in establishing notice,
    88
annual loss expectancy (ALE), 22–23, 23t
ANSI (American National Standards Institute),
    103
anti-intrusion card-access system, 36
appendix, business continuity plan, 254
arrested fall emergencies
  overview, 141–143

  preparedness, 142–143
  prevention and mitigation, 142
  recovery, 143
  response, 143
art museums, 36–37
ASIS standard
  Incident Command System (ICS), 108
  Organizational Resilience Standard, 102–103
  planning committee, 231
  Program Coordinator, 227
assess risks phase, FEMA, 115
asset data, developing, 10
assets
  identifying, national infrastructure
      protection plan, 41
  risk control and, 9
Association of Certified Fraud Examiners
    (ACFE), 45–46
assumptions, business continuity plan,
    250–251
auditing
  guidelines, 276–277
  procedures
    aids to surveys, 53–54
    fieldwork, 54–58
Aum Shinrikyo, 120, 153
Australia, emergency number, 252
Automated External Defibrillators (AED), 189

B
balance, report accuracy, 72
Barber, Robert L., 135
barriers, security checklist, 14–15
base flood, 175–176
BIA (business impact analysis)
  methodology
    data collection, 211–213
    overview, 204–214
    project planning, 205–211
  overview, 201–202, 203

BIA (business impact analysis) (Continued)
   questions for, 214–215
   resource questionnaires and forms
      customers, 217
      data analysis, 218–219
      employees and consultants, 215–216
      equipment, 217–218
      forms and supplies, 218
      internal and external contacts, 216–217
      overview, 215–221
      presentation of the data, 219–221
      reanalysis, 221
      software and applications, 217
      vital records, 218
   versus risk analysis, 203–204
   sample table, 211t
biological attack
   overview, 152–156
   preparedness, 154–155
   prevention and mitigation, 154
   recovery, 156
   response, 155–156
Bio-terrorism Act, 39
biotoxins, 152
bomb incident management
   evacuation, 145–146
   overview, 143–152
   preparedness, 150–151
   prevention and mitigation, 149–150
   recovery, 152
   response, 151–152
   searches
      dirty bombs, 148–149
      overview, 146–149
      package bombs, 147
      suicide bombs, 147–148
      suspicious object, 147
   threat evaluation, 144–145
British Incident Management System, 109
Bronze Commander, British Incident
     Management System, 109
Burgess method, 80
business continuity planning
   crisis management plan and, 262
   decentralizing facilities, 122

EOC and, 110
importance of, 224–225
mitigation, 113–114
overview, 223–224
planning process, 225–226
preservation and recovery of records, 126
project management
   conducting BIA, 232–233
   conducting risk identification and
     mitigation inspections, 231–232
   defining scope and planning
     methodology, 228–231
   exercising plan, 242–244
   identifying critical functions, 233
   identifying planning coordinator,
     227
   maintaining plan, 244–245
   obtaining management support and
     resources, 228
   overview, 226–245
   recovery strategies, 233–240
   setting up recovery teams, 240–241
   training recovery teams, 242
   role of senior management, 205–211
Business Continuity Planning Manager, 227
business continuity vendors, 259, 260
business impact analysis (BIA). *See* BIA
     (business impact analysis)
business premises extension, K & R coverage,
     95

C
CAD (computer-aided design) analysis, 34
call centers, 124, 125, 236–237
call trees, 215, 243, 255
Canada
   Fujita tornado scale, 192, 193t
   National Building Code, 161
Cannistraro, Vince, 147
Cardoza, Barry, 212–213
CARVER + Shock method, 39
casualty protection, insurance, 93–94
catastrophic events, planning for, 101
Cawood, James S., 33
CC&Rs (codes, covenants, and restrictions), 37

CCTV (closed-circuit television) cameras, 31–32, 82, 279–280
cellular communications, 124–125, 236–237
CEM (Comprehensive Emergency Management), 101–103, 130–131
CEO (Chief Executive Officer)
    business continuity planning and, 205–206
    crisis management team, 264
CERT (Citizens Emergency Response Team), 131, 141
CFO (Chief Financial Officer), 205–206, 207, 224
Charter for Organization, 61–62
chemical attack
    overview, 152–156
    preparedness, 154–155
    prevention and mitigation, 154
    recovery, 156
    response, 155–156
Chernobyl, 7–8
Chief Executive Officer (CEO)
    business continuity planning and, 205–206
    crisis management team, 264
Chief Financial Officer (CFO), 205–206, 207, 224
chief operations officer, crisis management team, 264
Chubb Insurance Company, 94–95
Citizens Emergency Response Team (CERT), 131, 141
civil disturbances
    defined, 7
    overview, 156–158
    preparedness, 157
    prevention and mitigation, 157
    recovery, 157–158
    response, 157
Ckonjevic, Maxialo, 260
claims reimbursement, BIA, 219
clarity, survey report, 72–73
client/server architectures, 125
closed-circuit television (CCTV) cameras, 31–32, 82, 279–280
CMP (crisis management plan), kidnap and ransom, 312

CMT (crisis management team), 240, 260, 264–266
codes, covenants, and restrictions (CC&Rs), 37
cold sites, 235–236
comments, security checklist, 20
communication
    alternate, 123–125
    security checklist, 18
    telecommunications, 236–237
    while conducting searches, 146–147
Comprehensive Emergency Management (CEM), 101–103, 130–131
computer-aided design (CAD) analysis, 34
conciseness, survey report, 73–74
conclusions, survey report, 77–78
conditioned power, 123
confidentiality, business continuity plan, 254
confined space emergency
    overview, 158–160
    preparedness, 158–159
    prevention and mitigation, 158
    recovery, 160
    response, 159–160
conflicts of interest, 7
consequence, NIPP framework, 41–42
consultants, security
    business continuity planning, 230
    outside
        assisting with program set up, 282–283
        versus in-house, 279–281
        justifying cost, 282
        reasons to use, 281–283
    security proposals
        alarm systems, 285
        communications, 285
        construction of security facilities, 285
        cost, 286–288
        evaluation of, 288
        guard force, 284
        losses, 284
        management, 286
        objectives, 284
        organization, 284
        overview, 283–288
        personnel security, 284

consultants, security (Continued)
    physical security conditions, 285
    procurement, 286
    regulations and procedures, 284
    security and fire safety hardware, 285
    security hardware, 285
    surveillance, 285
    utilities, 285
contact phone numbers, 257
continuity coordinator, 241
control, assessing criticality process, 30
controlled property, 86
corporate team, 240–241
cost valuation (impact), event, 21–23
cost/benefit analysis
    building redundancy into system, 36–37
    overview, 33
    security countermeasure, 37–38
    system design engineering
        cost, 34
        delay, 35–36
        overview, 33–36
        reliability, 34–35
cost/benefit ratio, 30, 114, 129–130
countermeasure, security, 37–38
Cox, Jack E., 135
crime insurance, 92–94
crime prediction
    establishing notice, 87–89
    external crime, 81–85
    inadequate security, 85–87
    internal crime, 80–81
    overview, 79–80
Criminal Intent workplace violence, 196
criminality, defined, 7
crisis management
    crisis management team (CMT), 240, 260,
        264–266
    initial contact, 266–267
    media control, 270–271
    overview, 259–260
    plan activation, 263–264
    plan documentation, 262–263
    preventive security, 268–269
    ransom considerations, 267–268

    suggestions for kidnapped individuals, 269
    threat identification, 261–262
crisis management plan (CMP), kidnap and
    ransom, 312
crisis management team (CMT), 240, 260,
    264–266
critical functions, identifying, 233
Crush syndrome, 142, 169
Customer / Client Related workplace violence,
    196
customer list, 257
customers resource questionnaire form, 217

D
data analysis resource questionnaire form,
    218–219
data systems
    backup, 125–126
    business impact analysis (BIA), 211–213
    recovery, 237–238
decision matrix, 31–32
declaration fee, hot site, 235–236
deductibles, insurance, 94
defense costs, fees, and judgments section,
    K & R coverage, 95
delay, system design engineering, 35–36
Department of Homeland Security (DHS)
    establishing priorities based on risk
        assessments, 43
    measuring effectiveness, 43
    national critical assets database, 41
    standards for PS-Prep, 104
dirty bombs, 148–149
disaster levels, 251–252
Disaster Mitigation Act of 2000, 115
disaster recovery planning. See business
    continuity planning
disclosure, crisis management, 270
displacement costs, 129
distribution, business continuity plan,
    253–254
diverse routing
    communications, 124
    microwave transmissions, 125
    telecommunications, 236–237

document scanning, 127

domestic violence, "spillover" workplace violence, 196

duplicate systems, 238

E

earthquakes
  overview, 160–164
  preparedness, 162–163
  prevention and mitigation, 161–162
  recovery, 164
  response, 163–164

Eisenhower, Dwight, 223

electronic plans, 253–254

electronics systems, 34–35

elimination principle, risk management, 121

EMAP (Emergency Management Accreditation Program), 104

Emergency Action Guide, 165

emergency management
  Comprehensive Emergency Management (CEM), 101–103
  Emergency Operations Center (EOC), 110–112
  Incident Command System (ICS), 104–109
    ASIS and NFPA 1600, 108
    British Incident Management System, 109
    example, 108
    finance and administration, 108
    incident commander, 106–107
    logistics, 108
    operations, 107
    planning and intelligence, 107
  Multi-Agency Coordination System (MACS), 109
  National Incident Management System (NIMS), 104
  overview, 101
  Private Sector Preparedness Accreditation and Certification Program (PS-Prep), 104
  standards, 103
  Unified Command, 109–110

Emergency Management Accreditation Program (EMAP), 104

Emergency Operations Center (EOC), 104–105, 106, 110-112, 256

emergency planning, security checklist, 19–20

emergency procedures
  arrested fall emergencies
    overview, 141–143
    preparedness, 142–143
    prevention and mitigation, 142
    recovery, 143
    response, 143
  bomb incident management
    evacuation, 145–146
    overview, 143–152
    preparedness, 150–151
    prevention and mitigation, 149–150
    recovery, 152
    response, 151–152
    searches, 146–149
    threat evaluation, 144–145
  chemical or biological attack
    overview, 152–156
    preparedness, 154–155
    prevention and mitigation, 154
    recovery, 156
    response, 155–156
  civil disturbance
    overview, 156–158
    preparedness, 157
    prevention and mitigation, 157
    recovery, 157–158
    response, 157
  confined space emergency
    overview, 158–160
    preparedness, 158–159
    prevention and mitigation, 158
    recovery, 160
    response, 159–160
  earthquake
    overview, 160–164
    preparedness, 162–163
    prevention and mitigation, 161–162
    recovery, 164
    response, 163–164
  evacuation planning
    overview, 164–169

emergency procedures (Continued)
   preparedness, 167–168
   prevention and mitigation, 166–167
   recovery, 169
   response, 168–169
  excavation or trench collapse
   overview, 169–171
   preparedness, 170
   prevention and mitigation, 169–170
   recovery, 171
   response, 170–171
  fires
   overview, 171–174
   preparedness, 172–173
   prevention and mitigation, 171–172
   recovery, 174
   response, 173–174
  floods and heavy rain
   overview, 174–179
   preparedness, 177–178
   prevention and mitigation, 176–177
   recovery, 178–179
   response, 178
  hazardous materials incidents
   overview, 179–182
   preparedness, 180–181
   prevention and mitigation, 180
   recovery, 182
   response, 181–182
  hurricanes
   overview, 182–186
   preparedness, 185–186
   prevention and mitigation, 185
   recovery, 186
   response, 186
  lightning
   overview, 186–188
   preparedness, 187–188
   prevention and mitigation, 187
   recovery, 188
   response, 188
  overview, 141–199
  serious injury or illness
   overview, 188–190
   preparedness, 189–190
   prevention and mitigation, 189
   recovery, 190
   response, 190
  structural collapse
   overview, 190–192
   preparedness, 191
   prevention and mitigation, 191
   recovery, 192
   response, 191–192
  tornados
   overview, 192–196
   preparedness, 194–195
   prevention and mitigation, 194
   recovery, 195–196
   response, 195
  workplace violence
   overview, 196–199
   preparedness, 197–198
   prevention and mitigation, 197
   recovery, 199
   response, 198–199
Emergency Response Team (ERT)
  advertising, 141
  conducting regular drills, 141
  determining equipment and resource
    needs, 140
  developing training programs, 141
  duties and responsibilities, 140
  ICS, 104–105
  management acceptance and support,
    139–140
  overview, 139–141
  planning, 140
  recruiting team members, 140
employee abuse, 47
employees and consultants resource
    questionnaire form, 215–216
engineering controls, mitigation, 121
Enhanced Fujita tornado scale, 192, 193–194t
entry and movement, control of, 14
entry rescue, confined space emergency, 158
EOC (Emergency Operations Center), 104–
    105, 106, 110–112, 256
EPA (Environmental Protection Agency), 39,
    179

equipment
  ERT, 140
  rental, business continuity planning, 239
  resource listings, 256–257
  security checklist, 18–19
equipment resource questionnaire form,
    217–218
ERT (Emergency Response Team). *See*
    Emergency Response Team (ERT)
establishing notice, 87–89
evacuation
  bomb incident management, 145–146
  overview, 164–169
  preparedness, 167–168
  prevention and mitigation, 166–167
  recovery, 169
  response, 168–169
excavation. *See* trench collapse
executive kidnappings, 262
executive protection program outline, 96–98
exercising business continuity plan, 254
external crime
  predicting, 81–85
  purpose of analyses, 81
extortion, 316. *See also* crisis management
extraordinary expenses, estimating, 212–213
Exxon Valdez, 260
eye of storm, hurricanes, 182

F
Failure Mode and Effects Analysis (FMEA),
    118
fault zones, 160
FBI (Federal Bureau of Investigation), 83–85
FDA (Food and Drug Administration), 39
fear of litigation, 85
Federal Bureau of Investigation (FBI), 83–85
FEMA (Federal Emergency Management
    Agency)
  assess risks phase, 115
  Flood Insurance Rate Map (FIRM) program,
      175–176
  flooding deaths and property damage, 174
  mitigation, 113
  multihazard functional planning, 248

fences, security checklist, 14–15
Fidelity bonds, 93
fieldwork
  analyzing, 56
  evaluating, 57–58
  investigating, 56–57
  observing, 55–56
  overview, 54–58
  questioning, 56
  verifying, 56
finance and administration section, ICS, 108
financial Information, preliminary survey, 62
findings, survey report, 77
fires
  overview, 171–174
  preparedness, 172–173
  prevention and mitigation, 171–172
  recovery, 174
  response, 173–174
FIRM (Flood Insurance Rate Map) program,
    175–176
first responders, suicide bombs, 148
Fisher, Robert J., 91
flash floods, 175
Flesch, Rudolf, 69
Flood Insurance Rate Map (FIRM) program,
    175–176
floods
  overview, 174–179
  preparedness, 177–178
  prevention and mitigation, 176–177
  recovery, 178–179
  response, 178
floor wardens, building evacuations,
    165–166
flow of information, EOC, 111
flowcharts
  formal, 65f
  informal, 67t
  obtaining information, 63
  standard symbols, 64f
FMEA (Failure Mode and Effects Analysis),
    118, 118
Food and Drug Administration (FDA), 39
Food Safety and Inspection Service (FSIS), 39

forecasting
  earthquakes, 161
  floods, 175
  hurricanes, 182–185
Forgery bonds, 93
format, survey report
  body of report
    findings, 77
    introduction (foreword), 76
    purpose, 76
    scope, 76
    title, 76
  cover letter, 75–76
  overview, 75–78
  statement of opinion (conclusions), 77–78
forms, resource. *See* resource questionnaires
    and forms
forms and supplies resource questionnaire
    form, 218
fraud
  cost of, 45–46
  distribution across sectors, 46–47
frequency of occurrence
  cost valuation and, 21–23
  estimating, 27
FSIS (Food Safety and Inspection Service),
    39
Fujita tornado scale, 192, 193t
functional annexes, 248–249
functional testing, 274–275
funnel cloud, 192

G
gates, security checklist, 14–15
generators, 123
Geographical Information System (GIS)
    software, 117–118
Gold Commander, British Incident
    Management System, 109
governmental EOC, 110
graduated activation, 251
Great Sumatra earthquake, 160
Green, Gion, 91
guide and procedures, auditing

aids to surveys, 53–54
fieldwork, 54–58
  analyzing, 56
  evaluating, 57–58
  investigating, 56–57
  observing, 55–56
  questioning, 56
  verifying, 56
guide words, HAZOP study, 118

H
hardware consultants. *See* consultants,
    security
Hayden, Charles, 38, 288
Hazard and Operability (HAZOP), 118
hazard identification
  BIA and, 204
  cause and effect, 119–120
  checklists, 117
  experts, 118–119
  Hazards United States (HAZUS), 117–118
  history, 116–117
  inspections, 117
  methodology, 120
  overview, 116–120
  process analysis, 118
hazardous materials incidents
  overview, 179–182
  preparedness, 180–181
  prevention and mitigation, 180
  recovery, 182
  response, 181–182
hazardous waste operations (hazwoper),
    139–140
hazards
  defined, 3
  manmade, 113
HAZOP (Hazard and Operability), 118
hazwoper (hazardous waste operations),
    139–140
heavy rain. *See* floods
high-availability solutions, 234
high-rise building evacuations, 164–166
home preparedness, 131

Homeland Security Presidential Directive-7
    (HSPD-7), 40
Homeland Security Presidential Directives, 130
Hoover, J. Edgar, 53
hostage negotiators, 265–266
hot sites, 235–236
HSPD-7 (Homeland Security Presidential
    Directive-7), 40
Hurricane Floyd, 182
Hurricane Mitch, 182–185
hurricanes
    overview, 182–186
    preparedness, 185–186
    prevention and mitigation, 185
    recovery, 186
    response, 186

I

IAP (Incident Action Plans), 107
ICS (Incident Command System), Incident
    Command System (ICS)
IEDs (improvised explosive devices), 144, 148
illness. *See* injury or illness
impact, event, 21–23
improvised explosive devices (IEDs), 144, 148
Incident Action Plans (IAP), 107
Incident Command System (ICS)
    ASIS and NFPA 1600, 108
    British Incident Management System, 109
    developing Incident Action Plans, 140
    example, 108
    finance and administration section, 108
    incident commanders
        information officer, 107
        liaison officer, 107
        overview, 106–107
        safety officer, 107
    logistics section, 108
    operations section, 107
    overview, 104–109
    planning and intelligence section, 107
independent consultant. *See* consultants,
    security
independent events, 25

India, Union Carbide plant, 7–8, 259
industrial disasters, 7
informal flowcharts, 67t
information flow, EOC, 111
information officer, ICS, 107
information technology (IT), 233–234
initial interview, preliminary survey, 60–61
Initial Public Offering (IPO), 230
injury or illness
    overview, 188–190
    preparedness, 189–190
    prevention and mitigation, 189
    recovery, 190
    response, 190
insurance
    crime, 92–94
    crisis management definition, 259
    "key man" life insurance, 268
    kidnap and ransom (K & R), 94–98, 317
    overview, 91
    risk control, 92
    risk management, 91–92
Insurance Company of North America, 94–95
intentional acts, 6
interim reports, 74
internal and external contacts resource
    questionnaire form, 216–217
internal crime
    predicting, 80–81
    theft, 47
internal facilitator, 229
International Organization for
    Standardization (ISO), 103
interviews
    for BIA, 213
    versus interrogations, 56
introduction section
    security systems specification, 324
    survey report, 76
IPO (Initial Public Offering), 230
Irish Republican Army, 119–120
ISO (International Organization for
    Standardization), 103
IT (information technology), 233–234

J

Jacobs, Herbert H., 276
Janson, Jojm J., 29
justifying preparedness, 132–133

K

K & R (kidnap and ransom). *See* kidnap and
        ransom (K & R)
Kennedy, John F., 201
"key man" life insurance, 268
key-and-lock system, cost, 34
keys, security checklist, 16–17
KI (potassium iodide), 149
kidnap and ransom (K & R). *See also* crisis
        management
    contingency plan
        analysis, 316
        coordinator, 310–311
        counterdemands and proposals, 317
        crisis management center, 311
        crisis management plan (CMP), 312
        crisis management program, 312
        crisis management team composition,
            309–310
        extortion demands, 316
        insurance, 317
        overview, 309
        policy regarding ransom, 309
        presentation of demands, 316–317
        resources of verification, 314
        response, 317
        threateners, 313–314
        verification of validity of threat, 313
    insurance, 94–98, 317
Kingsford Manufacturing Company's charcoal
        plant, 113–114
Kobe earthquake, 237

L

Laplace, Marquis de, 24–25, 26
Lass, A. H., 69
law enforcement, role in handling extortion
        demands, 266, 267
Laye, John, 113
letter bombs, 144

Lewthwaite, Samantha, 39
liability cases, 85–86
liaison officer, ICS, 107
life cycle cost, 34
lighting, security checklist, 15–16
lightning
    overview, 186–188
    preparedness, 187–188
    prevention and mitigation, 187
    recovery, 188
    response, 188
Lloyds of London Underwriters, 94–95
locks, security checklist, 16–17
logistics section, ICS, 108
Loma Prieta earthquake, California, 124–125
loss potential
    assessing criticality or severity, 30
    decision matrix, 31–32
Loss to Sales (Profit) Ratio, 47, 48t

M

MACS (Multi-Agency Coordination System),
        109
maintenance, business continuity plan, 254
Major Incident Procedure Manual, 109
management audit techniques. *See* auditing;
        preliminary survey
Management by Objectives (MBO), 105, 108
management memorandum document, 60–61
manmade hazards, 113
manpower consultants. *See* consultants,
        security
manpower costs, 34, 35–36
manual methods, business continuity
        planning, 238
Massive Damage category, tornados, 192
Matters of Special Interest, preliminary survey,
        62
Maximum Tolerable Downtime (MTD), 202
MBO (Management by Objectives), 105, 108
measurements, fieldwork, 54–55
Mercantile open-stock policy, 93
Mercantile safe-burglary policy, 93
methodology
    BIA

data collection, 211–213
    overview, 204–214
    project planning, 205–211
  business continuity planning, 228–231
  hazard identification, 120
microfiche, 127
microwave transmissions, 125, 236–237
Minor Damage category, tornados, 192
mitigation
  arrested fall emergencies, 142
  bomb incident, 149–150
  chemical or biological attack, 154
  civil disturbance, 157
  confined space emergency, 158
  cost-effectiveness, 129–130
  earthquakes, 161–162
  evacuation planning, 166–167
  fires, 171–172
  floods, 176–177
  hazard identification
    cause and effect, 119–120
    checklists, 117
    experts, 118–119
    Hazards United States (HAZUS),
        117–118
    history, 116–117
    inspections, 117
    methodology, 120
    overview, 116–120
    process analysis, 118
  hazardous materials incidents, 180
  hurricanes, 185
  injury or illness, 189
  lightning, 187
  Organizational Resilience Standard, ASIS,
      102–103
  overview, 113–130
  specific mitigation
    alternate communications, 123–125
    alternate power sources, 123
    data backup, 125–126
    facilities salvage and restoration,
        128–129
    overview, 122–129
    policies and procedures, 125

    records management, 126–127
  strategies
    administrative controls, 121–122
    engineering controls, 121
    overview, 120–122
    redundancies/divergence, 122
    regulatory controls, 121
    risk management, 121
    separation of hazards, 122
    service agreements, 122
  structural collapse, 191
  tornados, 194
  trench collapse, 169–170
  workplace violence, 197
Mitigation phase, CEM, 102
mobile satellite transmission, 236–237
Modified Mercalli scale, 160
"Mom and Pop" stores, 47
Momboisse, Raymond M., 45
monitoring safeguards
  audit guidelines, 276–277
  avoiding predictable failure, 275–276
  developing plan of action, 277–278
  overview, 273
  scientific method, 274
  testing existing system, 273–274
  types of testing, 274–275
MTD (Maximum Tolerable Downtime), 202
Mueller, Robert, 147
Multi-Agency Coordination System (MACS),
    109
Multi-Hazard Functional Planning, 101,
    248–249
Murphy's Law, 275
museums, security, 36–37
mutual aid agreements, 132

N
National Building Code, Canada, 161
National Fire Prevention Association's (NFPA)
    1600 standard. See NFPA 1600 standard
National Incident Management System
    (NIMS), 104
National Incident-Based Reporting System
    (NIBRS), 83–85

national infrastructure protection plan, 40–44
    assessing risk based on consequences,
        vulnerabilities and threats, 41–43
    consequence, 42
    threat, 43
    vulnerability, 42–43
    establishing priorities based on risk
        assessments, 43
    identifying assets, systems, networks, and
        functions, 41
    implementing protection programs and
        resiliency strategies, 43
    measuring effectiveness, 43
    setting goals and objectives, 41
National Infrastructure Protection Plan (NIPP)
    Framework, 40, 41–42
National Oceanic and Atmospheric
    Administration (NOAA), 116–117, 175
National Preparedness goals, United States, 130
National Response Plan (NRP), 104
natural catastrophes, 7
natural hazards, 6
negative findings, survey report, 77
negotiations, fees, and expenses section, K & R
    coverage, 95
negotiations, kidnapping, 265–266
Neighborhood Emergency Response Teams
    (NERT), 141
nerve agents, 152
NFPA 1600 standard, 103
    constructing emergency planning program,
        136–137
    Incident Command System (ICS), 108
    planning committee, 231
    Program Coordinator, 227–228
NIBRS (National Incident-Based Reporting
    System), 83–85
NIMS (National Incident Management
    System), 104
NIPP (National Infrastructure Protection Plan)
    Framework, 40, 41–42
NOAA (National Oceanic and Atmospheric
    Administration), 116–117, 175
non-entry rescue, confined space emergency,
    158

nonstructural hazards, 117
Norona, Charles Seoane, 259
notice, establishing, 87–89
NRC (Nuclear Regulatory Commission), 7–8
NRP (National Response Plan), 104
nuclear accidents, 7–8
Nuclear Regulatory Commission (NRC), 7–8

O
objectives, business continuity plan, 250
obtaining information
    flowcharting, 63
    physical observation, 63
    sources of information, 62–63
    what information to obtain, 61–62
Occupational Safety and Health
    Administration (OSHA), 58–59,
    136–137
Office of Food Security and Emergency
    Preparedness (OFSEP), 39
offsite storage facilities, data backup, 126, 127
OFSEP (Office of Food Security and
    Emergency Preparedness), 39
Oklahoma City bombing, 144
1 percent flood, 175–176
100-year flood, 175–176
on-site storage, 127
opening conference (initial interview),
    preliminary survey, 60
Operating Instructions, preliminary survey, 62
operational loss, in BIA, 220
Operational Risk Management, 39
operations section, ICS, 107
oral questions, 56
order-of-magnitude expressions, risk, 32
organization, security checklist, 13–14
Organizational Resilience Standard, ASIS,
    102–103
orthostatic intolerance, 142
OSHA (Occupational Safety and Health
    Administration), 58–59, 136–137
outage tolerances
    BIA, 201, 202
    determining, 219
outline, security survey report, 69–71

P

package bombs, 147

panic, 138–139

paper plans, 253–254

PDCA (Plan, Do, Check, Act), 103

performance testing, 274–275

perils, defined, 3

Perrier bottling plant, 259

personal assets extension, K & R coverage, 95

personal fall protection, 142

personal preparedness, 131

personal property, 18–19

personnel screening, security checklist, 20

physical observation, obtaining information, 63

PIO (public information officer), 107

pitch, survey report, 75

Plan, Do, Check, Act (PDCA), 103

plan documentation, business continuity

  activation procedures and authority, 251–252

  alternate locations and allocations, 252

  appendix, 254

  assumptions, 250–251

  confidentiality, 254

  emergency telephone numbers, 252

  exercising, 254

  maintenance, 254

  Multihazard functional planning, 248–249

  objectives, 250

  overview, 247

  pertinent information, 253

  plan distribution, 253–254

  policy, 250

  recovery priorities or recovery time objectives (RTOs), 253

  required elements, 247–248

  scope, 250

  table of contents, 250

  team recovery plans

    blank forms, 257

    information from the basic plan, 255

    overview, 254–257

    overview of the team plan, 255

    resource listings, 256–257

    scripted continuity instructions, 256

    team member contact list, 255

    training, 254

planning and intelligence section, ICS, 107

planning coordinator, business continuity planning, 227

policy

  administrative controls, 121–122

  business continuity plan, 250

  data system backup, 125

  security checklist, 13

Policy on Critical Infrastructure Protection (Presidential Directive 63), 39–40

positive findings, survey report, 77

positive opinions, 78

potassium iodide (KI), 149

Preliminary Hazard Analysis (PrHA), 118

preliminary survey

  definition and purpose, 58–60

  initial interview, 60–61

  obtaining information

    flowcharting, 63

    physical observation, 63

    sources of information, 62–63

    what information to obtain, 61–62

premiums, crime insurance, 93

preparedness

  arrested fall emergencies, 142–143

  bomb incident, 150–151

  chemical or biological attack, 154–155

  civil disturbance, 157

  confined space emergency, 158–159

  earthquakes, 162–163

  emergency supplies, 131–132

  evacuation planning, 167–168

  fires, 172–173

  floods, 177–178

  hazardous materials incidents, 180–181

  home and personal, 131

  hurricanes, 185–186

  injury or illness, 189–190

  justification, 132–133

  lightning, 187–188

  overview, 130–133

  public–private partnerships, 132

preparedness (Continued)
  structural collapse, 191
  tornados, 194–195
  trench collapse, 170
  vendor relations, 132
  workplace violence, 197–198
Preparedness phase, CEM, 102
Presidential Directive 63 (Policy on Critical
    Infrastructure Protection), 39
prevention
  arrested fall emergencies, 142
  assessing criticality process, 30
  bomb incident, 149–150
  chemical or biological attack, 154
  civil disturbance, 157
  confined space emergency, 158
  earthquakes, 161–162
  evacuation planning, 166–167
  fires, 171–172
  floods, 176–177
  hazardous materials incidents, 180
  hurricanes, 185
  injury or illness, 189
  lightning, 187
  structural collapse, 191
  tornados, 194
  trench collapse, 169–170
  workplace violence, 197
PrHA (Preliminary Hazard Analysis), 118
prioritizing loss potential
  assessing criticality or severity, 30
  decision matrix, 31–32
  overview, 29–30
Private Sector Preparedness Accreditation and
    Certification Program (PS-Prep), 104
proactive response, crisis management,
    263–264
probability
  defined, 24
  of occurrence, estimating, 7
  risk measurement
    principles of, 23–25
    security and, 25–27
Problem Areas, preliminary survey, 62
procedures

administrative controls, 121–122
  data system backup, 125
professional judgment, 57
Profit (Loss to Sales) Ratio, 47, 48t
program, security checklist, 13
Program Coordinator, 227
project management, business continuity
    planning
  conducting BIA, 232–233
  conducting risk identification and
    mitigation inspections, 231–232
  critical functions, identifying, 233
  defining scope and planning methodology,
    228–231
  exercising plan, 242–244
  maintaining plan, 244–245
  obtaining management support and
    resources, 228
  overview, 226–245
  planning coordinator, identifying, 227
  recovery strategies
    data systems, 237–238
    equipment rental, 239
    hot, cold, and warm sites, 235–236
    overview, 233–240
    purchasing materials from competitors,
      237
    reallocating resources, 239
    reciprocal agreements, 239
    relocation, 236
    rescheduling production, 239
    reverting to manual methods, 238
    service-level or quick-ship agreements,
      239–240
    telecommunications, 236–237
    third-party manufacturing, 237
    virtual manufacturing, 238
    workforce management, 238–239
    working at home, 236
  setting up recovery teams, 240–241
  training recovery teams, 242
project planning, BIA, 205–211
property control, security checklist, 18–19
property damage coverage extension, K & R
    coverage, 95

property protection, insurance, 93–94
Proposal Pricing Worksheet, 287t
proposals, security
  alarm systems, 285
  communications, 285
  construction of security facilities, 285
  cost, 286–288
  evaluation of, 288
  guard force, 284
  losses, 284
  management, 286
  objectives, 284
  organization, 284
  overview, 283–288
  personnel security, 284
  physical security conditions, 285
  procurement, 286
  regulations and procedures, 284
  security and fire safety hardware, 285
  security hardware, 285
  surveillance, 285
  utilities, 285
proprietary alarm system, 36
protection consulting, 280
proximate cause, 86
PS-Prep (Private Sector Preparedness
    Accreditation and Certification
    Program), 104
public information officer (PIO), 107
public–private partnerships, 132
purchasing materials from competitors,
    237
pure risk, defined, 3
purpose, survey report, 76

Q
quantifying loss potential
  assessing criticality or severity, 30
  decision matrix, 31–32
  overview, 29–30
Quantitative Risk Analysis, 40
questionnaires, resource. See also resource
    questionnaires and forms
"quick fit" device, generator, 123
quick-ship agreements, 239–240

R
radiological dispersal device (RDD), 148, 149
RAID (redundant array of inexpensive) drives,
    122
rain, heavy. See floods
ransom, 309. See also crisis management
RCRA (Resource Conservation and Recovery
    Act), 179
RDD (radiological dispersal device), 148,
    149
reallocating resources, business continuity
    planning, 239
recidivism rate, 83–85
reciprocal agreements, business continuity
    planning, 239
records management, 126–127
recovery
  arrested fall emergencies, 143
  assessing criticality process, 30
  bomb incident, 152
  chemical or biological attack, 156
  civil disturbance, 157–158
  confined space emergency, 160
  earthquakes, 164
  evacuation planning, 169
  fires, 174
  floods, 178–179
  hazardous materials incidents, 182
  hurricanes, 186
  injury or illness, 190
  lightning, 188
  Organizational Resilience Standard, ASIS,
    102-103
  structural collapse, 192
  tornados, 195–196
  trench collapse, 171
  workplace violence, 199
Recovery phase, CEM, 102, 114–115
recovery point objectives (RPOs)
  BIA, 202
  determining, 219
recovery strategies, business continuity
    planning
  data systems, 237–238
  equipment rental, 239

hot, cold, and warm sites, 235–236
overview, 233–240
purchasing materials from competitors, 237
reallocating resources, 239
reciprocal agreements, 239
relocation, 236
rescheduling production, 239
reverting to manual methods, 238
service-level or quick-ship agreements, 239–240
telecommunications, 236–237
third-party manufacturing, 237
virtual manufacturing, 238
workforce management, 238–239
working at home, 236
recovery time objectives (RTOs), BIA, 201, 202
redundancy
building into system, 36–37
mitigation, 122
redundant array of inexpensive (RAID) drives, 122
regression testing, 274–275
regulatory controls, mitigation, 121
relocation, business, 236
replacement cost
key-and-lock system, 34
versus restoration, 128
request for proposal (RFP), 286
rescheduling production, business continuity planning, 239
Resource Conservation and Recovery Act (RCRA), 179
resource questionnaires and forms
customers, 217
data analysis, 218–219
Employee Resource, 216t
employees and consultants, 215–216
equipment, 217–218
Equipment Resource, 218t
forms and supplies, 218
internal and external contacts, 216–217
Internal/External Contacts, 216t
overview, 215–221
presentation of the data, 219–221

reanalysis, 221
software and applications, 217
Software Resource, 217t
vital records, 218
Vital Records Resource, 219t
Response phase, CEM, 102
response planning
emergency procedures
arrested fall emergencies, 141–143
bomb incident management, 143–152
chemical or biological attack, 152–156
civil disturbance, 156–158
confined space emergency, 158–160
earthquake, 160–164
evacuation planning, 164–169
excavation or trench collapse, 169–171
fires, 171–174
floods and heavy rain, 174–179
hazardous materials incidents, 179–182
hurricanes, 182–186
lightning, 186–188
overview, 141–199
serious injury or illness, 188–190
structural collapse, 190–192
tornados, 192–196
workplace violence, 196–199
Emergency Response Team (ERT)
advertising, 141
conducting regular drills, 141
determining equipment and resource needs, 140
developing training programs, 141
duties and responsibilities, 140
management acceptance and support, 139–140
overview, 139–141
planning, 140
recruiting team members, 140
overview, 135, 136
plans, 136–139
restoration, 128–129, 226
resumption, 226
reward extension, K & R coverage, 95
RFP (request for proposal), 286
Richter scale, 160

RIMS (Risk Insurance Management Society), 94
risk. *See also* risk analysis; risk assessment; risk
    measurement
  defined, 3–4
  risk exposure assessment, 6–8
risk analysis
  versus business impact analysis (BIA), 204
  defined, 4
  national infrastructure protection plan
    assessing risk based on consequences,
      vulnerabilities and threats, 41–43
    establishing priorities based on risk
      assessments, 43
    identifying assets, systems, networks, and
      functions, 41
    implementing protection programs and
      resiliency strategies, 43
    measuring effectiveness, 43
    overview, 40–44
    setting goals and objectives, 41
  overview, 39–40
risk assessment
  defined, 4–6
  management and, 4–6
risk avoidance, 121
risk control, insurance and, 92
risk exposure assessment, 6–8
Risk Insurance Management Society (RIMS),
    94
risk management
  insurance and, 91–92
  mitigation, 121
risk manager, 207–211
risk measurement
  cost valuation and frequency of occurrence,
    21–23
  estimating frequency of occurrence, 27
  overview, 21
  probability
    principles of, 23–25
    security and, 25–27
Risk Priority Number (RPN), 118
robbery, 198–199
robbery-homicide, 196
RPN (Risk Priority Number), 118

RPOs (recovery point objectives). *See* recovery
    time objectives (RTOs), BIA
RTOs (recovery time objectives), BIA, 201, 202

S
safety officer, ICS, 107
safety testing, 274–275
Saffir-Simpson Hurricane Scale, 183–184t
salvage, facilities, 128–129
sarin, 153
satellite transmission, 125
scenario planning, 119, 232
scheduled testing, 276
scientific method, security testing, 274
scope
  business continuity plan, 250
  survey report, 76
scope of work section, security systems
    specification, 324
searches, bomb
  dirty bombs, 148–149
  overview, 146–149
  package bombs, 147
  suicide bombs, 147–148
  suspicious object, 147
security checklist
  alarms, 17–18
  barriers, 14–15
  comments, 20
  communications, 18
  control of entry and movement, 14
  emergency planning, 19–20
  lighting, 15–16
  locks and keys, 16–17
  organization, 13–14
  overview, 12–20
  personnel screening, 20
  policy and program, 13
  property control, 18–19
Security Consulting (Sennewald), 52
security countermeasure, 37–38
security managers, 224
security proposals. *See* proposals, security
security survey
  attitude of business toward security, 48–49

security survey (Continued)
    need for security professional, 50
    needed by whom, 46–47
    overview, 45
    selling security, 50–52
    what's accomplished by, 49–50
    why needed, 45–46
    work sheets
        Annex A: hospital surveys, 303–306
        Annex B: university and college surveys, 306–307
        cafeteria, 292
        classified operations, 293
        company store, 292–293
        credit union, 292
        custodial service, 292
        employees, 291–292
        funds on hand, 293
        general questions before starting survey, 291
        overview, 291
        reference materials, 303
        theft experience, 293–302
security systems specifications. See specification, security systems
security-related losses, 91
self rescue, confined space emergency, 158
self-insured principle, 4
Sennewald, Charles A., 9, 45, 50, 52, 279
separation of hazards, 122
service-level agreements
    business continuity planning, 239–240
    mitigation, 122
short-term outage, equipment, 217–218
Silver Commander, British Incident Management System, 109
simulating plan, 242–244
slant, survey report, 75
smoke sensors, 36
software and applications resource questionnaire form, 217
SOPs (standard operating procedures), 249
Soviet Union, Chernobyl, 7–8
specific mitigation
    alternate communications, 123–125
    alternate power sources, 123

data backup, 125–126
facilities salvage and restoration, 128–129
overview, 122–129
policies and procedures, data system backup, 125
records management, 126–127
specification, security systems
    integrated electronic security system example
        introduction section, 325
        scope of work section, 325
        system requirements section, 325–326
        user requirements section, 326–327
    introduction section, 324
    overview, 323–325
    scope of work section, 324
    system requirements section, 324
    user requirements section, 324
speculative risk, 3
standard flowchart symbols, 64f
standard operating procedures (SOPs), 249
standards. See also ASIS standard; NFPA 1600 standard
    emergency management, 103–104
    regulatory controls, 121
statement of opinion, survey report, 77–78
statistical sampling, 276
steering committees, 231
storm surge, 182
stress testing, 274–275
structural collapse
    overview, 190–192
    preparedness, 191
    prevention and mitigation, 191
    recovery, 192
    response, 191–192
substitution principle, risk management, 121
suicide bombs, 147–148
suitcase bombs, 148
survey report
    criteria
        accuracy, 72
        clarity, 72–73
        conciseness, 73–74
        overview, 72–75
        slant or pitch, 75

timeliness, 74–75
format
    body of report, 76–77
    cover letter, 75–76
    overview, 75–78
    statement of opinion (conclusions),
        77–78
    overview, 69
    writing, 69–72
suspension trauma, 142
suspicious objects, 147
synergistic effect concept, 33
system design engineering, 33–36
    cost, 34
    delay, 35–36
    reliability, 34–35
system requirements section, security systems
        specification, 324

T
table of contents, business continuity plan,
        250
tabletop simulation, 243
team recovery plans
    basic plan information, 255
    blank forms, 257
    contact list, 255
    overview, 254–257
    resource listings, 256–257
    scripted continuity instructions, 256
technical advisors
    crisis management team, 264
    emergency response teams (ERTs), 107
technology
    HAZUS, 117–118
    information technology (IT), 233–234
tectonic plates, 160
telecommunications, 236–237
telephone threats, 145
templates, business continuity planning, 230
terrorism, 7, 27, 119–120
testing security. *See* monitoring safeguards
texting, 124–125
Théorie Analytique des Probabilités, 24–25, 26
third-party crime, 85–86
third-party manufacturing, 237

threat identification. *See* vulnerability and
        threat identification
threats
    defined, 6
    development of occurrence rates and
        probabilities, 6
    evaluation, bomb, 144–145
    NIPP framework, 41–42
3D policy, crime insurance, 93
timeliness, survey report, 74–75
title, survey report, 76
tools, 18–19
tornado warning, 192–194
tornado watch, 192–194
tornados
    overview, 192–196
    preparedness, 194–195
    prevention and mitigation, 194
    recovery, 195–196
    response, 195
training, business continuity plan, 254
training programs, ERT, 141
transferring risk, 92
transit extension, K & R coverage, 95
traumatic rhabdomyolysis, 142, 169
trench collapse
    overview, 169–171
    preparedness, 170
    prevention and mitigation, 169–170
    recovery, 171
    response, 170–171
Tucker, Eugene, 247

U
UCRs (Uniform Crime Reports), 80, 81
uncontrolled property, 86
Unified Command, 109–110
Uniform Building Code, United States, 161
Uniform Crime Reports (UCRs), 80, 81
uninterruptible power supplies (UPS), 122, 123
Union Carbide plant, 7–8, 259
United Kingdom, emergency number, 252
United States
    emergency number, 252
    Enhanced Fujita tornado scale, 192,
        193–194t

EPA and RCRA, 179
National Preparedness goals, 130
Nuclear Regulatory Commission (NRC), 7–8
Uniform Building Code, 161
United States National Response Framework, 102–103
UPS (uninterruptible power supplies), 122, 123
U.S. Department of Agriculture (USDA), 39–40
U.S. Geological Survey (USGS), 116–117
user requirements section, security systems specification, 324

V
vandalism, 144
vehicle-borne improvised explosive devices (VBIEDs), 144
vendor relations, 132
vesicants, 152
violence, workplace
  overview, 196–199
  preparedness, 197–198
  prevention and mitigation, 197
  recovery, 199
  response
    active shooter, 199
    robbery, 198–199
violent crimes, 83–85
virtual EOCs, 110
virtual manufacturing, 238
vital records resource questionnaire form, 218
VSAT (Vulnerability Self Assessment Tool), 39–40
vulnerability, NIPP framework, 41–42
vulnerability and threat identification
  problems of, 11–12
  risk identification, 9–11
  security checklist
    alarms, 17–18
    barriers, 14–15
    comments, 20
    communications, 18
    control of entry and movement, 14
    emergency planning, 19–20
    lighting, 15–16

  locks and keys, 16–17
  organization, 13–14
  overview, 12–20
  personnel screening, 20
  policy and program, 13
  property control, 18–19
vulnerability assessment, defined, 40
vulnerability reduction, 113
Vulnerability Self Assessment Tool (VSAT), 39–40

W
walls, security checklist, 14–15
warm sites, 235–236
waterspout, 192
Work Recovery Time (WRT), 202
work sheets, security survey
  Annex A: hospital surveys, 303–306
  Annex B: university and college surveys, 306–307
  cafeteria, 292
  classified operations, 293
  company store, 292–293
  credit union, 292
  custodial service, 292
  employees, 291–292
  funds on hand, 293
  general questions before starting survey, 291
  overview, 291
  reference materials, 303
  theft experience, 293–302
Worker versus Worker workplace violence, 196
workforce management, business continuity planning, 238–239
working at home, 236
workplace violence
  overview, 196–199
  preparedness, 197–198
  prevention and mitigation, 197
  recovery, 199
  response
    active shooter, 199
    robbery, 198–199
writing, survey report, 69–72
WRT (Work Recovery Time), 202

Printed and bound by CPI Group (UK) Ltd, Croydon, CR0 4YY

03/10/2024

01040314-0001